IET POWER AND ENERGY SERIES 75

Power Distribution Automation

Other volumes in this series:

Volume 1	**Power Circuit Breaker Theory and Design** C.H. Flurscheim (Editor)
Volume 4	**Industrial Microwave Heating** A.C. Metaxas and R.J. Meredith
Volume 7	**Insulators for High Voltages** J.S.T. Looms
Volume 8	**Variable Frequency AC Motor Drive Systems** D. Finney
Volume 10	**SF_6 Switchgear** H.M. Ryan and G.R. Jones
Volume 11	**Conduction and Induction Heating** E.J. Davies
Volume 13	**Statistical Techniques for High Voltage Engineering** W. Hauschild and W. Mosch
Volume 14	**Uninterruptible Power Supplies** J. Platts and J.D. St Aubyn (Editors)
Volume 15	**Digital Protection for Power Systems** A.T. Johns and S.K. Salman
Volume 16	**Electricity Economics and Planning** T.W. Berrie
Volume 18	**Vacuum Switchgear** A. Greenwood
Volume 19	**Electrical Safety: A guide to causes and prevention of hazards** J. Maxwell Adams
Volume 21	**Electricity Distribution Network Design, 2nd Edition** E. Lakervi and E.J. Holmes
Volume 22	**Artificial Intelligence Techniques in Power Systems** K. Warwick, A.O. Ekwue and R. Aggarwal (Editors)
Volume 24	**Power System Commissioning and Maintenance Practice** K. Harker
Volume 25	**Engineers' Handbook of Industrial Microwave Heating** R.J. Meredith
Volume 26	**Small Electric Motors** H. Moczala *et al.*
Volume 27	**AC-DC Power System Analysis** J. Arrillaga and B.C. Smith
Volume 29	**High Voltage Direct Current Transmission, 2nd Edition** J. Arrillaga
Volume 30	**Flexible AC Transmission Systems (FACTS)** Y-H. Song (Editor)
Volume 31	**Embedded Generation** N. Jenkins *et al.*
Volume 32	**High Voltage Engineering and Testing, 2nd Edition** H.M. Ryan (Editor)
Volume 33	**Overvoltage Protection of Low-Voltage Systems, Revised Edition** P. Hasse
Volume 36	**Voltage Quality in Electrical Power Systems** J. Schlabbach *et al.*
Volume 37	**Electrical Steels for Rotating Machines** P. Beckley
Volume 38	**The Electric Car: Development and future of battery, hybrid and fuel-cell cars** M. Westbrook
Volume 39	**Power Systems Electromagnetic Transients Simulation** J. Arrillaga and N. Watson
Volume 40	**Advances in High Voltage Engineering** M. Haddad and D. Warne
Volume 41	**Electrical Operation of Electrostatic Precipitators** K. Parker
Volume 43	**Thermal Power Plant Simulation and Control** D. Flynn
Volume 44	**Economic Evaluation of Projects in the Electricity Supply Industry** H. Khatib
Volume 45	**Propulsion Systems for Hybrid Vehicles** J. Miller
Volume 46	**Distribution Switchgear** S. Stewart
Volume 47	**Protection of Electricity Distribution Networks, 2nd Edition** J. Gers and E. Holmes
Volume 48	**Wood Pole Overhead Lines** B. Wareing
Volume 49	**Electric Fuses, 3rd Edition** A. Wright and G. Newbery
Volume 50	**Wind Power Integration: Connection and system operational aspects** B. Fox *et al.*
Volume 51	**Short Circuit Currents** J. Schlabbach
Volume 52	**Nuclear Power** J. Wood
Volume 53	**Condition Assessment of High Voltage Insulation in Power System Equipment** R.E. James and Q. Su
Volume 55	**Local Energy: Distributed generation of heat and power** J. Wood
Volume 56	**Condition Monitoring of Rotating Electrical Machines** P. Tavner, L. Ran, J. Penman and H. Sedding
Volume 57	**The Control Techniques Drives and Controls Handbook, 2nd Edition** B. Drury
Volume 58	**Lightning Protection** V. Cooray (Editor)
Volume 59	**Ultracapacitor Applications** J.M. Miller
Volume 62	**Lightning Electromagnetics** V. Cooray
Volume 63	**Energy Storage for Power Systems, 2nd Edition** A. Ter-Gazarian
Volume 65	**Protection of Electricity Distribution Networks, 3rd Edition** J. Gers
Volume 66	**High Voltage Engineering Testing, 3rd Edition** H. Ryan (Editor)
Volume 67	**Multicore Simulation of Power System Transients** F.M. Uriate
Volume 68	**Distribution System Analysis and Automation** J. Gers
Volume 69	**The Lightening Flash, 2nd Edition** V. Cooray (Editor)
Volume 70	**Economic Evaluation of Projects in the Electricity Supply Industry, 3rd Edition** H. Khatib
Volume 74	**Power Electronic Converters and Systems: Frontiers and applications** A. M. Trzynadlowski (Editor)
Volume 75	**Power Distribution Automation** B. Das (Editor)
Volume 76	**Power System Stability: Modelling, analysis and control** B. Om P. Malik
Volume 78	**Numerical Analysis of Power System Transients and Dynamics** A. Ametani (Editor)
Volume 79	**Vehicle-to-Grid: Linking electric vehicles to the smart grid** J. Lu and J. Hossain (Editors)
Volume 905	**Power System Protection, 4 volumes**

Power Distribution Automation

Edited by
Biswarup Das

The Institution of Engineering and Technology

Published by The Institution of Engineering and Technology, London, United Kingdom

The Institution of Engineering and Technology is registered as a Charity in England & Wales (no. 211014) and Scotland (no. SC038698).

First published 2016

The Institution of Engineering and Technology
Michael Faraday House
Six Hills Way, Stevenage
Herts, SG1 2AY, United Kingdom

www.theiet.org

British Library Cataloguing in Publication Data
A catalogue record for this product is available from the British Library

ISBN 978-1-84919-828-8 (hardback)
ISBN 978-1-84919-829-5 (PDF)

Typeset in India by MPS Limited
Printed in the UK by CPI Group (UK) Ltd, Croydon

Contents

Chapter 1
Communication systems for distribution automation

Debashis Ghosh[1]

1.1 Introduction

Sophisticated communication systems form the backbone of a reliable and efficient distribution automation system [1, 2]. In the context of distribution automation, communication systems are necessary for transmitting the information regarding the health of different components and the status of the switches in the distribution network to a central monitoring and control station for round-the-clock "unmanned" monitoring and sensing of the distribution system. Further, the same communication network also sends the control signals to different controllable devices (such as switchable capacitors, reactors, tap changers of the transformers, etc.) in order to improve the operational efficiency of the distribution system.

Communication or, more specifically, *telecommunication* refers to the transmission of messages over a long distance. People have been communicating with each other over long distance since the days in the BC era – the way they communicate has changed significantly over time, thanks to the advancements in technology [3, 4]. Starting from the prehistoric age of smoke signals and drumbeats, telecommunication systems evolved continuously with the use of non-electrical means, such as carrier pigeons, semaphore, acoustic phonograph, etc. until electrical telecommunication systems, viz. telegraphy and telephony, started to appear in 1830s. Following the discovery of electromagnetic waves (*radio* waves) that can propagate through free space, wireless communication systems started to develop in the last decade of the nineteenth century. Revolution in communication technology started with the invention of wireless telegraphy, radio, television (TV), videophone, fiber-optic communication, etc. during the first half of the last century. With the advent of digital technology aided by the advances in computer and very-large-scale integration (VLSI) technology, the past few decades witnessed a tremendous growth in communication technology. Today we are in the *communication age* where smart communication gadgets have

[1] Department of Electronics and Communication Engineering, Indian Institute of Technology Roorkee, Roorkee, Uttarakhand, India

penetrated our daily lives so much that our lives these days are not only difficult without communication but maybe even impossible.

1.2 Importance of communication technology in distribution automation

Essentially, a distribution automation is used to monitor, control, coordinate and operate distribution system components from remote locations in real time. A distribution automation system involves integration of computer and communication technologies, in addition to electric devices. Several important quantities related to the health condition of the distribution system are sensed at different points in the distribution network, and then transmitted to the monitoring stations in control rooms. The information so received is processed, decisions are made, following which supervisory control commands are sent to actuators. Thus, the smooth functioning of a distributed automation system relies on the proper coordination among various elements in the system. It is necessary to have a clear picture of the operating status and the overall health of the distribution system, monitor critical data and exchange information among various elements in the system so as to automate transmission and distribution processes. The data from sensors to monitoring stations and set commands from control rooms to actuators need to be sent via wired or wireless communication links. Hence, a data communication network capable of supporting these functions of the system is essential and forms an integral part of any distributed automation system.

However, there is no particular communication technology suitable for the purpose. The choice of a communication system and/or protocol largely depends on the situation, nature and functions of the distribution system. The basic parameters that are generally used to assess the suitability of a communication system are:

- **Bandwidth**, the capacity or the maximum throughput of a communication link, defined as the maximum number of bits or bytes per unit time that the link can handle, depends on the bandwidth of the communication link; higher the bandwidth, larger is the capacity.
- **Availability**, defined as the percent uptime of the link over a given period of time.
- **Latency**, defined as the delay in sending data packets through the link.
- **Jitter**, defined as the variation in latency of the packets transmitted through the link.

As discussed above, a reliable and efficient data communication system plays a major role in distributed automation. Therefore, the purpose of this chapter is to introduce basic concepts, terminology and the working principles of data communication systems for better understanding of the functioning of a distributed automation system. Although the primary focus is data communication, this chapter presents a brief treatment of general digital communication system that supports data communications. Further, analog communication technology is reviewed to provide a background for the introduction of digital communication technology. However, the scope of this

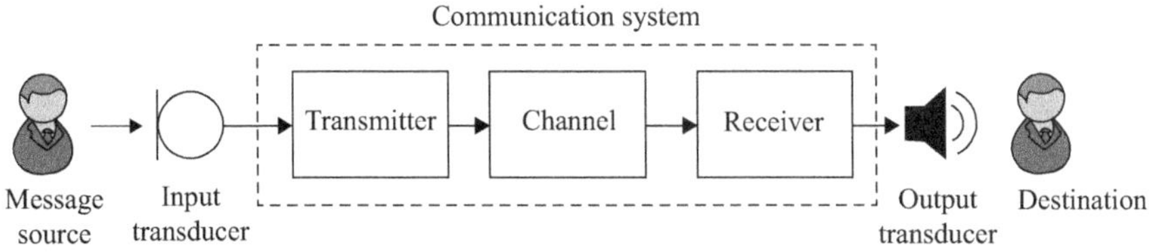

Figure 1.1 Model of a communication system

chapter is limited only to the basic concepts. Several books such as References 4–10, to name a few, cover communication theory in detail. The reader is encouraged to refer any of these books to obtain a thorough knowledge on this subject.

1.3 Communication system model

A typical communication system may be modeled as illustrated in Figure 1.1. The function of a communication system, as said above, is to transmit information-bearing signal from a *source* to a *destination*. The signal generated by the source is generally non-electrical in nature (e.g., human voice, TV picture, etc.) that needs to be converted to an electrical waveform, called the message signal, by an *input transducer*.

The *transmitter* modifies the message signal into a form suitable for transmission over the channel. A communication system may be either analog or digital depending on the transmitted message type. An analog signal is a continuous function of time whose value varies over a continuous range. On the other hand, a digital signal is defined only at discrete times with its amplitude taking on a finite set of discrete values. Communication systems were primarily developed for voice transmission and later, with the invention of TV system, for image and video transmission. Real-world voice, image and video signals are analog in nature that are continuous both in time and amplitude. Accordingly, since the early days of telecommunication, analog system has been the primary technology platform. Voice digitization and transmission started to evolve in the late 1950s although there is no fixed date that marks the transition from analog technology to digital technology. Digital communication technology has gradually taken over from analog technology during the second half of the twentieth century and the trend still continues. Following are some of the advantages of digital communication over analog systems that has contributed to its growth:

1. Immunity to noise – since the amplitude of a digital signal takes on a finite set of discrete values, any small change in the amplitude due to additive noise can be detected and compensated for to recover the original digital waveform.
2. Easy multiplexing – message signals from multiple sources can be transmitted simultaneously using time division multiplexing (TDM).
3. Data integration – voice, video and text messages can be merged and transmitted over a common digital transmission system.

4. Ease of signaling – control signals can be readily incorporated into a digital transmission system by inserting special control codes into the message channel.
5. Error detection and correction capability.
6. Easy processing such as data compression, encryption, etc.
7. Use of modern technology – relatively less expensive and high-performance digital circuits, advances in VLSI technology designed specifically for communication purposes such as encoding/decoding, multiplexing/demultiplexing, general-purpose digital signal processors and application-specific integrated circuits favor digital systems over analog systems.

Nevertheless, digital communication technology has certain disadvantages too, as listed below:

1. Generally, more bandwidth is required for transmission.
2. There is a need for proper bit, character and frame synchronization.
3. Extra processing is involved – analog to digital conversion in the transmitter and the reverse in the receiver.

In nature, signals are generally neither discrete in time nor discrete in amplitude. In a digital communication system, the transmitter converts an analog message signal to its digital form via sampling and quantization. Sampling accomplishes time-discretization of the input analog signal while amplitude-discretization is achieved in the quantization process. The digitized message signal is subsequently binary encoded to produce a sequence of binary digits (*bits*). Digital data may also arise directly in cases where the source of information is inherently discrete in nature. For example, text messages from a digital computer is composed of discrete ASCII characters[1], each character represented by eight bits or one *byte*. In case of distributed automation, the data acquisition devices, such as the remote terminal units, measure several electrical parameters, viz. the root mean square (RMS) values of voltage, current, frequency, power, etc., and other various quantities such as switch status, temperature, etc. at the transformers and feeders, and outputs the measured values in binary format. Also, the commands and instructions issued by the central control unit are binary coded. Thus, the information generated and shared by the various components in a distribution system is inherently digital.

The first ever electrical communication system, i.e., telegraphy, developed in the first half of the nineteenth century may be regarded as a digital system of communication in which characters in a text message are represented and transmitted using variable-length *Morse code* (consisting of sequences of *dots* and *dashes*), later replaced by fixed-length *Baudot code* (consisting of sequences of *marks* and *spaces*, each of length 5). However, it was not until the second half of the twentieth century that the age of modern digital communications actually began. It started with the work of Harry Nyquist when he investigated on finding the maximum rate at which

[1] ASCII stands for *American Standard Code for Information Interchange*, a code for representing characters as numbers.

binary-coded telegraphic signals in the form of mark and space can be transmitted without causing intersymbol interference (ISI) [11, 12]. His work led to the *Nyquist theory* which states that to represent an analog signal by its sampled version, the sampling frequency should be at least twice the highest frequency contained in the analog signal. Nyquist's result was later stated precisely by Shannon and is known as *Shannon's sampling theorem* [13, 14]. Shannon also formulated the problem of reliable transmission in statistical terms and determined the maximum transmission bit-rate that a channel can support, called the *channel capacity* [15]. His works eventually laid the foundations of modern digital communication.

Depending on the range of frequencies used for transmission, the communication system may also be classified as either baseband or passband systems. In baseband transmission, the transmitted signal occupies frequencies from zero up to a maximum that depends on the message signal. In passband transmission, on the other hand, the transmitted signal occupies frequencies around the frequency of a high-frequency carrier signal. Almost all real-world sources generate signals that have frequencies close to zero, such as the human voice (20 Hz–3.4 kHz) and TV picture (0 Hz–5.5 MHz). Accordingly, the message signals are generally baseband signals. In an analog baseband transmission system, the message signal is transmitted as it is without any frequency translation. An example of baseband communication is the local telephone system. In passband communication, as in case of long-distance telephone call via satellite link, the baseband message signal is frequency shifted to a higher frequency and then transmitted over the channel. This shifting in frequency is achieved by means of *modulation*. Modulation is the process in which certain parameter of a carrier signal, such as amplitude, frequency or phase, is varied in accordance to the message signal. Section 1.4 describes various modulation schemes applied to analog messages.

In digital baseband transmission, the binary sequence of information is transmitted directly in the form of electrical pulses. The process of coding binary digits into electrical pulses is called *line coding*. The simplest form of line code is *on–off* signaling in which a **1** is represented by a positive pulse and a **0** is represented by no pulse. Another commonly used line code is the *non-return-to-zero* (NRZ) polar format that uses a positive pulse to represent bit **1** and negative pulse to represent bit **0**. Some other popular line coding formats are Manchester coding, bipolar coding or alternate mark inversion, etc. All these signaling formats are used for transmission of the binary encoded digital message directly over a baseband channel. However, when it is required to transmit the digital message over a band-pass channel, the incoming data are modulated onto a carrier wave that maps the input binary sequence of information into signal waveforms by switching or keying the amplitude, frequency or phase of the carrier waveform in accordance with the incoming bits. We describe some of these digital modulation techniques in Section 1.5. Figures 1.2 and 1.3 illustrate the functional diagram of an analog and a digital communication systems, respectively.

In Figure 1.2, the modulator in the transmitter performs the task of modulation that translates the input baseband signal to an intermediate frequency (IF), as mentioned above. This is followed by up-conversion of the modulated signal from the IF to the desired channel frequency which is generally higher than the IF and is in the radio frequency (RF) or microwave range. Finally, the signal is amplified and transmitted

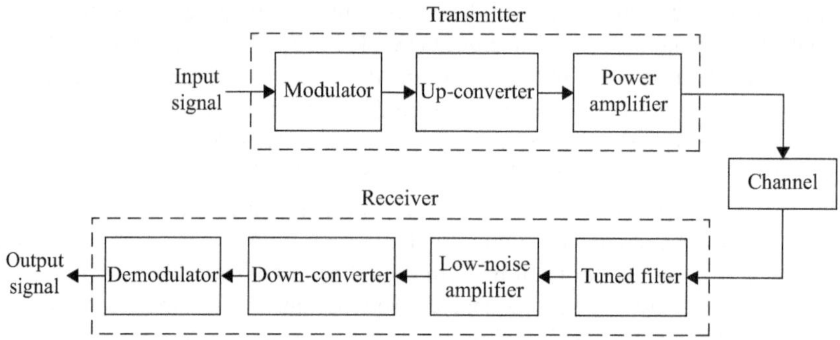

Figure 1.2 Basic elements of an analog communication system

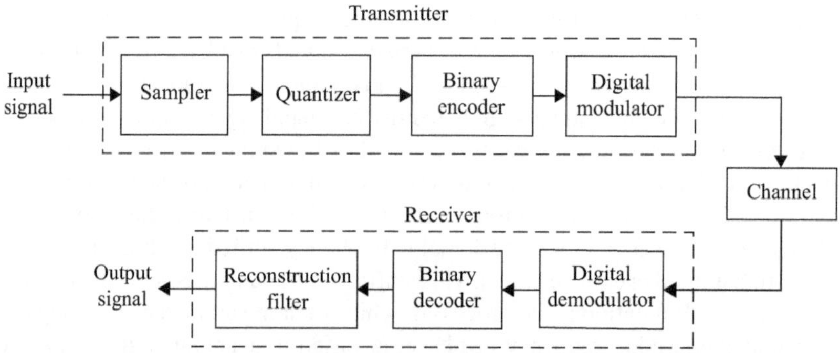

Figure 1.3 Basic elements of a digital communication system

over the channel. The power amplifier provides the power necessary for the signal to reach the remote destination. In a digital communication system, as depicted in Figure 1.3, the input time-continuous signal is sampled and then quantized in amplitude to achieve time-discretization and amplitude-discretization, respectively. The quantized samples are subsequently binary encoded to generate a bit-stream that is transmitted over the channel. While the bits are transmitted in the form of line-coded pulses in case of baseband transmission, they are modulated onto a carrier wave via modulation in case of passband transmission, as described above.

The *channel* is the physical medium over which the transmitter output is sent to the destination. Channels may be classified into two categories – wire and wireless. The wire channels are guided media in the sense that the propagation of the message signal is confined to a specific path formed by the solid physical medium connecting the transmitter and the receiver. Twisted-pair transmission line, coaxial cable, fiber-optic cable, waveguide, etc. are some examples of wire channels. In wireless transmission, on the other hand, the transmitter radiates the message signal in the form of electromagnetic waves into air, water, vacuum or free space. This signal, thus, is available to anyone who has a device (viz. antenna) capable of receiving it from the medium. Accordingly, wireless channels are regarded as unguided transmission media. A detailed description of some of these transmission media is given in

Section 1.7. During the course of transmission through a channel, the transmitted signal gets distorted due to channel noise, interference and/or non-linearity in the frequency response of the channel. The main challenge in designing a communication system lies in careful design of the transmitter and receiver so as to minimize the effect of noise and other channel distortion.

The signal transmitted over the channel is received at the destination end by the *receiver* that converts the received signal to its original form by reversing all modifications made by the transmitter. A band-pass filter at the front-end of the receiver, tuned to the desired channel frequency, extracts out the desired message signal from the channel. A low-noise amplifier following the tuned filter amplifies the received weak signal embedded in noise. Down-conversion of passband signal to baseband signal is accomplished by *demodulation*, which is the reverse of the modulation process used in the transmitter. In analog transmission system, the output of the demodulator is the desired baseband message signal. In a digital system, the demodulator output is the binary encoded digitized message signal. Therefore, decoding followed by signal reconstruction is carried out in subsequent stages to obtain the desired time-continuous message signal. According to *Shannon's sampling theorem*, portions of the signal waveform in between the samples can be recovered exactly by passing the sampled signal through a low-pass filter. However, since the quantization process is irreversible, the input to the reconstruction filter is not the original sampled message signal but its quantized version. Consequently, the reconstructed signal is not exactly same as the original message signal but an approximation to it. The difference between the original signal and the reconstructed signal is a measure of the distortion due to quantization introduced by the digital communication system.

1.4 Continuous-wave modulation techniques

As said above, the basic purpose of modulation is to translate baseband message signal to a higher frequency band, as necessary for passband. This frequency translation is particularly useful for simultaneous transmission of a number of message signals over a common channel using *frequency division multiplexing* (FDM) described in Section 1.6. This section presents an overview of the different modulation techniques used in analog communication systems, viz. amplitude modulation (AM), phase modulation (PM) and frequency modulation (FM). For a detailed study on these topics, the reader may refer Reference 16, in addition to the books [4, 5, 8, 9] mentioned earlier.

1.4.1 Amplitude modulation

AM is defined as the process in which the amplitude of the carrier wave is varied linearly with the baseband message signal.

Consider a baseband message signal (modulating wave) $m(t)$, as shown in Figure 1.4(a), to be transmitted over a channel using amplitude modulation of a sinusoidal carrier wave $c(t)$, as shown in Figure 1.4(b), defined as

$$c(t) = A_c \cos 2\pi f_c t \tag{1.1}$$

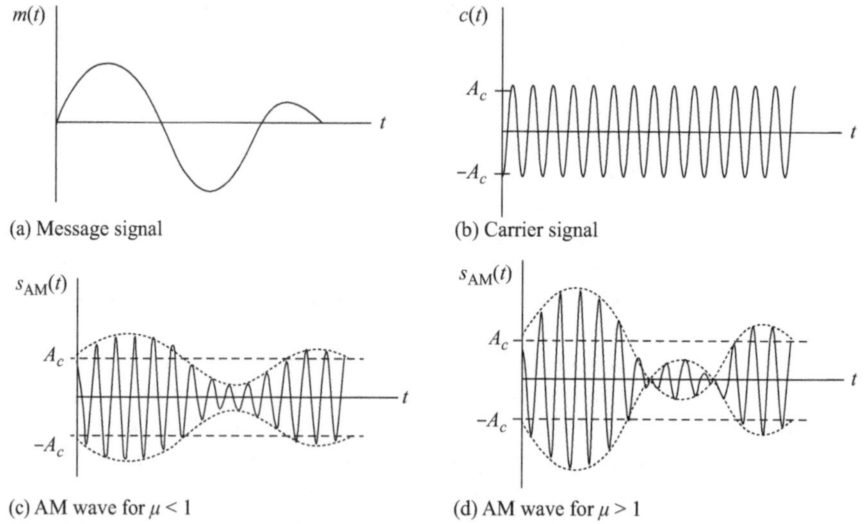

Figure 1.4 Amplitude modulation: message signal in (a) modulates the carrier wave in (b) with small amplitude sensitivity (less than 100% modulation) in (c) and large amplitude sensitivity (more than 100% modulation) in (d)

where A_c is the carrier amplitude and f_c is the carrier frequency. The carrier frequency is generally chosen much greater than the highest frequency component of the message signal, i.e., $f_c \gg W$, where W is the highest frequency component of the message signal $m(t)$. In AM, the amplitude of $c(t)$ is varied in accordance to the message signal $m(t)$ as

$$s_{\mathrm{AM}}(t) = A_c\,[1 + k_a m(t)] \cos 2\pi f_c t \tag{1.2}$$

where k_a is a constant referred to as *amplitude sensitivity* of the modulator. The *modulation index* μ, or the percentage modulation when expressed numerically as percentage, is defined as $\mu = \left|k_a m_p\right|$, where m_p is the peak amplitude (positive or negative) of the message signal $m(t)$. For example, in case of single-tone modulation with $m(t) = A_m \cos 2\pi f_m t$, the modulation index is given as

$$\mu = k_a A_m = \frac{A_{\max} - A_{\min}}{A_{\max} + A_{\min}} \tag{1.3}$$

where $A_{\max} = A_c\,[1 + k_a A_m]$ and $A_{\min} = A_c\,[1 - k_a A_m]$ are, respectively, the maximum and the minimum values of the envelope of the AM wave $s_{\mathrm{AM}}(t)$ given by $A_c\,[1 + k_a m(t)]$. Generally, for a given message signal $m(t)$, the constant k_a is chosen such that $0 < \mu \leq 1$. Consequently, we have $|k_a m(t)| \leq 1$ for all t which ensures that the envelope of the modulated wave is always non-negative and hence has the same shape as the message signal $m(t)$, as depicted in Figure 1.4(c). For large k_a that makes $\mu > 1$ or $|k_a m(t)| > 1$ for some t, the carrier wave is *over-modulated* (i.e., percentage

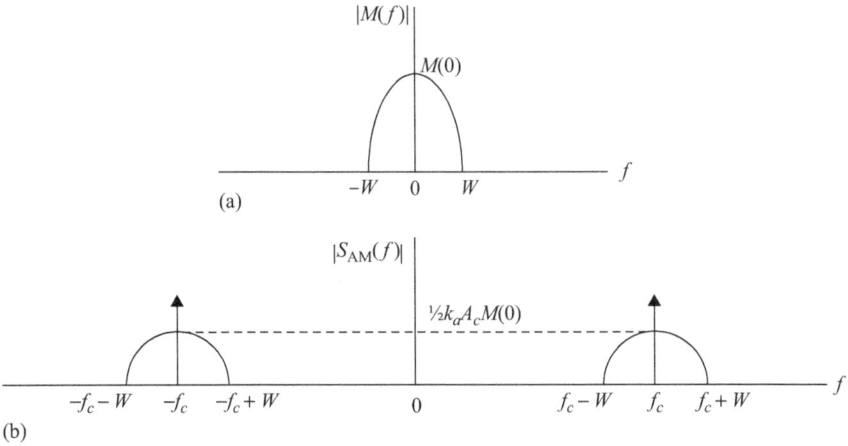

Figure 1.5 Magnitude spectra of the modulating baseband message signal and the corresponding AM wave. (a) Magnitude spectrum of baseband message signal and (b) magnitude spectrum of AM wave

modulation more than 100%) resulting in envelope distortion. This is illustrated in Figure 1.4(d).

From (1.2), the spectrum of the AM wave $s_{AM}(t)$ is obtained as

$$S_{AM}(f) = \frac{A_c}{2}\left[\delta(f+f_c) + \delta(f-f_c)\right] + \frac{k_a A_c}{2}\left[M(f+f_c) + M(f-f_c)\right] \quad (1.4)$$

where $M(f)$ is the spectrum of the baseband message signal $m(t)$. This is illustrated in Figure 1.5 in which Figure 1.5(a) shows the magnitude spectrum $|M(f)|$ of the message signal $m(t)$ and Figure 1.5(b) shows the magnitude spectrum $|S_{AM}(f)|$ of the AM wave $s_{AM}(t)$.

A notable advantage with AM described above is its simple demodulation using envelope detector but at the cost of excess transmission power. From (1.2), we have $s_{AM}(t) = c(t) + k_a m(t)c(t)$. That is, the AM wave comprises of two components – carrier signal modulated by the message signal and the carrier signal alone. Since this carrier signal component is independent of the message signal, its transmission does not convey any message but causes waste of power. To overcome this shortcoming, the carrier component may be suppressed resulting in *double-sideband suppressed carrier* (DSBSC) modulation. The DSBSC-modulated wave, hence, is given as

$$s_{DSB}(t) = A_c m(t)\cos 2\pi f_c t$$

$$\rightleftharpoons S_{DSB}(f) = \frac{A_c}{2}\left[M(f+f_c) + M(f-f_c)\right] \quad (1.5)$$

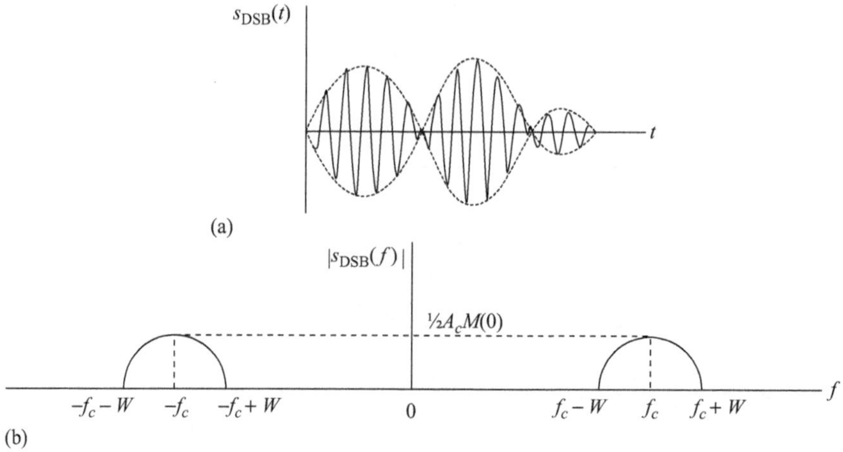

Figure 1.6 Double-sideband suppressed carrier modulation: DSBSC-modulated wave and its magnitude spectrum. (a) DSBSC wave and (b) magnitude spectrum of the DSBSC wave

and depicted in Figure 1.6; Figure 1.6(a) shows the DSBSC-modulated wave and Figure 1.6(b) shows the corresponding magnitude spectrum of the same. Demodulation of the DSBSC-modulated wave can be done by multiplying the modulated signal with another carrier signal $A'_c \cos 2\pi f_c t$ that has *exactly* the same frequency and phase as the carrier signal in the modulator. The output of the multiplier is

$$\begin{aligned} v(t) &= s_{DSB}(t) \times A'_c \cos 2\pi f_c t \\ &= A_c A'_c m(t) \cos^2 2\pi f_c t \\ &= \frac{1}{2} A_c A'_c m(t) \left[1 + \cos 4\pi f_c t\right] \\ &= \frac{1}{2} A_c A'_c m(t) + \frac{1}{2} A_c A'_c m(t) \cos 4\pi f_c t \end{aligned} \tag{1.6}$$

From (1.6), we see that the second term represents a DSBSC-modulated wave with a carrier frequency $2f_c$ that corresponds to a passband signal centered around $2f_c$, whereas the first term is proportional to the baseband message signal. Hence, the desired message signal can be obtained by low-pass filtering the product signal $v(t)$.

As we observe in Figures 1.5 and 1.6, both AM and DSBSC modulations require a transmission bandwidth $2W$ which is equal to twice the message bandwidth W; one half of the transmission bandwidth, $f_c - W \leq f \leq f_c$, is occupied by the lower sideband of the modulated wave and the other half, $f_c \leq f \leq f_c + W$, is occupied by the upper sideband. However, these two sidebands are uniquely related to each other and hence transmitting both the sidebands is a waste of bandwidth as well as transmission power. Accordingly, another variant of the AM may be used in which only one of the two sidebands is transmitted with the carrier component suppressed. Such

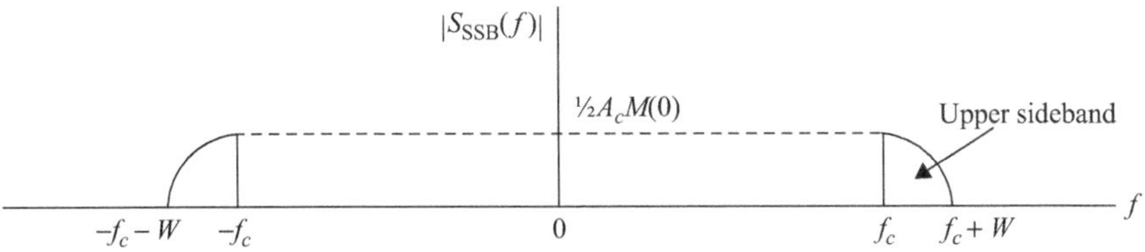

Figure 1.7 Magnitude spectrum of SSB wave

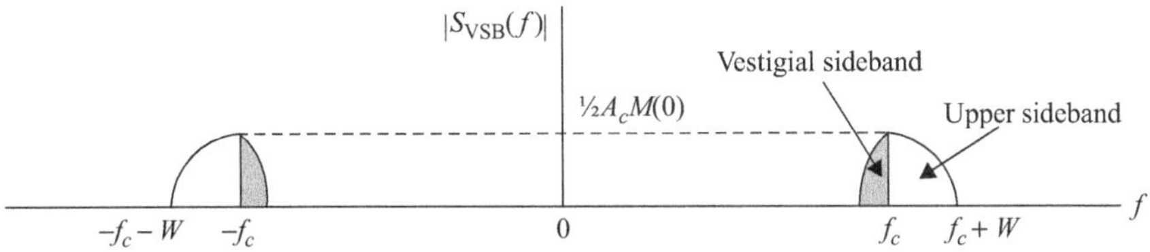

Figure 1.8 Magnitude spectrum of VSB wave

a modulation scheme is referred to as *single sideband* (SSB) modulation. Figure 1.7 shows the spectrum of an SSB wave with the upper sideband transmitted. The major disadvantage with the SSB system is its cost and complexity.

Since it is generally difficult and expensive to separate out the two sidebands precisely, as necessary in SSB modulation, a compromise between the DSBSC and the SSB schemes is adopted, known as the *vestigial sideband* (VSB) modulation. In this scheme, one of the two sidebands is transmitted as in SSB modulation while, unlike SSB modulation, the other sideband is partially suppressed. This is illustrated in Figure 1.8 that shows the spectrum of a VSB wave with the upper sideband transmitted along with a trace of the lower sideband.

1.4.2 Angle modulation

In AM, the message is impressed upon the carrier signal by varying its amplitude. So, any change in amplitude during transmission will result in distortion in the message recovered after demodulation in the receiver. Channel noise is generally additive in nature and hence, an AM signal is highly prone to noise. This may be taken care of by modulating the angle (frequency or phase) of the carrier signal in accordance with the message signal while the amplitude of the carrier signal is maintained constant. Angle modulation schemes provide good immunity to noise but at the cost of increased bandwidth – better noise immunity can be achieved by increasing the transmission bandwidth. AM and its variants do not provide such trade-off between noise immunity and bandwidth.

There are two forms of angle modulation – PM and FM. As the names suggest, the phase angle and the instantaneous frequency of the carrier signal are varied linearly in accordance with the message signal in these two modulation schemes, respectively. Accordingly, the phase-modulated and the frequency-modulated waves are given as

$$s_{\mathrm{PM}}(t) = A_c \cos\left[2\pi f_c t + k_p m(t)\right] \tag{1.7}$$

$$s_{\mathrm{FM}}(t) = A_c \cos\left[2\pi \left(f_c + k_f m(t)\right) t\right] \tag{1.8}$$

where k_p and k_f represent the *phase sensitivity* and the *frequency sensitivity* of the modulators, respectively. From (1.7) and (1.8), the instantaneous frequency of the PM wave and the instantaneous angle of the FM wave can be derived, respectively, as

$$f_{\mathrm{PM}}(t) = \frac{1}{2\pi}\frac{d}{dt}\left(2\pi f_c t + k_p m(t)\right) = f_c + \frac{k_p}{2\pi}\frac{d}{dt}m(t) \tag{1.9}$$

$$\theta_{\mathrm{FM}}(t) = 2\pi \int_{-\infty}^{t} \left(f_c + k_f m(\tau)\right) d\tau = 2\pi f_c t + 2\pi k_f \int_{-\infty}^{t} m(\tau)\, d\tau \tag{1.10}$$

Using these, we may now rewrite the expressions for PM and FM waves as

$$s_{\mathrm{PM}}(t) = A_c \cos\left[2\pi \left(f_c + \frac{k_p}{2\pi}\frac{d}{dt}m(t)\right) t\right] \tag{1.11}$$

$$s_{\mathrm{FM}}(t) = A_c \cos\left[2\pi f_c t + 2\pi k_f \int_{-\infty}^{t} m(\tau) d\tau\right] \tag{1.12}$$

Thus, the PM wave $s_{\mathrm{PM}}(t)$ is expressed in a form similar to the FM wave in (1.8) and the FM wave $s_{\mathrm{FM}}(t)$ in a form similar to the PM wave in (1.7). This shows the relationship between the two angle modulation schemes. Accordingly, it is possible to generate frequency-modulated wave using a phase modulator and vice versa, as illustrated in Figures 1.9(a) and 1.9(b), respectively.

Consider the case of single-tone sinusoidal modulation. Figure 1.10 shows how the phase angle and the instantaneous frequency of the carrier signal varies due to PM and FM, respectively, when modulated using a single-tone sinusoidal wave. In this figure, Figure 1.10(c) and Figure 1.10(d) are the phase-modulated and the frequency-modulated waves, respectively, that are generated when the carrier signal in Figure 1.10(b) is modulated by the sinusoidal message signal in Figure 1.10(a).

For the modulating message signal $m(t) = A_m \cos 2\pi f_m t$, the maximum departure of the instantaneous frequency in the FM wave from its carrier frequency f_c is $\Delta f = k_f A_m$. This quantity Δf is called the *frequency deviation*. The ratio of the frequency deviation to the modulating frequency is called the *modulation index*, denoted as

$$\beta = \frac{\Delta f}{f_m} = \frac{k_f A_m}{f_m} \tag{1.13}$$

Following (1.12), the FM wave in this case is obtained as

$$s_{\mathrm{FM}}(t) = A_c \cos\left[2\pi f_c t + \beta \sin 2\pi f_m t\right] \tag{1.14}$$

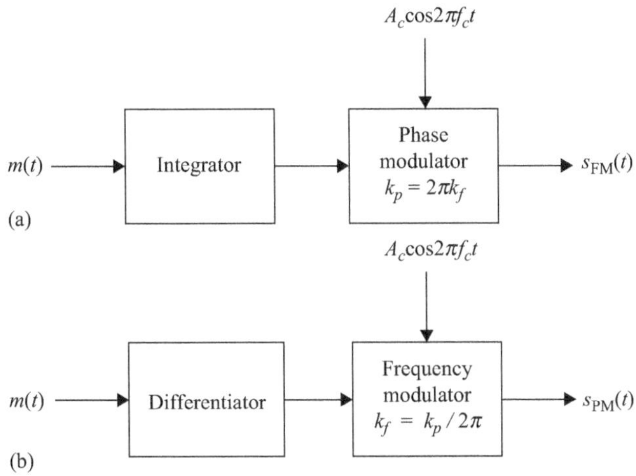

Figure 1.9 Scheme for generating frequency-modulated wave using a phase modulator and vice-versa. (a) FM generation by phase modulator and (b) PM generation by frequency modulator

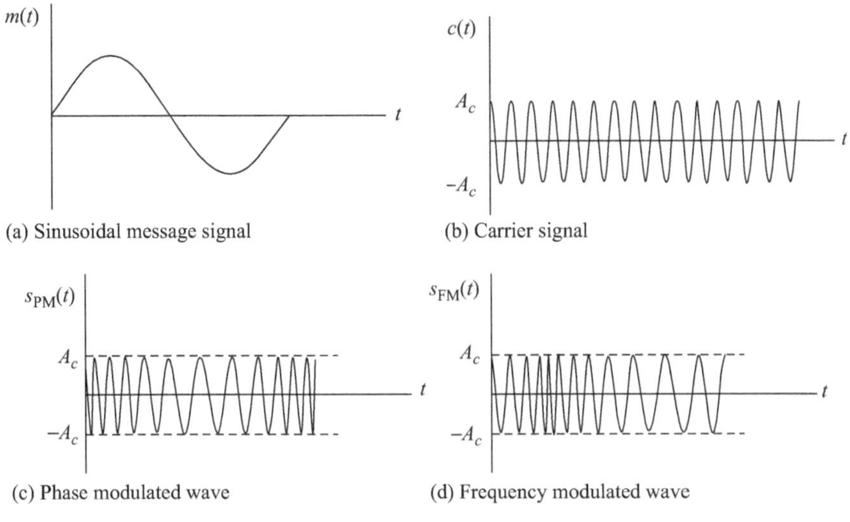

Figure 1.10 Angle modulation: message signal in (a) modulates the carrier wave in (b) by phase modulation in (c) and by frequency modulation in (d)

The modulation index β, therefore, equals to the maximum departure of the instantaneous angle in the FM wave from the angle $2\pi f_c t$ of the unmodulated carrier signal. Hence, the modulation index may also be termed as *phase deviation* of the FM wave.

FM may be either narrow-band when β is small or wide-band when β is large. Expanding the relation given in (1.14) and assuming β to be very small, the narrow-band FM wave may be approximated as

$$s_{FM}(t) = A_c \cos 2\pi f_c t - \beta A_c \sin(2\pi f_c t) \sin(2\pi f_m t) \tag{1.15}$$

which is somewhat like a double-sideband modulated wave with carrier signal. In case of wide-band FM, the relation given in (1.14) may be rewritten as

$$s_{\mathrm{FM}}(t) = \Re\,[\tilde{s}(t)\exp(j2\pi f_c t)] \tag{1.16}$$

where $\Re(x)$ denotes the real part of a complex quantity x and $\tilde{s}(t)$ is the complex envelope of the FM wave $s_{\mathrm{FM}}(t)$, defined as

$$\tilde{s}(t) = A_c \exp(j\beta \sin(2\pi f_m t)) = A_c \sum_{n=-\infty}^{\infty} J_n(\beta)\exp(j2\pi n f_m t) \tag{1.17}$$

where $J_n(\beta)$ is the nth order *Bessel function* of the first kind with argument β. Substituting $\tilde{s}(t)$ in (1.16), we get

$$\begin{aligned} s_{\mathrm{FM}}(t) &= A_c \sum_{n=-\infty}^{\infty} J_n(\beta)\cos[2\pi(f_c + nf_m)t] \\ &\rightleftharpoons S_{\mathrm{FM}}(f) = \frac{A_c}{2}\sum_{n=-\infty}^{\infty} J_n(\beta)\,[\delta(f + f_c + nf_m) + \delta(f - f_c - nf_m)] \end{aligned} \tag{1.18}$$

The above expression for single-tone wide-band FM wave may be generalized to multi-tone wide-band FM as

$$s_{\mathrm{FM}}(t) = A_c \sum_{n_1=-\infty}^{\infty} \cdots \sum_{n_K=-\infty}^{\infty} J_{n_1}(\beta_1)\cdots J_{n_K}(\beta_K)\cos[2\pi(f_c + n_1 f_1 + \cdots + n_K f_K)t] \tag{1.19}$$

where K is the number of tone frequencies $f_1, \ldots, f_K$ present in the modulating message signal $m(t)$, with corresponding modulation indices $\beta_1, \ldots, \beta_K$, respectively. Thus, the resultant FM wave contains a carrier component and an infinite set of side-frequencies located symmetrically on either side of the carrier, i.e., frequency components $(f_c + n_1 f_1 + \cdots + n_K f_K)$, where $n_1, \ldots, n_K = 0, \pm 1, \pm 2, \ldots$. This implies that, theoretically, the bandwidth required to transmit FM wave is infinite. However, in practice, significant portion of the FM wave is limited to a finite number of side-frequencies. In case of single-tone sinusoidal modulation, the transmission bandwidth is approximated using *Carson's rule*, given as

$$B \simeq 2(f_m + \Delta f) = 2f_m\,(1 + \beta) = 2\Delta f\left(1 + \frac{1}{\beta}\right) \tag{1.20}$$

In case of an arbitrary modulating wave with highest frequency component W, the transmission bandwidth is approximated as

$$B \simeq 2\Delta f\left(1 + \frac{1}{D}\right) \tag{1.21}$$

where the frequency deviation Δf corresponds to the maximum amplitude m_p of the modulating signal $m(t)$, i.e., $\Delta f = k_f m_p$, and D denotes the *deviation ratio* defined as the ratio of the frequency deviation Δf to the highest modulation frequency W, i.e., $D = \Delta f / W$.

1.5 Digital modulation techniques

Digital modulation or *digital signaling* is the process of mapping digital data, which is usually in the form of a sequence of **0**s and **1**s, to signals suitable for transmission over a transmission channel. Real channels are generally characterized as bandlimited waveform channels. If binary data (bits) are transmitted directly over these channels in the form of electrical pulses, the frequency components constituting the pulses are differently attenuated and delayed resulting in *dispersion* of the pulses over an interval longer than the pulse duration. This in turn results in overlapping of successive pulses into one another, a phenomenon known as *intersymbol interference* (ISI). Methods to counter the ISI problem include appropriate pulse shaping, correlative coding or use of an equalizer in the receiver.

Alternatively, digital data may be transmitted in the form of band-pass signals with frequency components restricted within the frequency band of the channel. For example, bits **0** and **1** may be transmitted as waveforms $s_0(t)$ and $s_1(t)$, respectively. This mapping of bits to waveforms is accomplished by the modulation process in which a sinusoidal carrier wave is modulated in accordance with the modulating wave consisting of the binary data stream. The most common digital modulation techniques involve switching or *keying* the amplitude, phase or frequency of the carrier in accordance with the incoming bits. Accordingly, there are three basic digital modulation schemes – amplitude shift keying (ASK), phase shift keying (PSK) and frequency shift keying (FSK), as described in the following sections. These basic schemes may also be combined and/or multiple shifts be used, instead of binary shifts, to transmit more number of bits per symbol. Below we describe two such schemes, viz. quadriphase-shift keying (QPSK) that uses four phase shifts and quadrature amplitude modulation (QAM) that combines AM with PM. More information on several other digital modulation schemes are available in References 4–9.

1.5.1 Amplitude shift keying

In ASK method of modulation, the pair of signals $s_0(t)$ and $s_1(t)$ used to represent bits **0** and **1**, respectively, are defined as

$$\begin{aligned} s_0(t) &= 0, & 0 \le t \le T_b \\ s_1(t) &= A_c \cos 2\pi f_c t, & 0 \le t \le T_b \end{aligned} \tag{1.22}$$

where T_b is the bit duration and f_c is the frequency of the carrier signal. In order to ensure an integral number of cycles of the carrier wave within a bit period T_b, the carrier frequency f_c is generally chosen as integral multiple of the bit frequency, i.e.,

$f_c = N/T_b$, for some positive integer N. If E_b denotes the transmitted signal energy per bit, then

$$\begin{aligned} E_b &= \frac{1}{2}A_c^2 T_b \\ \text{or} \quad A_c &= \sqrt{\frac{2E_b}{T_b}} \end{aligned} \tag{1.23}$$

Equation (1.22) may also be written in the form of a DSBSC-modulated wave given in (1.5). That is,

$$s_{\text{ASK}}(t) = A_c m(t) \cos 2\pi f_c t \tag{1.24}$$

where

$$m(t) = \begin{cases} 0, & 0 \le t \le T_b, \quad \text{for input bit} = \mathbf{0} \\ 1, & 0 \le t \le T_b, \quad \text{for input bit} = \mathbf{1} \end{cases} \tag{1.25}$$

This implies that ASK is essentially DSBSC modulation in which the amplitude of the carrier signal $c(t) = A_c \cos 2\pi f_c t$ is modulated by an on–off signal $m(t)$ representing the binary sequence in unipolar format. Hence, this modulation scheme is also referred to as *on–off keying* (OOK). Figure 1.11(c) shows the result of ASK modulation of the sinusoidal carrier signal in Figure 1.11(b) by the arbitrary sequence of bits in Figure 1.11(a).

The signal constellation diagram for the ASK signaling system is shown in Figure 1.12. *Signal constellation* diagram is the graphical representation of all possible signals in an N-dimensional signal space spanned by N number of basis functions. In this particular case of ASK, there is only one basis function of unit energy as follows:

$$\phi_1(t) = \sqrt{\frac{2}{T_b}} \cos 2\pi f_c t, \qquad 0 \le t \le T_b \tag{1.26}$$

Consequently, the two possible signals in ASK are given in terms of $\phi_1(t)$ as $s_0(t) = 0$ and $s_1(t) = \sqrt{E_b}\phi_1(t)$, as shown graphically in Figure 1.12.

1.5.2 Binary phase-shift keying

In PSK, the phase of the carrier signal is varied in accordance with the input digital data. In the case of binary data, the phase of the carrier is shifted by 0° and 180° corresponding to bits **1** and **0**, respectively. Accordingly, the pair of signals $s_0(t)$ and $s_1(t)$ are defined as

$$\begin{aligned} s_0(t) &= A_c \cos(2\pi f_c t + \pi) = -\sqrt{\frac{2E_b}{T_b}} \cos 2\pi f_c t, \quad 0 \le t \le T_b \\ s_1(t) &= A_c \cos 2\pi f_c t = \sqrt{\frac{2E_b}{T_b}} \cos 2\pi f_c t, \qquad 0 \le t \le T_b \end{aligned} \tag{1.27}$$

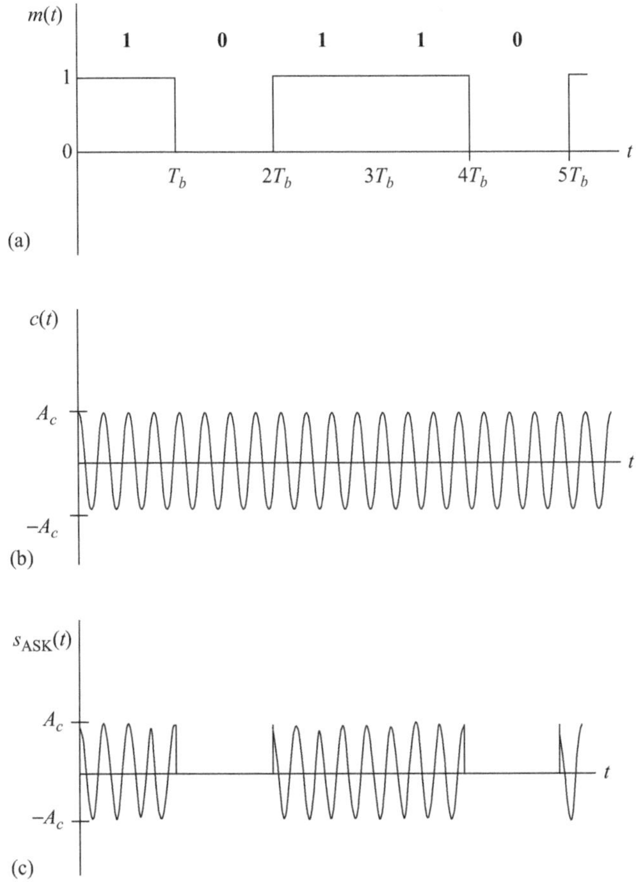

Figure 1.11 An example of ASK modulation. (a) Digital message and corresponding unipolar signaling, (b) carrier signal and (c) ASK modulated wave

It may be noted that, like ASK modulation scheme, binary phase-shift keying (BPSK) is also equivalent to DSBSC modulation but with NRZ polar binary signal as the modulating wave $m(t)$, i.e.,

$$m(t) = \begin{cases} -1, & 0 \leq t \leq T_b, \quad \text{for input bit} = \mathbf{0} \\ +1, & 0 \leq t \leq T_b, \quad \text{for input bit} = \mathbf{1} \end{cases} \tag{1.28}$$

Example of BPSK modulation, for the digital message considered in the ASK example above, and the corresponding signal-space diagram are depicted in Figures 1.13 and 1.14, respectively. Figure 1.13(b) shows the BPSK modulated wave obtained when the carrier signal in Figure 1.11(b) is modulated by the digital message signal in NRZ format shown in Figure 1.13(a). Here also there is only one basis function of unit

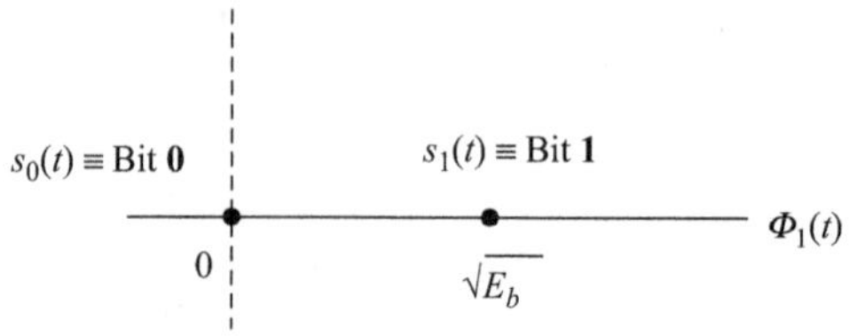

Figure 1.12 Signal constellation diagram for ASK modulation

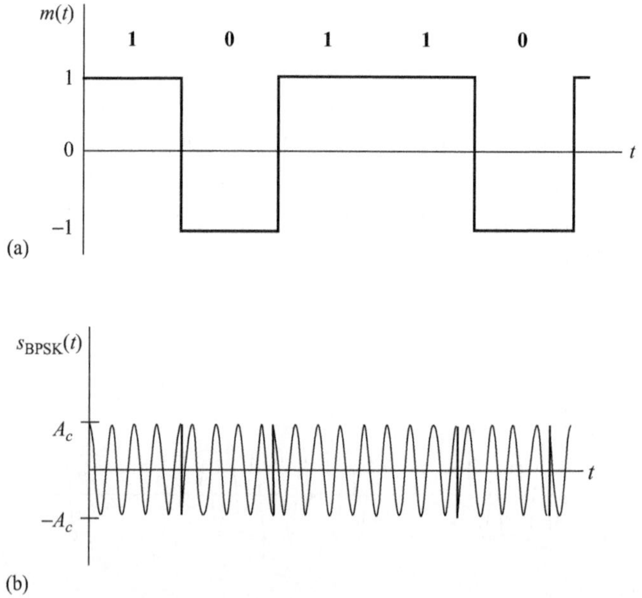

Figure 1.13 An example of BPSK modulation. (a) Digital message and corresponding NRZ polar signaling and (b) BPSK modulated wave

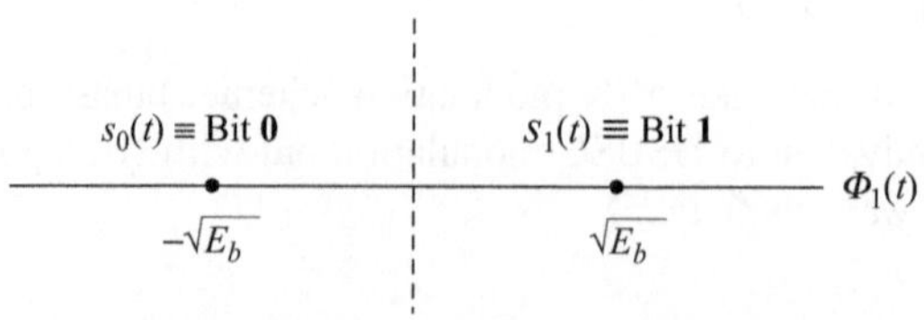

Figure 1.14 Signal constellation diagram for BPSK modulation

energy which is same as that given in (1.26) for the case of ASK above, and the two possible signals in BPSK are given as $s_0(t) = -\sqrt{E_b}\phi_1(t)$ and $s_1(t) = \sqrt{E_b}\phi_1(t)$.

1.5.3 Binary frequency-shift keying

FSK is that form of digital modulation in which digital data are transmitted by varying the frequency of the carrier waveform. For transmission of binary data, the pair of signals $s_0(t)$ and $s_1(t)$ corresponding to transmission of bits **0** and **1**, respectively, are defined as

$$\begin{aligned} s_0(t) &= A_c \cos 2\pi f_0 t = \sqrt{\frac{2E_b}{T_b}} \cos 2\pi f_0 t, \quad 0 \le t \le T_b \\ s_1(t) &= A_c \cos 2\pi f_1 t = \sqrt{\frac{2E_b}{T_b}} \cos 2\pi f_1 t, \quad 0 \le t \le T_b \end{aligned} \tag{1.29}$$

As in the cases of ASK and BPSK above, binary frequency-shift keying (BFSK) may be conceived as FM modulation in which the frequency of the carrier signal $c(t) = A_c \cos 2\pi f_c t$ is modulated by a signal $m(t)$ representing the binary sequence in NRZ polar format, as defined in (1.28). Accordingly, following (1.8), we may write the BFSK wave as

$$\begin{aligned} s_{\text{BFSK}}(t) &= A_c \cos\left[2\pi\left(f_c + m(t)\frac{\Delta f}{2}\right)t\right] \\ &= \begin{cases} A_c \cos\left[2\pi\left(f_c - \frac{\Delta f}{2}\right)t\right], & 0 \le t \le T_b, \quad \text{for input bit} = \mathbf{0} \\ A_c \cos\left[2\pi\left(f_c + \frac{\Delta f}{2}\right)t\right], & 0 \le t \le T_b, \quad \text{for input bit} = \mathbf{1} \end{cases} \end{aligned} \tag{1.30}$$

Therefore, the two transmitting frequencies in BFSK, given in (1.29), are $f_0 = f_c - \Delta f/2$ and $f_1 = f_c + \Delta f/2$, with frequency separation $\Delta f = f_1 - f_0$. An example of BFSK modulation is illustrated in Figure 1.15. Figure 1.15(b) shows the BFSK modulated wave obtained when the carrier signal in Figure 1.11(b) is modulated by the digital message signal in NRZ format shown in Figure 1.15(a). The BFSK modulation described here may be represented graphically in a two-dimensional signal space spanned by the pair of orthonormal basis functions

$$\begin{aligned} \phi_1(t) &= \sqrt{\frac{2}{T_b}} \cos 2\pi f_0 t, \qquad 0 \le t \le T_b \\ \text{and} \quad \phi_2(t) &= \sqrt{\frac{2}{T_b}} \cos 2\pi f_1 t, \qquad 0 \le t \le T_b \end{aligned} \tag{1.31}$$

The two transmitted signals in BFSK are, hence, given as $s_0(t) = \sqrt{E_b}\phi_1(t)$ and $s_1(t) = \sqrt{E_b}\phi_2(t)$. These are shown in Figure 1.16 by two message points defined, respectively, by the signal vectors

$$\begin{aligned} \mathbf{s}_0 &= \begin{bmatrix} \sqrt{E_b} \\ 0 \end{bmatrix} \\ \text{and} \quad \mathbf{s}_1 &= \begin{bmatrix} 0 \\ \sqrt{E_b} \end{bmatrix} \end{aligned} \tag{1.32}$$

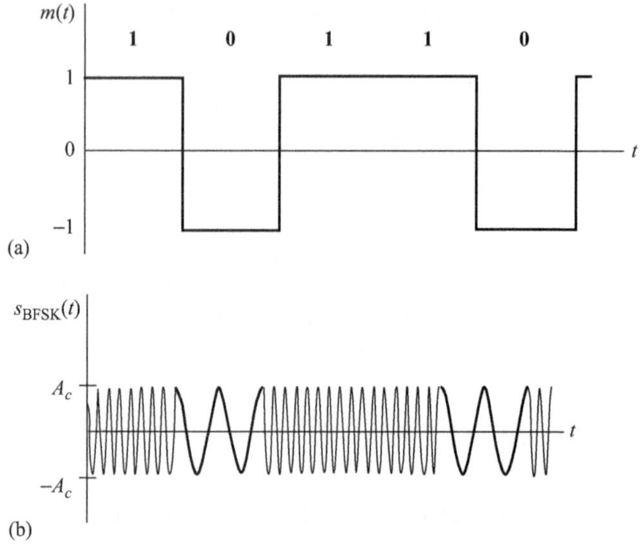

Figure 1.15 An example of BFSK modulation. (a) Digital message and corresponding NRZ polar signaling and (b) BFSK modulated wave

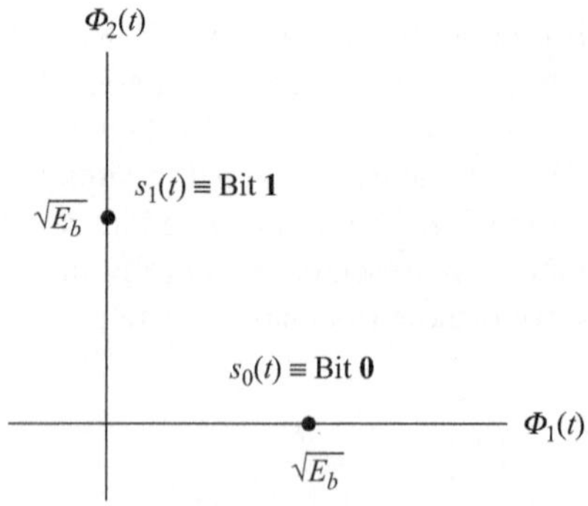

Figure 1.16 Signal constellation diagram for BFSK modulation

It may be shown that the two basis functions defined in (1.31) are orthogonal if and only if the frequency separation $\Delta f = k/2T_b$ for some positive integer k.

1.5.4 Continuous-phase frequency-shift keying

The BFSK scheme described above may be implemented by having transmitted signal waveforms generated from two oscillators tuned to frequencies f_0 and f_1 and selecting one of these two oscillators according to the bit to be transmitted. However, such abrupt switching from one oscillator to another results in large spectral side lobes thereby requiring large transmission bandwidth. This problem of spectral side lobes may be avoided if the frequency of the carrier wave is changed (modulated) continuously in accordance to the information bits. The resulting modulated signal is phase continuous and hence, is called continuous-phase frequency-shift keying (CPFSK). The CPFSK wave, hence, may be represented as

$$s_{\text{CPFSK}}(t) = \sqrt{\frac{2E_b}{T_b}} \cos[2\pi f_c t + \theta(t)] \tag{1.33}$$

The phase $\theta(t)$ decreases or increases linearly with time during every bit period corresponding to transmission of bits **0** and **1**, respectively; the phase $\theta(t)$ of the CPFSK wave during transmission of the nth bit in the binary sequence is given as

$$\theta(t) = \begin{cases} \theta(t_{n-1}) - 2\pi \dfrac{\Delta f}{2}(t - t_{n-1}), & t_{n-1} \le t \le t_n, \quad \text{if } n\text{th bit} = \mathbf{0} \\ \theta(t_{n-1}) + 2\pi \dfrac{\Delta f}{2}(t - t_{n-1}), & t_{n-1} \le t \le t_n, \quad \text{if } n\text{th bit} = \mathbf{1} \end{cases} \tag{1.34}$$

where $t_{n-1} = (n-1)T_b$ and $t_n = nT_b$ denote the time instants at which the nth bit period starts and ends, respectively. The phase $\theta(t_{n-1})$ at the start of the nth bit period depends on the past history of the modulation process. Accordingly, the pair of CPFSK signals $s_0(t)$ and $s_1(t)$ corresponding to the transmission of bits **0** and **1**, respectively, are defined as

$$\begin{aligned} s_0(t) &= \sqrt{\frac{2E_b}{T_b}} \cos\left[2\pi\left(f_c - \frac{\Delta f}{2}\right)t + \theta(0)\right], \quad 0 \le t \le T_b \\ s_1(t) &= \sqrt{\frac{2E_b}{T_b}} \cos\left[2\pi\left(f_c + \frac{\Delta f}{2}\right)t + \theta(0)\right], \quad 0 \le t \le T_b \end{aligned} \tag{1.35}$$

where $\theta(0)$ is the phase of the CPFSK wave at the end of the previous bit period that depends on the sequence of bits transmitted so far. Thus, while in case of normal BFSK the phase of the carrier wave in every bit period starts from zero and then decreases/increases by $\pi \Delta f T_b$ radians, that in case of CPFSK decreases/increases by the same amount but starting from $\theta(0)$ radians. This ensures that the phase $\theta(t)$ in CPFSK wave is a continuous function of time. Consequently, the CPFSK modulated wave $s_{\text{CPFSK}}(t)$ is also continuous at all time instants.

Defining parameter $h = T_b\Delta f$ as the *deviation ratio*, the change in phase over a bit period is calculated as $\pm\pi h$ radians; the phase of the CPFSK signal decreases (increases) by πh radians over one bit period when bit **0** (bit **1**) is transmitted. Since the minimum frequency separation Δf that guarantees orthogonality of the two FSK signals is $\Delta f = 1/2T_b$, the minimum value of h to be taken is 1/2. The CPFSK scheme with $h = 1/2$ is known as *minimum shift keying* (MSK) [17, 18]. MSK, and more generally CPFSK, is a special case of continuous-phase modulation (CPM) scheme that offers both power and bandwidth efficiency while maintaining constant amplitude and good error performance [19].

The modulation schemes described above are binary modulation in the sense that the modulator maps one of the two possible symbols (**0** or **1**) into a waveform

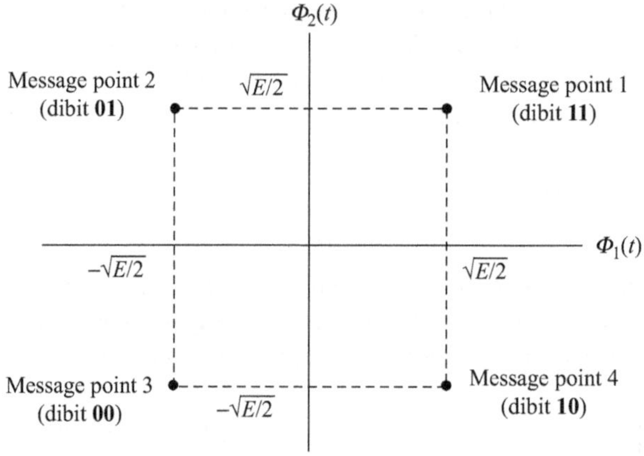

Figure 1.17 Signal constellation diagram for QPSK modulation

($s_0(t)$ or $s_1(t)$). Thus, in these schemes, only one bit is transmitted at a time. In order to use the channel bandwidth more efficiently, however, it is possible to design M-ary modulators that can transmit B bits at a time by mapping $M = 2^B$ possible symbols (B-bit sequences) into M distinct signals $s_i(t)$, $i = 0, 1, \ldots, M - 1$. This is called *M-ary modulation*. Two popularly used M-ary modulation schemes are as follows:

- **Quadriphase-shift keying (QPSK):** This is one of the frequently used multi-phase signaling scheme in which $M = 4$. The scheme uses phase shifts of $\pi/4$, $3\pi/4$, $5\pi/4$ and $7\pi/4$, as shown in the signal-space diagram of Figure 1.17, to transmit two bits (*dibits*) at a time thereby doubling the transmission rate without increasing the bandwidth.
- **Quadrature amplitude modulation (QAM):** This scheme, also known as *amplitude phase keying*, is a combination of AM with phase-shift keying. The carrier signal experiences phase shift of $\pi/2$ to produce a pair of quadrature carriers $\phi_I(t)$ (in-phase carrier) and $\phi_Q(t)$ (quadrature-phase carrier), defined as

$$\phi_I(t) = \sqrt{\frac{2}{T_b}} \cos 2\pi f_c t, \qquad 0 \leq t \leq T_b$$

$$\text{and} \quad \phi_Q(t) = \sqrt{\frac{2}{T_b}} \sin 2\pi f_c t, \qquad 0 \leq t \leq T_b \tag{1.36}$$

A QAM signal is a combination of these quadrature carriers weighted by amplitudes $A_I\sqrt{E_b}$ and $A_Q\sqrt{E_b}$. That is, a QAM signal may be represented as

$$s_i(t) = A_I\sqrt{E_b}\phi_I(t) + A_Q\sqrt{E_b}\phi_Q(t), \qquad i = 0, 1, \ldots, M - 1 \tag{1.37}$$

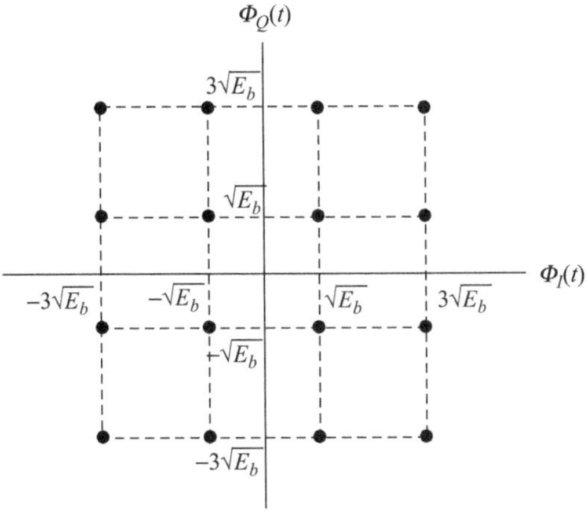

Figure 1.18 Signal constellation diagram for QAM-16 modulation

where $I = 0, 1, \ldots, M_1 - 1$ and $Q = 0, 1, \ldots, M_2 - 1$. That means, A_I takes M_1 discrete values and A_Q takes M_2 discrete values, such that $M = M_1 \times M_2$. Thus, QAM signals are two-dimensional signals represented by the signal vectors

$$\mathbf{s}_i = \begin{bmatrix} A_I\sqrt{E_b} \\ A_Q\sqrt{E_b} \end{bmatrix}, \quad i = 0, 1, \ldots, M - 1 \tag{1.38}$$

Figure 1.18 shows the constellation diagram for a 16-ary QAM signaling scheme (QAM-16). Here, each of the amplitudes A_I and A_Q takes four discrete values, resulting in a total 16 different signaling waveforms, as seen in the constellation diagram.

1.6 Multiplexing

The modulation schemes described above are good for transmission of one message signal over a single channel. However, economy and maximum utilization of the communication networks demand sharing of channels among multiple signals via multiplexing. As a consequence, several multiplexing schemes have been developed over the years, as described below.

1.6.1 Frequency division multiplexing

The inherent bandwidth of a transmission medium is generally much greater than that needed for a single signal. FDM can be employed to take advantage of this in sending multiple signals over a single medium [20]. In this scheme, the available bandwidth of the transmission medium is divided into a number of narrower frequency bands

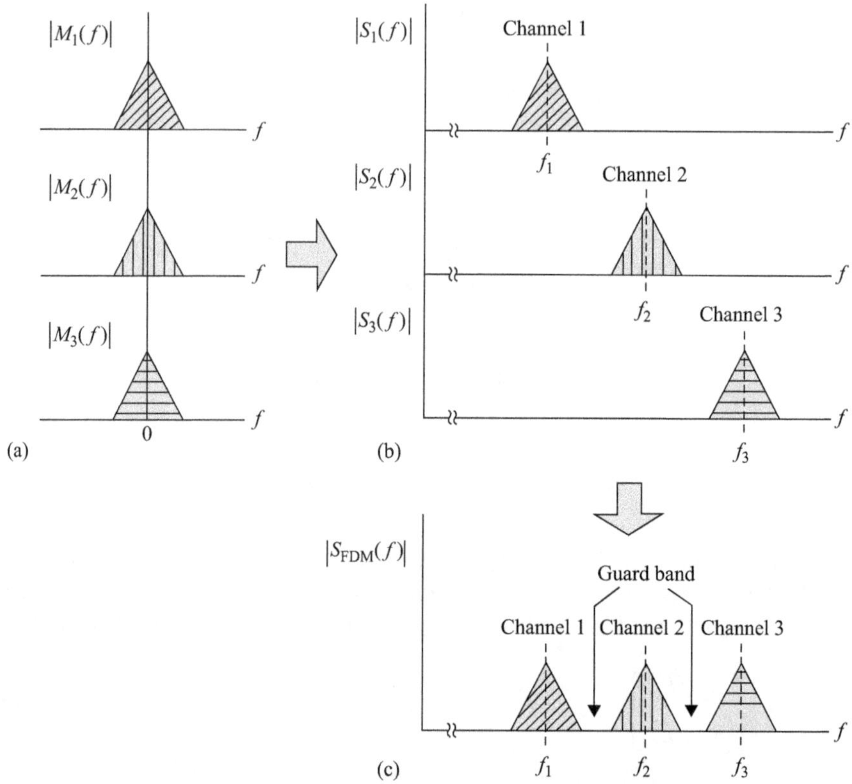

Figure 1.19 Frequency division multiplexing (FDM). (a) Magnitude spectra of input baseband message signals $m_1(t)$, $m_2(t)$ and $m_3(t)$. (b) Magnitude spectra of frequency-translated message signals $s_1(t)$, $s_2(t)$ and $s_3(t)$. (c) Magnitude spectrum of the composite multiplexed signal $s_{FDM}(t)$

(subbands), as depicted in Figure 1.19. Each of these subbands is allocated to each user or transmitting station. A carrier signal of appropriate frequency is modulated by the signal originated at a particular station so that the modulated wave is inserted within the subband allocated to that station. As an example, the magnitude spectra of the message signals originated from three transmitting stations, as given in Figure 1.19(a), are frequency translated and then combined to form the composite multiplexed signal, as shown in Figures 1.19(b) and 1.19(c), respectively. A sufficient amount of separation (guard band) between the subbands is allowed so as to avoid any interference between the adjacent channels. This scheme has been used in radio and TV broadcasting, in multiplexing calls over a telephone line and continues to be used in present day cellular networks and satellite communication. FDM is also utilized in fiber-optic transmission systems, where it is customarily referred to as wavelength division multiplexing (WDM). When multiple users share a physical communication channel using FDM, it is called *frequency division multiple access* (FDMA).

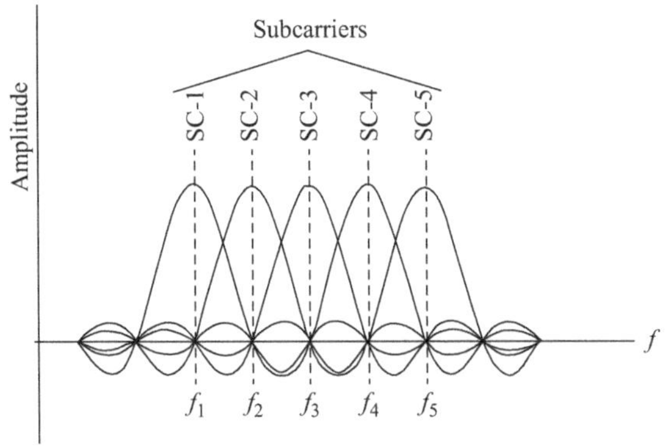

Figure 1.20 Spectrum of the subcarriers in orthogonal frequency division multiplexing (OFDM)

1.6.2 Orthogonal frequency division multiplexing

In digital data communication, it is possible to divide the transmission band more efficiently by inserting more number of closely packed subbands without any guard band in between the adjacent subbands; a scheme known as *orthogonal frequency division multiplexing* (OFDM) [21–24], and the scheme by which multiple users access a channel using OFDM is called *orthogonal frequency division multiple access* (OFDMA).

Suppose, $c_k(t) = \cos 2\pi f_k t$ be the sinusoidal carrier signal associated with the kth subband, where f_k is the center frequency of the subband. These subband carriers are referred to as subcarriers. If the frequency separation between adjacent subcarriers is taken as $\Delta f = 1/T_s$, where T_s is the symbol (data) duration, then the frequency separation between any pair of subcarriers $c_k(t)$ and $c_j(t)$ is an integral multiple of the symbol rate $1/T_s$, i.e., $f_k - f_j = n/T_s$ for some positive integer n. Consequently, this pair of subcarriers is mutually orthogonal over the symbol duration T_s, independent of the values of the phases ϕ_k and ϕ_j associated with them. That is,

$$\int_0^{T_s} \cos(2\pi f_k t + \phi_k) \cos\left(2\pi f_j t + \phi_j\right) dt = 0 \tag{1.39}$$

Thus, by virtue of orthogonality among the subcarriers, data can be detected on each of these subcarriers without any interference from other subcarriers.

A frequency domain understanding to the above is as follows. The frequency response of the kth subband carrying a data symbol of duration T_s is given by a *sinc* function centered at f_k with zero crossings at integral (both positive and negative) multiples of $1/T_s$. This is illustrated in Figure 1.20. Thus, signals carried by a subcarrier extend into other subbands. Nevertheless, as said above, the frequency separation between adjacent subcarriers is taken as $\Delta f = 1/T_s$ due to which the center frequency of every subband is located at the zero crossings of the frequency responses of all other subbands. This ensures that if the subbands are sampled precisely at their center frequencies, then data on each of these subcarriers can be detected without any interference from other subcarriers.

1.6.3 Time division multiplexing

An alternative to the FDM is the TDM in which the entire channel bandwidth is available to all users but periodically for a restricted interval of time [20]. In this scheme, samples and/or data symbols from several sources are time interleaved and then transmitted over a single communication channel. In line with FDMA and OFDMA schemes, TDM is also utilized for accessing a channel by multiple users, referred to as *time division multiple access* (TDMA).

1.6.4 Code division multiple access

Starting from the last decade in the twentieth century, code division multiple access (CDMA) has been widely deployed in cellular networks and continues to be the standard for mobile telephony in many countries. Not only cellular networks, it is also used in satellite and cable networks [23, 25]. This is a form of *spread spectrum* communication in which every user/station can use the entire channel bandwidth all the time. Here, each bit time T_b is divided into N smaller intervals called *chips*. Every user/station is assigned a unique N-bit code called the *chip sequence*. Chip sequences for all users are designed such that they are pairwise orthogonal. That is, the normalized inner product of two distinct sequences is zero while the normalized inner product of a sequence with itself is unity. For example, $(+1, +1, +1, +1)$, $(+1, -1, +1, -1)$, $(+1, +1, -1, -1)$ and $(+1, -1, -1, +1)$, where -1 corresponds to bit **0** and +1 corresponds to bit **1** (i.e., NRZ bipolar signaling format), form a set of four such sequences of length $N = 4$.

Digital data are transmitted by each user in the form of this chip sequence – the station transmits a data bit **1** by sending the N-bit chip sequence (code) while bit **0** is represented by the negative of the sequence. Thus, a narrowband data signal spreads out over an N times wider frequency band. When multiple stations are transmitting data, the bit-streams from all the transmitters are simply added and transmitted over the channel. This method of combining signals from multiple sources is called *code division multiplexing*. A receiver intending to listen to a particular transmitting station computes the normalized inner product between the received sequence and the chip sequence of the desired station.

As an example, suppose that data are transmitted from three stations A, B and C whose assigned chip sequences are $\mathbf{C}_A$, $\mathbf{C}_B$ and $\mathbf{C}_C$, respectively. The bit-stream transmitted from Station A over one bit period is

$$\mathbf{S}_A = \begin{cases} \overline{\mathbf{C}}_A, & \text{if data bit to be transmitted} = \mathbf{0} \\ \mathbf{C}_A, & \text{if data bit to be transmitted} = \mathbf{1} \end{cases} \tag{1.40}$$

In a similar manner, Station B and Station C transmit bit-streams $\mathbf{S}_B$ and $\mathbf{S}_C$, respectively, over the same bit interval. Therefore, any receiver in the network receives the combined sequence $(\mathbf{S}_A + \mathbf{S}_B + \mathbf{S}_C)$. In order to recover the data transmitted from a particular transmitting station, say Station B, a receiver computes the normalized inner product between the received sequence and the chip sequence $\mathbf{C}_B$ of the desired station. Due to pairwise orthogonality of the chip sequences, we have

$$\begin{aligned}(\mathbf{S}_A + \mathbf{S}_B + \mathbf{S}_C) \cdot \mathbf{C}_B &= \mathbf{S}_B \cdot \mathbf{C}_B \\ &= \begin{cases} \overline{\mathbf{C}}_B \cdot \mathbf{C}_B = -1, & \text{for data bit} = \mathbf{0} \\ \mathbf{C}_B \cdot \mathbf{C}_B = +1, & \text{for data bit} = \mathbf{1} \end{cases}\end{aligned} \tag{1.41}$$

Thus, from the calculated results, it is possible to determine the data bits transmitted from Station B while ignoring data transmitted from other stations.

1.7 Transmission media

The purpose of transmission media or the channel is to convey message signal from the transmitter to the destination. As mentioned in Section 1.3, transmission media are roughly grouped into two categories – wire (guided) and wireless (unguided) transmission media. We now briefly describe some of the transmission media used commonly in present day communication systems, especially in data communication. Further discussion on transmission media for digital communications is available in Reference 26.

1.7.1 Twisted pair cable

Twisted pair cable is one of the oldest and still the most common guided transmission medium, especially used for telephone communications and modern Ethernet networks. It can be used for transmitting both analog and digital information. It consists of two insulated copper wires twisted together in a helical form. Since two parallel wires constitute a fine antenna that radiates electromagnetic energy, the wires are twisted to prevent any radiation loss during transmission by nullifying the waves from different twists. Typically, multiple pairs of twisted cables are bundled together by wrapping them in a tough protective insulating sheath. Crosstalk interference between neighboring pairs in a bundle is reduced by having different twist lengths for different pairs.

In twisted pair cable, a message signal is usually carried as the difference in voltage between the two wires in the pair. Since both the wires in a pair are generally affected by same amount of external noise, the difference in voltage between the two wires is hardly affected by noise thereby providing better immunity to channel noise. Twisted pair cable is generally very cheap, readily available, flexible and lightweight. It is easy to work with and install. Thus, due to its adequate performance and other added advantages, twisted pair cables are widely used and are expected to remain so for several years ahead. However, twisted pair cable suffers from certain inherent drawbacks. It is highly susceptible to external interference from other channels, offers high attenuation and generally has a very narrow transmission bandwidth.

There are two basic types of twisted pair cables – unshielded and shielded. Unshielded twisted pair (UTP) is the ordinary wire and so is the cheapest of all the transmission media. It is easy to work with and install and, hence, used commonly for telephone networks and local area networks (LANs) in office buildings. UTP comes in several categories that are based on the number of twists in the wires, the diameter of the wires and the material used in the wires. Among all the different categories of UTP, Categories 3 and 5 have received the most attention. Category 3 cable with transmission bandwidth up to 16 MHz is used primarily for telephone connections. Category 5 uses the same connector as in Category 3 but with more number of twists per unit length providing less crosstalk and a better-quality signal over longer distances. Consequently, the transmission characteristics of Category 5 cable are specified up to 100 MHz making the cable more suitable for high-speed computer communication. Recently, Category 6 cable with more stringent specifications has been developed. Categories 5 and 6 are currently the most common Ethernet cables used.

A way to improve the performance of UTP so as to handle signals with higher bandwidth is to have a metallic shielding on every individual twisted-pair in a bundle as well as around the entire bundle but inside the protective plastic sheath. While the inner layer of shielding prevents any crosstalk between adjacent cable pairs, the outer layer of shielding makes the cable less susceptible to external interference. However, shielded twisted pair (STP) is more expensive and more difficult to work with than the UTP.

1.7.2 Coaxial cable

Coaxial cable is another commonly used guided transmission medium. It has better shielding and greater bandwidth compared to twisted pair cables. A coaxial cable consists of a hollow outer cylindrical conductor that surrounds a single inner wire conductor; the two conductors are electrically separated either by regularly spaced insulating rings or a layer of solid dielectric material between the two. The inner conductor consists of a stiff copper wire while the outer conductor is often made in the form of a closely woven braided mesh. A protective plastic sheath covers the outer conductor. A cutaway view of a coaxial cable is shown in Figure 1.21.

The construction and shielding of the coaxial cable give it a good combination of high bandwidth and excellent noise immunity. The outer conductor shields the inner conductor from picking up stray signal from outside. The transmission bandwidth of coaxial cable depends on the cable quality and length. Nowadays, coaxial cables with bandwidth up to few gigahertz are available.

Coaxial cable can be used to transmit both analog and digital signals. There are two kinds of coaxial cable; 50 Ω cable for digital transmission and 75 Ω cable for analog transmission. It is perhaps the most versatile transmission medium used in a wide variety of applications such as cable TV, long-distance telephony, short-range computer links, LAN, etc. In cable TV system, coaxial cable is used to accommodate

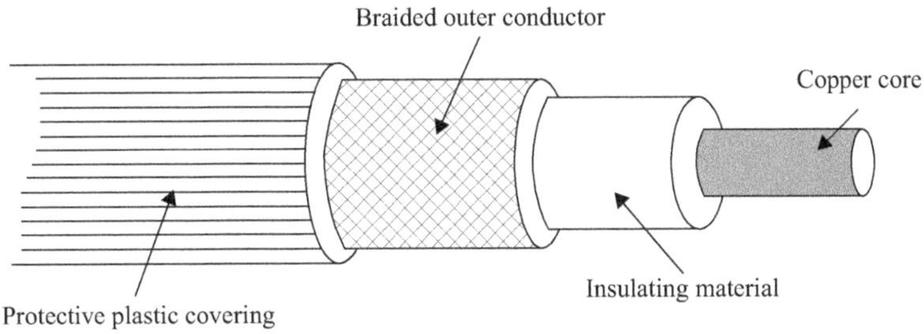

Figure 1.21 Construction of a coaxial cable – a cutaway view

hundreds of TV channels transmitted over a distance of few tens of kilometers. Coaxial cable can carry thousands of voice channels in long-distance telephone system.

1.7.3 Power line communication

Another common kind of guided transmission channel is the electrical power line used primarily for delivering electrical power to houses, and for distribution of power to various electrical outlets and appliances within houses. Power line has been used for many years by electricity companies for remote metering. It is also used for controlling devices in the home. In recent years, power line has also been used for high-speed digital data communication over LAN and Internet. The main advantage of power line communication is that no separate network setting is necessary for signaling and transmission of messages. The already existing infrastructure for electrical power distribution over a region can be utilized for the purpose. Any communication device (transmitter or receiver) needs power. So, whenever the device is plugged in, the device not only gets electric power supply but also sends and/or receives messages over the same electrical wiring. The high-frequency message signal is superimposed on the low-frequency power signal which is conveniently filtered out using a band-pass filter at the receiving end. However, power line is basically designed to distribute 50/60 Hz electrical signals and is generally incapable of supporting high-frequency message signals. As a result, message signals are severely attenuated in course of its transmission over the power line. Also, transient currents due to switching on and off create electrical noise over a wide range of frequencies thereby distorting the message signal to a large extent. Moreover, electrical wiring acts as an antenna, if not twisted properly, picking up external interference easily while radiating signals of its own.

1.7.4 Fiber-optic cable

In the race between computing technology and communication technology, the latter probably is the winner, thanks to the development of fiber-optic transmission system that has enabled high-speed transmission of digital data [27, 28]. Due to the high optical frequencies, fiber-optic cable has the capability to support data transmission of the order terabits per second (Tbps), but limited by the operating speed of transducers and other associated hardware. Nevertheless, data transmission at a rate as high as 100 gigabits per second (Gbps) can still be achieved in a fiber-optic system. Accordingly, fiber optics find applications in high-speed LANs, high-speed Internet access and so on. In this system of data transmission, digital bits are transmitted in the form of pulses of light. Conventionally, a pulse of light indicates bit **1** while bit **0** is indicated by the absence of any such light pulse. At present, three bands of light waves in the near-infrared part of the spectrum are most commonly used in fiber-optic transmission systems. These bands are centered at 850 nm, 1300 nm and 1550 nm (i.e., approximately at 350 THz, 230 THz and 190 THz, respectively) and are 25–30 THz wide.

The three key elements of a fiber-optic transmission system are the electrical-to-optical transducer in the transmitter, the fiber-optic cable and the optical-to-electrical transducer at the receiving end, in addition to other basic elements present in any digital communication system such as amplifiers, clock recovery circuit, binary symbol (bit) detection circuit, etc. The purpose of the electrical-to-optical transducer is to convert an electrical pulse corresponding to a message-bit **1** to a pulse of light (and none corresponding to message-bit **0**). This may be viewed as OOK modulation of a light wave (carrier signal) from an optical source by the input message bit-stream. Typically, a laser diode or a light-emitting diode (LED) is used for the purpose. Laser diodes have greater bandwidth, narrower spectrum, higher output power and hence, generally used in long-distance communication. On the other hand, LEDs are cheaper, more robust to environmental conditions and require simple interface circuitry. Hence, they are used when cost and robustness are more important than performance.

At the receiver, two types of photodetectors, viz. *p-i-n* diodes and avalanche photodiodes, are used for demodulation of the received modulated light beam. The process includes conversion of light pulses to electrical pulses followed by detection of data bits. *p-i-n* diodes are cheap and provide good performance, but operate at a lower range of wavelengths. A photodiode, on the other hand, can function at higher wavelengths and is more sensitive in detecting weak signals due to its inherent amplification. However, an avalanche photodiode needs high-supply voltage to operate, is very sensitive to temperature and has lower gain-bandwidth-product which restricts its use in very high data rate systems.

The transmission medium, i.e., the fiber-optic cable, is an ultra-thin fiber of glass. It consists of a glass core at the center through which the light propagates, surrounded by a glass cladding and a thin plastic jacket forming the outermost protective covering as shown in Figure 1.22. Although both the core and the cladding are made of glass that is transparent to light, they are designed such that the refractive index of the cladding material is lower than that of the core. As a result, light wave traveling through the core when impinges on the core-cladding boundary at an angle greater than the critical angle, it reflects back into the core due to the principle of total internal reflection of light. Thus, the light wave is reflected back and forth at the core-cladding boundary as it propagates along the fiber. This is illustrated in Figure 1.23. Since a beam of light is composed of a bundle of rays and different rays may strike the core-cladding boundary at different angles, these rays will reach the receiver via multiple paths with different

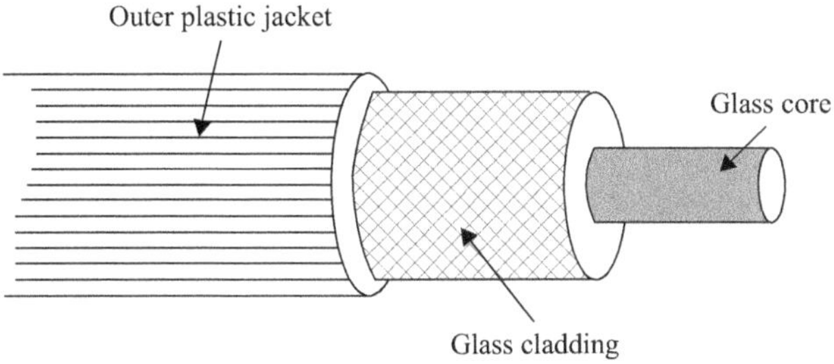

Figure 1.22 Construction of an optical fiber – a cutaway view

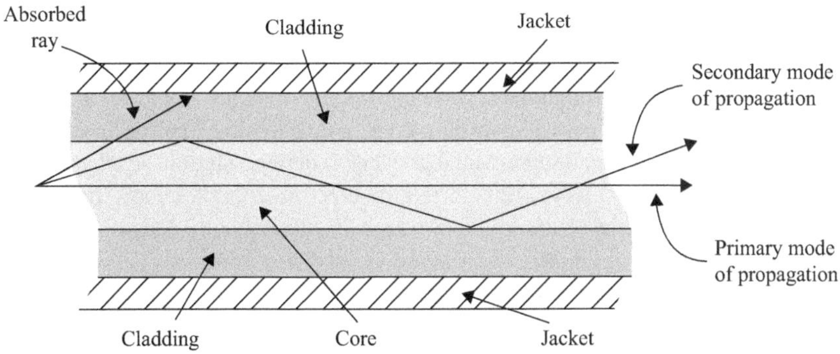

Figure 1.23 Propagation of light waves through an optical fiber

delays. That is, each ray has a different mode of propagation, and such a fiber-optic cable is referred to as *multimode fiber*. However, for fibers with diameter equal to only few wavelengths of the light, all the light rays can propagate in a straight line along the center of the core. Such optical fibers are referred to as *single-mode fibers*. Typically, a single-mode fiber has a core diameter of approximately 8 μm while that of a multimode fiber is approximately 50 μm. It may also be noted that in a multimode fiber some light rays may be incident on the core-cladding boundary at angles less than the critical angle. These rays, therefore, will leak through the cladding and finally absorbed by the opaque plastic jacket resulting in attenuation of the signal transmitted over the fiber. Another problem encountered during transmission through an optical fiber is the *chromatic dispersion* that occurs due to the differences in speeds at which different wavelengths composing the transmitted optical signal propagate. Chromatic dispersion results in spreading of the light pulses which in turn causes ISI.

Compared to copper wire, fiber-optic cable offers several advantages, as listed below:

1. Larger bandwidth,
2. less power requirement,
3. lighter in weight,
4. less susceptible to degradation,

5. no inter-fiber interference of light signals,
6. immune to electromagnetic interference.

However, optical communication systems suffer from certain disadvantages too. The transmitter and the receiver in an optical system are relatively complex requiring special interfacing devices, as discussed above. Further, fiber-optic cables can get damaged easily and are inherently unidirectional.

1.7.5 Wireless channel

While the above-described wireline channels can be used quite effectively between two fixed stations, they are hardly of any use in case of mobile transmitters and/or receivers. The only solution for providing connectivity in such cases is wireless channels. In this, the transmitter radiates electromagnetic wave through an antenna into the free space. This electromagnetic wave carries the message signal to be transmitted to one or many receivers which after traveling through the free space reaches the receiving station(s). A receiving antenna at every receiving station, irradiated by the received electromagnetic wave, transforms the electromagnetic energy into voltage across the two terminals of the antenna which is subsequently fed to the receiver circuit where the message signal is extracted via demodulation.

Electromagnetic waves may propagate through several possible paths in space depending on the frequency. Very low frequency (VLF) and low frequency (LF) waves with frequencies below 300 kHz are reflected from the surface of the Earth while traveling from the transmitter to the receiver. Thus, these waves are guided along the contour of the Earth's surface and are referred to as *ground waves*. However, the Earth's surface causes large attenuation to these waves and so cannot be used for long-distance communication. *Sky waves* are medium waves (0.3–3 MHz) and high-frequency waves (3–30 MHz) that are reflected from the ionosphere of the atmosphere. Very high frequency (VHF) waves in the range 30–300 MHz are the *tropospheric waves* that are reflected at different layers in the troposphere where abrupt changes in the effective dielectric of the troposphere occur. Electromagnetic waves with frequencies above 300 MHz and up to several hundreds of gigahertz, including the microwave band, can generally penetrate through the atmosphere. Hence, they are used for *line-of-sight* communication and/or satellite communication.

1.8 Data communication systems

In this section, we will study some of the different kinds of communication systems and networks that are used in transmitting data in distribution automation. They have different coverage areas (large or small) and use different technologies. Historically, communication systems and networks were primarily developed for analog transmission of voice, audio, image and video. Nowadays, with the advent of digital communication technology, many of these analog communication systems are being fully or partially replaced by digital systems. Also, except in broadcasting applications where analog communication is still used extensively, all the newly developed communication systems are digital systems. These digital transmission systems not only carry digitized voice, audio, image and video, but are also used for transmitting other digital messages, including those in case of distribution automation, by multiplexing these data along with the primary service data (voice, audio, image and/or video).

1.8.1 Telephone system

Following the invention of telephone by Alexander Graham Bell in 1876, there had been a tremendous demand for the same among the people who wished to talk to each other remotely. The demand prevailed for over a century and even till today resulting in the evolution of the system over the years. The need for one-to-one connectivity among all the subscribers necessitated in building central switching offices in towns and cities; credit goes to the Bell Telephone Company who opened the first switching office way back in 1878. Each subscriber's telephone set is connected to the *local central office* by a simple two-wire cable forming the *local loop*. All such local central offices within a larger geographical region are further connected to each other through a *toll office* located centrally within that region. The toll offices, in turn, are connected to each other through *intermediate switching offices*, thereby forming a hierarchical structure of interconnections. When two subscribers attached to the same local office wish to talk to each other, the switching mechanism within the office sets up a direct connection between the two local loops throughout the duration of the call enabling them to converse over the line. If the calling parties are residing within the same geographical region but not in the same locality, the connection between them is established through their respective local offices and the toll office connecting these two local offices. For long-distance call, a path is established through the intermediate switching office connecting the respective toll offices, local offices and the callers' telephone sets on either sides. In short, the three major components of a telephone system are – (i) *local loops* connecting the subscribers to the network, (ii) *trunks* connecting the switching offices and (iii) *switching offices* where calls are redirected from one trunk to another. A detailed description on the structure of telephony system is available in Reference 29.

While analog telephone system had been the predominant technology even till the last decade of the twentieth century, digital telephony has gradually taken over in most countries. The advent of computer technology, digital electronics and fiber-optic cables has made this possible. Telephone companies have replaced the analog switches with computer (software) controlled electronic switching systems. Intertoll trunk lines are being replaced by high-bandwidth fiber-optic cables, while various other transmission media still continue to exist, including low-bandwidth twisted pair cables in local loops, coaxial cables for long-distance trunk lines and microwave links. Owing to the high cost that the subscribers need to pay in converting local loops into digital link, the two-wire local loops still exist, except in some business houses where they have been replaced with optical fibers. Since baseband transmission of bits over low-bandwidth channel causes ISI, data bits are modulated onto time-continuous sinusoidal carriers and then transmitted over the local loops. A *modem* is used as an interface between the digital system (computer and/or data source) at the subscriber's end and the analog local loop. The modem, short form of *modulator–demodulator*, modulates the outgoing data bits onto a carrier and demodulates the incoming signal for data extraction. Unfortunately, the data transmission capacity of the telephone network is constrained by the low bandwidth and low signal-to-noise ratio of the local loops. Consequently, in order to achieve high-speed data transmission, telephone modems have been replaced by broadband digital subscriber lines (DSLs). There are many DSL methods that use either one or two twisted pair cables to serve the customers over the "last mile" local loops; a brief description on each of these DSL methods is given below.

1. **HDSL** (*high bit-rate digital subscriber line*): It uses two twisted pair cables, one for transmission and one for receiving data, to support data rate of 1.544 Mbps at full-duplex for a distance of about 12,000 ft.
2. **SDSL** (*symmetrical digital subscriber line*): It is a one-pair version of HDSL supporting full-duplex data transmission at a rate of 768 kbps in either direction. Data transmitted are separated from the data received by using a hybrid or echo canceller.
3. **ADSL** (*asymmetrical digital subscriber line*): It uses one twisted pair cable in which downstream and upstream data are multiplexed using FDM scheme. Two variations of ADSL are *G.DMT* and *G.Lite* that use "always on" (packet switching) ADSL technology. G.DMT uses discrete multitone (DMT) modulation with up to 256 carriers and up to 15 bits of data modulated on each carrier using QAM. Thus, G.DMT provides data rate of 6.1 Mbps downstream and 640 kbps upstream for a distance of about 12,000 ft. The upstream and downstream data carriers are placed in the 26–138 kHz and 138–1100 kHz bands, respectively, with a spacing of 4.3125 kHz between carriers. G.Lite, on the other hand, uses only 128 carriers with up to 8 bits of data modulated (QAM-256) on each carrier. G.Lite provides data rate of 1.5 Mbps downstream and 512 kbps upstream for a distance of about 18,000 ft.
4. **VDSL** (*very-high bit-rate digital subscriber line*): It uses one twisted pair cable that supports downstream data rate of 25 Mbps for a distance of up to 3000 ft or 51 Mbps for a distance of up to 1000 ft, and upstream data rate up to 3.2 Mbps.
5. **ISDN** (*integrated service digital network*): It uses one twisted pair cable to deliver only digital data and no voice data. There are two categories of ISDN – (i) narrowband ISDN (N-ISDN) that supports a data rate of 144 kbps in either direction for a distance of up to 18,000 ft and (ii) broadband ISDN (B-ISDN) that has a total data rate of 1.536 Mbps. Data from multiple channels are multiplexed and transmitted over ISDN using TDM.

Although telephone systems were primarily designed for voice transmission, telephone companies nowadays provide a multitude of other services over their public switched telephone networks (PSTNs). Modern digital telephone systems use TDM or packet data transmission for diverse applications such as fax, e-mails and data transfer, in addition to digitized voice communication. Digital services are provided to the users in any of the three ways as follows.

- **Dedicated leased service:** In this, the subscriber is connected to the network round-the-clock over a dedicated line that is leased out by the service provider to the subscriber on payment basis.
- **Circuit-switched service:** This service is available on a dial-up basis as and when the subscriber demands for it. When a subscriber's computer places a telephone call, the switching equipment within the telephone system establishes a physical path all the way from the caller's telephone to the receiver's telephone.
- **Packet-switched service:** In packet data transmission, data bits are transmitted in the form of packets as and when they are available without the need for establishing any dedicated path in advance. The link is "always-on" but used only when packets are exchanged. The packets are routed from the source to destination via routers that use a *store-and-forward* mechanism.

Further details on digital telephone system is beyond the scope of this chapter. For more information on this topic, see Reference 30.

1.8.2 Mobile phone network

Since the invention of the wireless communication systems, there had been a tremendous effort in providing wireless telephonic link even to persons who are on the move, walking or traveling from one place to another. However, electro-magnetic spectrum crunch prevents permanent allocation of a radio frequency (channel) for each and every subscriber within a geographical area. The cellular telephone system provides a solution to this. In this, a geographical area is divided into smaller regions called *cells*, each having a *base station*. The base station of every cell uses a set of frequencies, each frequency corresponding to a channel in that cell, that are shared by all mobile users within that cell via FDMA, TDMA or CDMA. The frequencies used in a cell are reused in other nearby cells, but not in its neighboring (adjacent) cells to avoid inter-cell interference. Thus, it is possible to cater to a large population within a given geographical area by dividing the area into numerous smaller-sized cells thereby permitting more frequency reuse. A user moving within a particular cell communicates wirelessly to the base station through his/her mobile telephone set using an available wireless link (channel). The base station is connected to a *mobile switching center* via telephone lines or microwave links, which in turn connects the caller to the called party. If the called party uses landline-based telephone, the switching center connects the mobile phone set of the caller party to the fixed telephone network via the local central office. If the called party is also mobile, the connection is established through the base station of the cell within which the called party is located using a wireless link available in that cell. The mobile switching center is also responsible for assigning a new transmission frequency (channel) to a mobile user whenever the user leaves a cell and enters a new adjacent cell, a process called *handoff*. This ensures uninterrupted conversation even when a mobile user travels from one cell to another.

Some of the basic advantages of the cellular telephonic system are as follows:

1. Large subscriber capacity by accommodating unlimited number of mobile users, theoretically, even for a limited number of channels.
2. Expandability of the network possible by splitting large congested cells into smaller cells.
3. Efficient use of the electromagnetic spectrum due to its ability to reuse allocated channel frequencies within a geographical area.
4. Reduced transmit power which is particularly important for battery-driven mobile devices, thereby reducing the cost of transmitters.
5. Reduced occurrences of multipath propagation due to short-range communications within a cell and between nearby cells.
6. Service to portable devices thereby allowing on-the-move communication.
7. High-quality voice and data communication at relatively low cost.

The cellular technology has, till now, gone through three distinct generations; popularly known as **1G**, **2G** and **3G**. These are described in brief below.

First-generation (1G) *system*:

The *Advanced Mobile Phone System* (AMPS), developed by AT&T Bell Laboratories in 1970s, is considered to be the *First Generation* (1G) of the modern mobile phone technology. It replaced the earlier *push-to-talk* system and the *Improved Mobile Telephone System* (IMTS) and was in use till the last decade. Every mobile telephone set contains a PROM (programmable read-only memory) or EPROM (erasable programmable read-only memory), known as *numeric assignment module* (NAM). The NAM stores the 10-digit telephone number or the *mobile identification number* (MIN) assigned to the subscriber using that mobile set, and a unique 32-bit identification number of the phone set given by the manufacturer, called *electronic service number* (ESN). Whenever the mobile is "on the air", it automatically sends its ESN and MIN to the base station and gets registered to the mobile switching center. Once registered, the switching center gains control over that phone and establishes connection with it as and when there is a "call". The AMPS is an analog system in which the analog voice and audio signals are frequency modulated, even though FSK is used for control signals. Four different sets of channels are used within a cell – data channels for carrying voice, control channels, paging channels for call alerts, and access channels for call setup and channel assignment. The channels are separated by FDM thereby allowing conversations between multiple pairs of users using the same wireless medium at the same time.

Soon after its commercial deployment in the USA in 1983, many other countries adopted the AMPS as well. The first commercial cellular system MCS (*Mobile Cellular System*) developed in 1979 in Japan was modified to MCS-L1 based on AMPS but with higher frequency and slightly smaller bandwidth. A system similar to AMPS, but also with higher frequency and smaller bandwidth, called TACS (*Total Access Communication System*), was introduced in 1985 in the United Kingdom and other European countries. Later, a narrowband version of AMPS (N-AMPS) with voice channels that are one-third of the regular AMPS was introduced in the year 1992. A variation of the TACS, called JTACS, was deployed in Japan in 1989, followed by N-TACS (narrowband-TACS) in 1993 with voice channels occupying half the bandwidth as that in JTACS. However, with growing popularity of digital communication systems and increasing demand for accommodating more mobile users, the analog 1G systems were gradually replaced by the digital second-generation systems.

Second-generation (2G) *system*:

The transition from analog cellular system to digital system marks the beginning of the second generation (2G) in mobile technology. Several different standards were developed among which the D-AMPS (digital-AMPS), GSM (*Global System for Mobile communications*) and the CDMA-based cellular system have been widely deployed. The D-AMPS is the digital upgradation of the first-generation AMPS. It uses both FDM and TDM – channels within a cell are separated by FDM while TDM is used to place multiple calls on the same channel. D-AMPS was designed to be compatible with AMPS so that both first-generation AMPS and second-generation D-AMPS can coexist and make a smooth transition from the analog generation to the digital generation. The GSM system also uses TDM combined with FDM in a manner similar to D-AMPS, whereas the third system is a completely different kind of system that is based on CDMA.

In an effort to develop a single 2G mobile standard throughout Europe, *Groupe Specialé Mobile* came up with the Global System for Mobile communications (GSM) [31] during the 1980s, although its first-field trial took place not before 1991. The European Telecommunication Standards Institute (ETSI) standardized this system which was subsequently adopted worldwide as the international mobile standard. Cell-based design, frequency reuse and the handoff features of the earlier 1G systems are retained in this system. The architecture is similar to that of AMPS with few modifications. The phone set in GSM no longer contains PROM or EPROM but a removable smart card with subscriber information, called *Subscriber Identity Module* (SIM). The SIM can be inserted into any phone set to activate and connect that phone set to the mobile network. The base stations are now connected to the mobile switching center through a *base station controller* that controls channel assignment and handoff. The base station uses the dedicated control channel for locating, registering and setting up of calls for mobiles within its range. There is also the common control channel that is split into three subchannels – paging channel for call alerts, random access channel through which a user requests for a slot in the dedicated control channel to set up a call, and access grant channel to announce whenever a slot is made available to a user. Finally, GSM employs a different mechanism for handoff, called *Mobile Assisted Handoff* (MAHO). Since a user transmits or sends data (digitized voice and audio) over a channel using TDM, it remains idle for most of the period. The user utilizes this idle period in assessing the quality of signal that it receives from other nearby base stations. This information is then passed to the base station controller which in turn determines the location of the user from this information. Based on the location of the mobile, as determined by the base station controller, the handoff is initiated whenever the mobile leaves a cell and enters a new cell. Thus, the mobile set also participates in the handoff process and hence the name.

Third-generation (3G) *system*:

As discussed above, 1G and 2G mobile technologies are only about analog and digital voice, respectively, with no provision for digital message data. Further, a world-wide compatible mobile system was still missing. Motivated by this, ETSI proposed the *Universal Mobile Telecommunication System* (UMTS). Soon, the International Telecommunications Union (ITU) developed the *International Mobile Telecommunications 2000* (IMT-2000) system which includes UMTS as a subset. IMT-2000 supports three different modes of operation, viz. *Wideband Code Division Multiple Access* (W-CDMA) in Europe and Japan, CDMA-2000 in the USA and some Asian countries, and *Time Division Code Division Multiple Access* (TD-CDMA) in China.

The basic services that the IMT-2000 provides are:

1. High-quality voice transmission,
2. messaging,
3. multimedia,
4. Internet access.

Additional services include video-conferencing, m-commerce, etc. All these services are available instantly everywhere that can be reached via satellite and/or terrestrial network. Thus, mobile technology is gradually penetrating every nook and corner of the world. The service can be used not only for voice communication but also for transmission of digital message data and multimedia data. Consequently, we have now stepped into a new era of mobile technology, popularly called 3G – the third generation of mobile technology.

Unlike the AMPS, GSM and most other earlier systems, the 3G technology uses CDMA. This provides us with the following advantages over the 1G and 2G systems:

1. Higher capacity which is particularly important in places with increasing tele-density.
2. Use of single frequency eliminates the need for complicated frequency planning.
3. Facilitates *soft handoff* in which the new base station takes control of the mobile before the previous base station gives up. This eliminates the possibility of unwanted call termination in case of fast-moving phone sets.

Although widespread deployment of 3G mobile networks is not yet complete, research in developing the next generation of mobile technology, in the name *Long Term Evolution* (LTE) or 4G, is underway [32, 33]. Some of the proposed features of LTE include high bandwidth, ubiquity (connectivity everywhere), seamless integration with other IP networks, adaptive resource and spectrum management, and high quality of service for multimedia.

1.8.3 Trunked radio system

A trunked radio system comprises of a number of mobile radios (users) with a central controller. All users must be enrolled to the system before they can participate in communicating within the system. The group of users in a trunked radio system is called a *talkgroup*. Every system uses a group of frequencies that is shared by this talkgroup, allowing multiple simultaneous conversations by FDM. Whenever a user wishes to communicate, the user presses a push-to-talk button that sends a digital message to the central controller requesting a channel assignment. If a channel is available, the system controller sends a message back to the user informing the user about the channel assignment. The system controller also broadcasts a similar message to other radios in the system so that they can tune to the designated channel if they desire so. After the channel assignment process is completed, voice conversation between two or more users in the system takes place.

In a trunked radio system, one out of all the channels provided in the system is dedicated for use as control channel. Exchange of control information between the users and the central controller is carried out over this channel. The remaining channels are available to the users for conversation. The central controller uses a dynamic channel assignment strategy to allocate these channels only to the demanding users. Since generally not many users in a system will require channel access at the same time, it is possible to accommodate a large number of users in a trunked radio system with limited number of channels even while guaranteeing channel availability to all its demanding users with high probability. A conventional (non-trunked) radio system, on the other hand, can only accommodate that many users as many channels are provided in the system. Thus, a trunked radio system offers greater benefits compared to the conventional radio system. Nevertheless, there are certain major disadvantages with trunked radio system as follows:

1. Trunked radio system depends on the smooth functioning of the central controller which in turn relies on computer software that controls the channel assignment task. So, any software as well as hardware problem with the central controller will affect all radio users in the system.
2. Trunked radio operation will be unavailable if the central controller becomes isolated from its remote mobile users. This may happen due to failure of the dedicated control channel link. In such a situation, the whole system will collapse.
3. Trunked radio system suffers from a significant delay in initiating conversation. This occurs due to the time spent in sending request for channel assignment by the user and subsequent channel assignment by the controller.

1.8.4 Cable TV network

In the past decade, the cable TV system has emerged as another major player for data transmission [34–36]. The idea of cable TV started sometime in the middle of the past century. This system, in the earlier days, consisted of a large antenna for picking up TV signals from the space, a *headend* to boost the received signal and then deliver this signal to every house in the locality via coaxial cable. Later, with increasing popularity, cable TV operators began to lay cables across cities connecting all local cable TV networks into a single network.

Starting from the beginning of this century, inter-city cables are replaced with fiber optics to support high-speed digital data transmission from the headend to the neighborhood of the customers, along with the TV signals. The data are multiplexed with the TV signals using FDM and subsequently distributed to all customers in the neighborhood via coaxial cables. In a manner similar to the TV signal transmission, cable TV networks initially could only be used for one-way downstream data transmission – from the headend to the customers. The present technology, however, has enabled upstream transmission as well. A modem at every user end demodulates the incoming signal to extract the downstream message data and modulates an RF wave with the upstream data for transmission over the coaxial cable to the outside world. Thus, cable TV system serves for data exchange between users connected to the network.

1.8.5 Satellite communication system

In astronomical term, a satellite is a celestial body. While the Moon is the only natural satellite of our Earth, advancement in space technology has led to the development of artificial satellites which are space vehicles launched by humans that orbit the Earth. Artificial satellites that provide communication and other related services to a variety of consumers is called a communication satellite or *comsat* in short. A communication satellite is essentially a microwave repeater in the sky that receives communication signals from a transmitting station on the Earth and relays it back to one or more receivers on the Earth. Over the last few years, the number of satellite communication systems has increased providing means for relaying of telephone signals, broadcasting of TV signals, communication links to remote locations and *direct-to-home* TV distribution via satellite. Satellite systems are also used for GPS (global positioning system) applications where the satellite signal is used to determine the user's exact location on the Earth.

A satellite communication system consists of one or more satellite space vehicles, a ground-based control station on the Earth, and a network of user Earth stations that communicates with each other via transmission and reception of communication signals through the satellite system. Communication satellites are generally placed in *geostationary orbit* such that their positions in the sky are fixed with respect to the Earth. This enables the antenna on the Earth to be pointed in fixed directions towards the satellites without the need for tracking the positions of the satellites. A transponder attached to the satellite receives the signal sent from the transmitting antenna on the Earth, amplifies the same and then relays back to the Earth. The uplink (Earth-to-satellite) and the downlink (satellite-to-Earth) use different frequencies to avoid any interference. The geographical area on the Earth's surface covered by the downward antenna beam is referred to as *footprint*. A communication satellite is generally equipped with multiple transponders, one transponder per channel, and multiple antennas to have a larger footprint area. Further reading on satellite communication is available in Reference 37.

Satellite communication technology has evolved tremendously in the past two decades. One major breakthrough in satellite communication had been the transition from analog to digital domain. This obviously has helped in utilizing the service of the communication satellites for digital data transmission. Satellite relays provide a channel for data communication, in addition to other services. The different methods used for multiple access of a single transponder by multiple uplink and downlink stations are:

1. Frequency division multiple access (FDMA)
2. Time division multiple access (TDMA)
3. Code division multiple access (CDMA)
4. Space division multiple access (SDMA) where narrow-beam antenna patterns are switched from one direction to another.

The system may use either *fixed-assigned multiple-access* (FAMA) or *demand-assigned multiple-access* (DAMA) schemes. In the former mode, the multiple access method used (FDMA, TDMA or CDMA) does not change. In the DAMA mode, the system adopts to either of the FDMA or TDMA methods of multiple access depending on the traffic condition. A multiple access method similar to TDMA, called ALOHA (an Hawaiian word, meaning "hello"), is also employed under low traffic condition [38]. In this method, multiple users send bursts of data (packets) as and when it is required to communicate. However, two or more packets from different users may overlap in time causing *collision*. In that case, each packet is retransmitted till it reaches the destination successfully.

However, while satellite communication systems are cost effective for broadcast purposes, they are less cost effective for point-to-point communication applications. The cost of broadcasting a message to a large crowd is not much compared to that in sending the message to a single recipient. From privacy point of view, satellite communication is highly insecure and hence, encryption is a must whenever data security is desired. Another drawback with satellite communication is the round-trip propagation delay which is typically of the order of hundreds of milliseconds (approximately 270 ms for a one-way transmission).

1.8.6 Wireless sensor network

Wireless sensor network is an emerging technology, generally deployed to monitor various aspects of the physical world such as monitoring the temperature, humidity and other weather conditions over a region, patient monitoring in a healthcare system, tracking of objects, monitoring performance of equipment, etc. Thus, it provides unprecedented opportunities for wide variety of applications such as disaster relief, border monitoring, medical care, surveillance in battlefield scenarios, fault diagnosis and many more. The network consists of a large number of sensor nodes that are somewhat like small computers fixed (or embedded) on the objects to be tracked or spatially distributed in the environment to be monitored. A sensor node is generally equipped with at least a power supply unit, sensing unit, processing unit to process the sensed data and a transmitter–receiver unit. The sensing unit consists of sensors that can sense temperature, vibration and other conditions, depending on the application, using various modalities, such as acoustic, seismic, thermal and infrared. The sensor nodes periodically sense the data, extract relevant information from the sensed data, process it and transmit it to a distant fusion center or sink. In order to realize it, peer-to-peer network technique may be used that allows direct communication between any two nodes. If two devices cannot communicate directly, other intermediate nodes relay data packets from the source node to the destination node using multi-hop routing. Due to the dense deployment of sensors, reliable multiple access techniques are necessary to allow multiple sensor nodes to transmit data simultaneously over the same transmission medium. For this purpose, multiplexing schemes commonly employed in wireless sensor networks are OFDM, its variants (OFDM in conjunction with TDMA or FDMA) and CDMA.

1.8.7 Wireless data networks

The desire to connect computers and/or laptops with the Internet wirelessly *anywhere–anytime* has led to the development of wireless LANs and wireless MANs, more popularly known as Wi-Fi and Wi-Max. However, compatibility among all such systems, developed by various companies, was an issue. Consequently, standardization of the systems was felt and the IEEE subsequently came up with the 802.11 and the 802.16 standards for Wi-Fi and Wi-Max systems, respectively.

Wi-Fi

The Wi-Fi is a *Wireless Local Area Network* (WLAN) that comprises the IEEE 802.11 family of networks. Starting from the beginning of this century, this family of WLANs has gained popularity over the last 15 years. Nowadays, most college and university campuses, office buildings, airport lounges, hotels and even private residences are provided with Wi-Fi facility where multiple computers housed in the same building are networked together and connected to the Internet via high-speed connection. Wi-Fi networks are even popular at homes for networking multiple computers and for connecting to the Internet as well.

The 802.11 standard defines an over-the-air interface between a wireless client and a base station or between two wireless clients. These wireless networks operate unlicensed in designated frequency bands allowing anyone to put any of these networks *on-the-air* without the need for any permission from the frequency allocation authority. The three major Wi-Fi standards in the IEEE 802.11 family are the 802.11a, 802.11b and 802.11g. 802.11a pertains to the WLAN standard that goes as fast as 54 Mbps in the 5 GHz band. 802.11b yields connection as fast as 11 Mbps in the 2.4 GHz band while 802.11g provides for 20+ Mbps transmission rate in the 2.4 GHz band. Table 1.1 gives a technical comparison among these three standards.

Table 1.1 Wi-Fi standards

	802.11a	**802.11b**	**802.11g**
Frequency band (GHz)	5.0	2.4	2.4
Maximum data rate (Mbps)	54	11	54
Channel bandwidth (MHz)	20	20	20
Radio technology	OFDM	DSSS*	OFDM

*Direct sequence spread spectrum.

Table 1.2 Wi-Max standards

	Licensed service	**Unlicensed service**
Frequency band (GHz)	2.5	5.5
Maximum data rate (Mbps)	108	108
Channel bandwidth (MHz)	20–40	20–40
Radio technology	OFDM	OFDM
Duplexing scheme	Frequency division	Time division

Wi-Max

The Wi-Max is a *Wireless Metropolitan Area Network* (WMAN) that is somewhat like a cellular telephone system, but designed only for data communication and connected to the Internet. Wi-Max standards are developed and approved by the IEEE 802.16 Working Group. The Wi-Max service may be licensed or unlicensed. Since unlicensed service provider need not to pay for the spectrum space and the license, the unlicensed service has the advantage of lower cost, but suffers from interference from other unlicensed systems. For licensed systems, the service provider has to buy spectral space and a license. It has the advantage of less interference and better *Non-Line of Sight* (NLOS) reception. Table 1.2 gives a technical comparison between the Wi-Max standards for licensed and unlicensed service.

1.8.8 Wireless mesh network

A wireless mesh network is a mesh network created through the connection of wireless access points installed at each user's location. In this system, the access points or the nodes are organized in a mesh topology forming a wireless ad hoc network. Wireless mesh networks generally consist of mesh clients, mesh routers and gateways. The mesh clients are wireless mobile devices such as laptops, mobile phones, etc. The mesh routers forward data to and from the gateways which may be connected to the Internet. Wireless mesh networks are implemented with various wireless technology including 802.11, 802.16, cellular technologies or combinations of more than one type. The networking infrastructure is decentralized and simplified where each network user forwards data as far as to the next node only. In a full mesh topology, every node communicates with every other node while in a partial mesh network, nodes communicate with all nearby nodes, but not distant nodes. Similar to wireless sensor networks, multiplexing schemes employed here are OFDM and its variants – OFDM in conjunction with TDMA, FDMA or CDMA. Some of the advantages of wireless mesh networks include:

1. Decreased need for Internet gateways,
2. collaborative backup ensuring data security in the event of disk failure,
3. dynamic route configuration,
4. lower power requirements,
5. increased reliability – each node connected to multiple nodes ensuring no broken link in the event of any node dropping out of the network.

1.8.9 Wireless automated meter reading system

With wireless communication becoming ubiquitous around the world nowadays, its application for gauging consumption of utilities by customers is rapidly becoming popular and essential. A wireless automatic meter reading system (WAMRS) is a system used to remote reading the consumption of energy. It consists of three main parts – the Sender (also called the Premises Unit) located inside the house of the consumer, the Central Point's Unit covering every geographical area and the Receiver (server) at the billing office of the electricity supply company. A digital energy meter at the consumer's premises continuously records the energy consumption by the customer. The primary objective of a WAMRS is to send periodical readings of the meter wirelessly to the billing office. Any of the available wireless technology may be used for the purpose. However, since the range of wireless coverage of each premises unit is generally limited, the information is not sent directly to the billing office but via a central point in that locality. These central points have long-range wireless transmitters that can deliver the meter-reading data over long distances to the billing office.

1.8.10 Advanced metering infrastructure

Apart from distribution automation system, communication networks also play a very vital role in advanced metering infrastructure (AMI). In an AMI, a smart meter is installed in the premises of a customer which collects time stamped measurement data. Subsequently, these measured data are sent to the AMI host system through communication channels for further processing and analysis in order to provide useful information to the service provider. However, in AMI the communication is carried out in both the directions (i.e., from the customer's smart meter to the utility and from the utility to the smart meter) so that necessary control signals or real-time price signals can be sent to the customer to help him/her to manage his/her electricity bill.

In an AMI, the communication system should have the following features [39]:

1. It should be able to send huge amount of data with appropriate confidentiality.
2. It should convey complete information about the customer's consumption pattern.
3. The precision and authenticity of the communicated data should be maintained.

4. It should be cost effective.
5. It should support future expansion.

Various communication technologies are used for AMI. These are [39]: (i) power line communication, (ii) optical fiber, (iii) cellular, (iv) Wi-Max, (v) General Packet Radio Service (GPRS), (vi) Internet, (vii) satellite, etc.

1.9 Conclusion

In this chapter, we discussed various issues related to communication systems and techniques. These include different modulation techniques, multiplexing schemes, transmission media and existing communication networks. Communication plays a crucial role in distribution automation and hence, sufficient knowledge in communication theory is necessary for proper understanding of this topic. Hope that this chapter will serve as a starting foundation in communication networks, as necessary for understanding the distribution automation system.

References

[1] V.C. Gungor and F.C. Lambert, "A survey on communication networks for electric system automation," *Computer Networks: The International Journal of Computer and Telecommunications Networking*, vol. 50, no. 7, pp. 877–897, May 2006.

[2] H. Chen and N.-H. Yu, "A survey of communication technology in distribution network," *Proceedings of the 2012 China International Conference on Electricity Distribution (CICED)*, Shanghai, China, 10–14 Sep. 2012.

[3] S. Millman, ed., *A History of Engineering and Science in the Bell System: Communication Sciences (1925–1980)*, AT&T Bell Laboratories, Murray Hill, NJ, 1984.

[4] L.W. Couch, II, *Digital and Analog Communication Systems*, 7th edn., Pearson/Prentice Hall, Upper Saddle River, NJ, 2007.

[5] B.P. Lathi, *Modern Digital and Analog Communication Systems*, 3rd edn., Oxford University Press, Oxford, 1998.

[6] B. Sklar, *Digital Communications*, 2nd edn., Prentice Hall, Upper Saddle River, NJ, 2001.

[7] J.G. Proakis and M. Salehi, *Digital Communications*, 5th edn., McGraw-Hill, New York, NY, 2008.

[8] H. Taub, D.L. Schilling and G. Saha, *Principles of Communication Systems*, 3rd edn., Tata-McGraw Hill, Noida, India, 2008.

[9] S. Haykin and M. Moher, *Communication Systems*, 5th edn., John Wiley & Sons, New York, NY, 2009.

[10] A.S. Tanenbaum and D.J. Wetherall, *Computer Networks*, 5th edn., Pearson/Prentice Hall, Upper Saddle River, NJ, 2011.

[11] H. Nyquist, "Certain factors affecting telegraph speed," *Bell System Technical Journal*, vol. 3, no. 2, pp. 324–346, Apr. 1924.
[12] H. Nyquist, "Certain topics in telegraph transmission," *Transactions of the AIEE*, vol. 47, pp. 617–644, Feb. 1928.
[13] C.E. Shannon, "A mathematical theory of communication," *Bell System Technical Journal*, vol. 27, no. 3, pp. 379–423, Jul. 1948.
[14] C.E. Shannon, "A mathematical theory of communication," *Bell System Technical Journal*, vol. 27, no. 4, pp. 623–656, Oct. 1948.
[15] C.E. Shannon, "Communication in the presence of noise," *Proceedings of the IRE*, vol. 37, no. 1, pp. 10–21, Jan. 1949.
[16] *Transmission Systems for Communications*, 5th edn., Bell Telephone Laboratories, Murray Hill, NJ, 1982.
[17] M.L. Doelz and E.H. Heald, *Minimum Shift Data Communication System*, U.S. Patent No. 2917417, Mar. 28, 1961.
[18] S. Pasupathy, "Minimum shift keying: A spectrally efficient modulation," *IEEE Communications Magazine*, vol. 17, no. 4, pp. 14–22, Jul. 1979.
[19] C.-E. Sundberg, "Continuous phase modulation," *IEEE Communications Magazine*, vol. 24, no. 4, pp. 25–38, Apr. 1986.
[20] W. Stallings, "Chapter 8: Multiplexing," in *Data and Computer Communications*, 10th edn., Pearson, Noida, India, 2013.
[21] S. Weinstein and P. Ebert, "Data transmission by frequency-division multiplexing using the discrete Fourier transform," *IEEE Transactions on Communication Technology*, vol. 19, no. 5, pp. 628–634, Oct. 1971.
[22] R. Prasad, *OFDM for Wireless Communications Systems*, Artech House, Norwood, MA, 2004.
[23] H. Schulze and C. Lueders, *Theory and Applications of OFDM and CDMA: Wideband Wireless Communications*, John Wiley & Sons, New York, NY, 2005.
[24] J. Armstrong, *OFDM: Principles and Applications of Orthogonal Frequency Division Multiplexing*, John Wiley & Sons, New York, NY, 2007.
[25] S. Hara and R. Prasad, "Overview of multicarrier CDMA," *IEEE Communications Magazine*, vol. 35, no. 12, pp. 126–133, Dec. 1997.
[26] D.N. Hatfield, "Chapter 1: Transmission media," in *Data Communications, Networks and Systems*, T.C. Bartee, ed., Howard W. Sams & Co., Indianapolis, IN, 1985.
[27] T. Li, "Advances in optical fiber communications: An historical perspective," *IEEE Journal on Selected Areas in Communications*, vol. SAC-1, no. 3, pp. 356–372, Apr. 1983.
[28] J. Hecht, *Understanding Fiber Optics*, 5th edn., Prentice Hall, Englewood Cliffs, NJ, 2006.
[29] W. Fraser, *Telecommunications*, MacDonald, London, 1978.
[30] J.C. Bellamy, *Digital Telephony*, 3rd. edn., John Wiley & Sons, New York, NY, 2000.
[31] M. Rahnema, "Overview of the GSM system and protocol architecture," *IEEE Communications Magazine*, vol. 31, no. 4, pp. 92–100, Apr. 1993.

[32] D. Astely, E. Dahlman, A. Furuskar, Y. Jading, M. Lindstrom and S. Parkvall, "LTE: The evolution of mobile broadband," *IEEE Communications Magazine*, vol. 47, no. 4, pp. 44–51, Apr. 2009.

[33] A. Larmo, M. Lindstrom, M. Meyer, G. Pelletier, J. Torsner and B. Wiemann, "The LTE link-layer design," *IEEE Communications Magazine*, vol. 47, no. 4, pp. 52–59, Apr. 2009.

[34] D. Fellows and D. Jones, "DOCSIS cable modem technology," *IEEE Communications Magazine*, vol. 39, no. 3, pp. 202–209, Mar. 2001.

[35] A. Dutta-Roy, "An overview of cable modem technology and market perspectives," *IEEE Communications Magazine*, vol. 39, pp. 81–88, Jun. 2001.

[36] G. Donaldson and D. Jones, "Cable television broadband network architectures," *IEEE Communications Magazine*, vol. 39, no. 6, pp. 122–126, Jun. 2001.

[37] T. Pratt, C.W. Bostian and J.E. Allnutt, *Satellite Communications*, 2nd edn., John Wiley & Sons, New York, NY, 2003.

[38] S.S. Lam, "Satellite packet communications, multiple access protocols and performance," *IEEE Transactions on Communications*, vol. 27, no. 10, pp. 1456–1466, Oct. 1979.

[39] R.R. Mohassel, A. Fung, F. Mohammadi and K. Raahemifar, "A survey on advanced metering infrastructure," *International Journal of Electrical Power and Energy Systems*, vol. 63, pp. 473–484, 2014.

Chapter 2
Load flow analysis

Biswarup Das[1]

In an automated distribution system, several tasks, such as network reconfiguration (for loss reduction, load balancing, service restoration), volt-var control, etc., are undertaken regularly to improve the performance of the system. Further, these tasks should be accomplished without violating any system constraints such as bus voltage magnitudes, line power flow limits, etc. These constraints are checked at each step of the algorithms (for accomplishing these tasks) through load flow analysis. Therefore, an efficient load flow analysis is an integral part of an effective distribution automation system. In this chapter, load flow analysis methods for balanced and unbalanced distribution system are discussed. In this regard, three different cases are considered in this chapter as follows:

- Load flow analysis of balanced radial distribution system.
- Load flow analysis of unbalanced radial distribution system.
- Load flow analysis of balanced weakly meshed distribution system.

It is to be noted that, in the foreseeable future, weakly meshed operation of low voltage, unbalanced distribution system is quite unlikely and therefore, this case is not considered in this chapter.

2.1 Load flow analysis of balanced radial distribution system

For load flow analysis of radial distribution system, the most popular technique is backward/forward load flow method [1]. In this method, at every iteration, utilising the most updated values of the node voltages, initially the load currents at every node are calculated. Subsequently, starting from the leaf nodes of the system and progressing towards the substation (root) node, the branch currents are calculated by applying the Kirchoff's current law. Finally, starting from the root node and moving towards the leaf nodes of the system, the node voltages are updated by applying the

[1] Department of Electrical Engineering, Indian Institute of Technology Roorkee, Roorkee, Uttarakhand, India

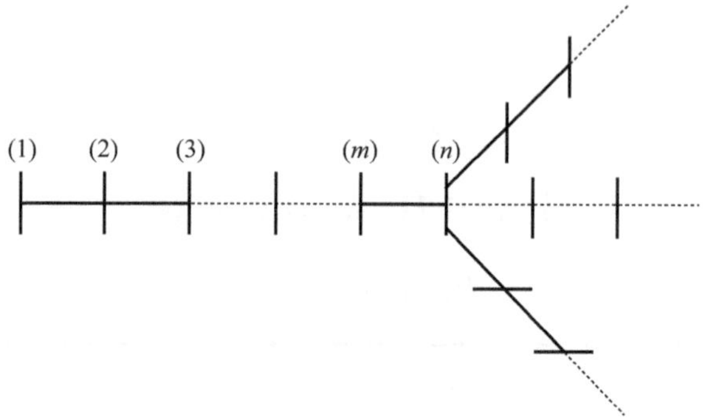

Figure 2.1 A sample radial distribution system

Kirchoff's voltage law (KVL). These three steps are repeated till the node voltages are converged.

The detailed steps of this method are described below with the help of the diagram of a sample radial distribution system shown in Figure 2.1. In the system shown in Figure 2.1, bus 1 is the substation bus (also called as root node) and as a result, its complex voltage is assumed to be known. Actually, the voltage magnitude of the root node is equal to the substation voltage while the angle of this bus voltage is taken to be zero, i.e. this bus (root node) acts as the reference node for calculating the angles of the other bus voltages.

For the algorithm described below, without any loss of generality it is assumed that the radial distribution system has N buses with bus 1 being the substation bus.

2.1.1 Detailed algorithm

Step 1: Initialise all the node voltages to $V_s\angle 0°$ (i.e. $\bar{V}_j^{(0)} = V_s\angle 0°$ for $j = 2, 3, \ldots, N$), where V_s denotes the substation voltage magnitude while $\bar{V}_j^{(0)}$ denotes the initial complex bus voltage of jth node.
Step 2: Initialise iteration count $k = 0$.
Step 3: Update iteration count $k = k + 1$.
Step 4: At iteration k, the load current $\bar{I}_j^{(k)}$ at node j is calculated as:

$$\bar{I}_j^{(k)} = \frac{P_{Lj} - jQ_{Lj}}{\left\{\bar{V}_j^{(k-1)}\right\}^*}; \quad j = 2, 3, \ldots, N \tag{2.1}$$

In (2.1), P_{Lj} and Q_{Lj} denote the real and reactive (inductive) power load at bus j, respectively, while $\{\bar{V}_j^{(k-1)}\}$ denotes the complex bus voltage of bus j corresponding to $(k - 1)$th iteration.
Step 5: This step is called backward sweep. In this step, initially the current in any branch directly connected to a leaf node is calculated to be equal to the load current at that leaf node. Subsequently, the current in any other branch is calculated as the sum

of the load current at the receiving end (of that branch) and the currents in the branches emanating from the receiving end (of that branch). For example, in Figure 2.1, the current in the branch between the buses m and n at iteration $k(\bar{I}_{mn}^{(k)})$ is given by,

$$\bar{I}_{mn}^{(k)} = \bar{I}_{n}^{(k)} + \sum \text{currents in the branches emanated from bus } n \tag{2.2}$$

By this process, the complex currents in all the branches are calculated.

Step 6: This step is called forward sweep. In this step, starting from the root node and moving towards the leaf nodes, the voltages at every node are updated from the knowledge of the voltage at the respective previous node. For example, in Figure 2.1, from the value of $\bar{V}_m^{(k)}$, the value of $\bar{V}_n^{(k)}$ is calculated as:

$$\bar{V}_n^{(k)} = \bar{V}_m^{(k)} - \bar{Z}_{mn}\bar{I}_{mn}^{(k)} \tag{2.3}$$

In (2.3), $\bar{Z}_{mn}$ is the impedance of the feeder section between buses m and n. By this process, the voltages at all the nodes are updated.

Step 7: Compute $e_i^{(k)} = |\bar{V}_i^{(k)} - \bar{V}_i^{(k-1)}|$ for all $i = 2, \ldots, N$.

Step 8: Compute $e^{(k)} = \max(e_2^{(k)}, e_3^{(k)}, \ldots, e_N^{(k)})$.

Step 9: If $e^{(k)} \leq \epsilon$ (tolerance limit), stop and print the solution. Else set $k = k + 1$ and go to Step 4.

2.1.2 Illustrative example

To illustrate the algorithm described above, a small radial distribution system shown in Figure 2.2 is considered. The data of this system are given in Table 2.1. In this system, bus 1 is the substation bus (root node) with a voltage of 11 kV. In Table 2.1, the symbols R and X denote the resistance and reactance of the feeder section between bus i and bus j, respectively. Further, the symbols P_L and Q_L denote the real and reactive power loads, respectively. For solving the load flow, initially all the bus voltages are initialised to $11\angle 0°$ kV. Subsequently, at each iteration k, following three steps are executed:

i. Calculation of load currents: $\bar{I}_i^{(k)} = \dfrac{P_{Li} - jQ_{Li}}{\{\bar{V}_i^{(k-1)}\}^*}$; $i = 2, 3, \ldots, 6$.

ii. Backward sweep: The branch currents are calculated as: $\bar{I}_{45}^{(k)} = \bar{I}_5^{(k)}$; $\bar{I}_{34}^{(k)} = \bar{I}_4^{(k)} + \bar{I}_{45}^{(k)}$; $\bar{I}_{36}^{(k)} = \bar{I}_6^{(k)}$; $\bar{I}_{23}^{(k)} = \bar{I}_3^{(k)} + \bar{I}_{34}^{(k)} + \bar{I}_{36}^{(k)}$; $\bar{I}_{12}^{(k)} = \bar{I}_2^{(k)} + \bar{I}_{23}^{(k)}$. It is to be noted that in this chapter, the symbol $\bar{I}_{ij}^{(k)}$ denotes the complex current flowing from bus i to bus j in iteration k.

iii. Forward sweep: The bus voltages are updated as: $\bar{V}_2^{(k)} = \bar{V}_1 - \bar{Z}_{12}\bar{I}_{12}^{(k)}$; $\bar{V}_3^{(k)} = \bar{V}_2^{(k)} - \bar{Z}_{23}\bar{I}_{23}^{(k)}$; $\bar{V}_4^{(k)} = \bar{V}_3^{(k)} - \bar{Z}_{34}\bar{I}_{34}^{(k)}$; $\bar{V}_5^{(k)} = \bar{V}_4^{(k)} - \bar{Z}_{45}\bar{I}_{45}^{(k)}$; $\bar{V}_6^{(k)} = \bar{V}_3^{(k)} - \bar{Z}_{36}\bar{I}_{36}^{(k)}$. Again, it is to be noted that the symbol $\bar{Z}_{ij}$ denotes the complex impedance of the branch connecting bus i and bus j.

After the voltages are updated, the convergence of the algorithm is checked following the procedure described in Steps 7–9 of the detailed algorithm given in

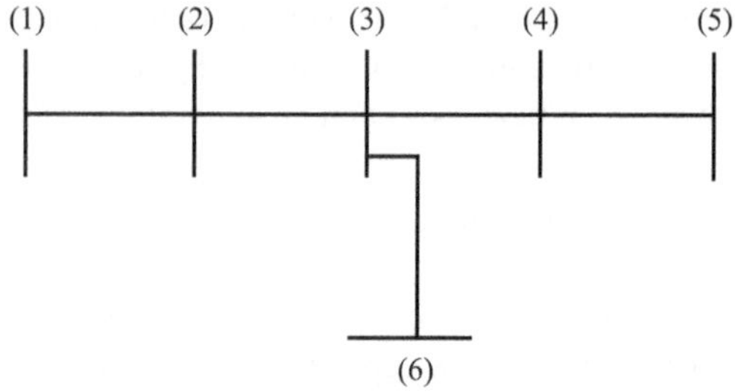

Figure 2.2 A small radial distribution system

Table 2.1 System data and results for the system of Figure 2.2

Line data				Load at bus j		V_{mag} at bus j	
Bus i	**Bus j**	**R (Ω)**	**X (Ω)**	**P_L (kW)**	**Q_L (kVAR)**	**V_{lf} (kV)**	**V_{sim} (kV)**
1	2	0.279	0.015	0	0	10.8654	10.8638
2	3	0.444	0.439	1572	174	10.6218	10.6181
3	4	0.864	0.751	1936	312	10.4178	10.4135
4	5	0.864	0.751	189	63	10.3976	10.3931
3	6	1.374	0.774	1336	112	10.4373	10.4333

Section 2.1.1. If the algorithm is not converged, then the iteration count is advanced and steps (i)–(iii) described above are repeated.

The voltage magnitudes of different buses obtained by the load flow algorithm described above are shown in the seventh column of Table 2.1 (denoted as V_{lf}). For computing the load flow solution, a tolerance of 10^{-12} (in kV) has been chosen and the algorithm converged in 11 iterations. Further, for comparing the accuracy of the load flow algorithm, the steady-state solution of the system shown in Figure 2.2 has also been computed through detailed time domain simulation using PSCAD/EMTDC software [2]. The solutions obtained from time domain simulation are taken as the benchmark for the purpose of comparison. The voltage magnitudes of different buses obtained from time domain simulation are also shown in the eighth column of Table 2.1 (denoted as V_{sim}). These two columns show that for this system, the load flow algorithm gives quite accurate result.

As a second example, a 31-bus radial distribution system described in Reference 3 is considered. The data of this system are given in Table 2.2. In this system, bus 1 is the substation bus (root node) with a voltage of 23 kV. For solving the load flow, initially all the bus voltages are initialised to $23\angle 0°$ kV. Subsequently, by following the steps of the algorithms, the load flow solution of this system has been computed. In this case, the algorithm converged in 17 iterations for a tolerance of 10^{-12} (in kV). The voltage magnitudes and angles (in °, denoted as θ_{lf}) at all the buses obtained after the solution are also shown in Table 2.2.

Table 2.2 System data and results for 31-bus radial distribution system

Line data				Load at bus j		Voltage at bus j	
Bus i	**Bus j**	**R (Ω)**	**X (Ω)**	**P_L (kW)**	**Q_L (kVAR)**	**V_{lf} (kV)**	**θ_{lf} (°)**
1	2	0.896	0.155	0	0	22.3154	0.2899
2	3	0.279	0.015	572	174	22.1447	0.4214
3	4	0.444	0.439	0	0	21.7990	0.0018
4	5	0.864	0.751	936	312	21.4050	−0.4061
5	6	0.864	0.751	0	0	21.0607	−0.7732
6	7	1.374	0.774	0	0	20.5589	−1.0234
7	8	1.374	0.774	0	0	20.2018	−1.2060
8	9	1.374	0.774	0	0	19.8448	−1.3952
9	10	1.374	0.774	189	63	19.6127	−1.5205
10	11	1.374	0.774	0	0	19.3964	−1.6397
11	12	1.374	0.774	336	112	19.1801	−1.7617
12	13	1.374	0.774	657	219	18.9926	−1.8695
13	14	1.374	0.774	783	261	18.8615	−1.9459
14	15	1.374	0.774	729	243	18.7982	−1.9831
9	16	0.864	0.751	477	159	19.7597	−1.4971
16	17	1.374	0.774	549	183	19.6745	−1.5449
17	18	1.374	0.774	477	159	19.6349	−1.5672
7	19	0.864	0.751	432	144	20.4716	−1.1242
19	20	0.864	0.751	672	224	20.4078	−1.1983
20	21	1.374	0.774	495	165	20.3681	−1.2199
7	22	0.864	0.751	207	69	20.5477	−1.0364
4	23	0.444	0.439	522	174	21.6635	−0.1713
23	24	0.444	0.439	1917	639	21.5423	−0.3278
24	25	0.864	0.751	0	0	21.4131	−0.4692
25	26	0.864	0.751	1116	372	21.2840	−0.6123
26	27	0.864	0.751	549	183	21.2134	−0.6911
27	28	1.374	0.774	792	264	21.1523	−0.7231
2	29	0.279	0.015	882	294	22.2815	0.3139
29	30	1.374	0.774	882	294	22.1513	0.2492
30	31	1.374	0.774	882	294	22.0861	0.2165

From Table 2.2, it is observed that at some load buses, the bus voltage is leading the substation voltage. This is due to significant amount of resistance of the distribution feeders, which can be explained as follows. From (2.3), $\bar{V}_n = \bar{V}_m - \bar{Z}_{mn}\bar{I}_{mn} = \bar{V}_m - (R_{mn} + jX_{mn})\bar{I}_{mn}$, where R_{mn} and X_{mn} are resistance and reactance of the feeder section between buses m and n, respectively. Now, if there is an active and reactive (inductive) load at bus n, then $\bar{I}_{mn}$ can be written as, $\bar{I}_{mn} = (I_{mnr} - jI_{mni})$, where I_{mnr} and I_{mni} are the real and imaginary part of $\bar{I}_{mn}$, respectively. Therefore, taking $\bar{V}_m$ as reference, $\bar{V}_n = V_m - (R_{mn} + jX_{mn})(I_{mnr} - jI_{mni}) = (V_m - R_{mn}I_{mnr} - X_{mn}I_{mni}) + j(X_{mn}I_{mnr} - R_{mn}I_{mni})$. Now, depending upon the values of R_{mn}, X_{mn}, I_{mnr} and I_{mni}, the quantity $(X_{mn}I_{mnr} - R_{mn}I_{mni})$ can be greater than zero thereby making $\bar{V}_n$ to lead $\bar{V}_m$. In this context, it is to be noted that if $R_{mn} \approx 0$, then $\bar{V}_n = V_m - jX_{mn}(I_{mnr} - jI_{mni}) = (V_m - X_{mn}I_{mni}) - jX_{mn}I_{mnr}$. Therefore,

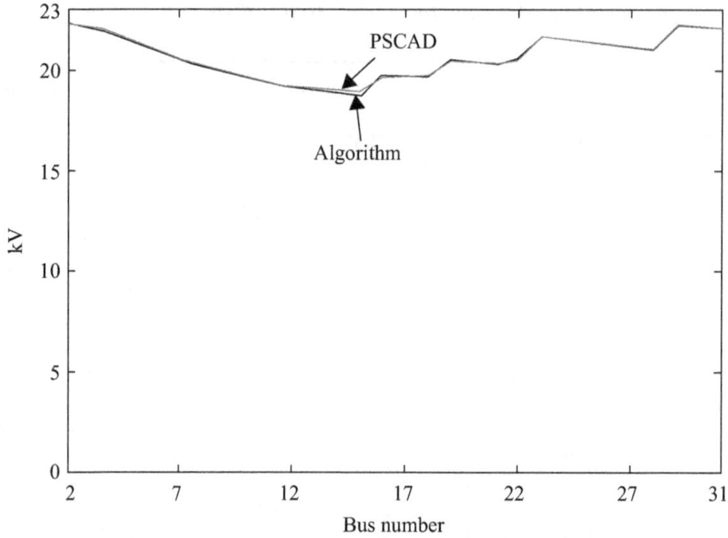

Figure 2.3 Voltage magnitude in 30-bus system

when the feeder resistance is negligible, $\bar{V}_n$ will always lag $\bar{V}_m$ for inductive load connected at bus n.

As in 6-bus system, the voltage magnitudes for the 30-bus system have also been computed using PSCAD/EMTDC and the results are shown in Figure 2.3 (along with the voltage magnitudes obtained by the algorithm). From this figure, it is again observed that the backward/forward sweep algorithm is quite accurate and as a result, the backward/forward load flow algorithm is considered as the standard algorithm for computing the steady-state solution of a balanced radial distribution system.

2.2 Load flow analysis of unbalanced radial distribution system

For the load flow solution of unbalanced radial distribution system, backward/forward load flow method [4] is again mostly used. Therefore, the basic procedure described in Section 2.1 is followed for this case also. However, as the system is unbalanced, in this case, the three-phase voltages and currents are considered. Also, while in the balanced system, a feeder section between bus i and bus j is represented by a series impedance $\bar{Z}_{ij}$, in an unbalanced system, a three-phase feeder is represented by a (3×3) matrix $\mathbf{\bar{Z}_{ij}}$. With these two basic differences, the detailed algorithm can be described as follows.

2.2.1 Detailed algorithm for unbalanced system

Step 1: Initialise all the node voltages to the substation bus voltage. Therefore, $\bar{V}_{ja}^{(0)} = V_s\angle 0°$, $\bar{V}_{jb}^{(0)} = V_s\angle{-120°}$ and $\bar{V}_{jc}^{(0)} = V_s\angle 120°$ for $j = 2, 3, \ldots, N$, where $\bar{V}_{jp}^{(0)}$

denotes the initial complex bus voltage of jth node corresponding to pth phase $(p=a,b,c)$.

Step 2: Initialise iteration count $k=0$.

Step 3: Update iteration count $k=k+1$.

Step 4: At iteration k, the load current $\bar{I}_{jp}^{(k)}$ at node j corresponding to phase p $(p=a,b,c)$ is calculated as:

$$\begin{bmatrix}\bar{I}_{ja}^{(k)} & \bar{I}_{jb}^{(k)} & \bar{I}_{jc}^{(k)}\end{bmatrix}^T = \begin{bmatrix}\dfrac{P_{Lja}-jQ_{Lja}}{\left\{\bar{V}_{ja}^{(k-1)}\right\}^*} & \dfrac{P_{Ljb}-jQ_{Ljb}}{\left\{\bar{V}_{jb}^{(k-1)}\right\}^*} & \dfrac{P_{Ljc}-jQ_{Ljc}}{\left\{\bar{V}_{jc}^{(k-1)}\right\}^*}\end{bmatrix}^T; \; j=2,3,\ldots,N \quad (2.4)$$

In (2.4), P_{Ljp} and Q_{Ljp} denote the real and reactive (inductive) power load at bus j corresponding to phase p $(p=a,b,c)$, respectively, while $\{\bar{V}_{jp}^{(k-1)}\}$ denotes the complex bus voltage of bus j at phase p corresponding to $(k-1)$th iteration.

Step 5: In this step (known as backward sweep), initially the three-phase current in any branch directly connected to a leaf node is calculated to be equal to the three-phase load current at that leaf node. Subsequently, the three-phase current in any other branch is calculated as the sum of the three-phase load current at the receiving end (of that branch) and the three-phase currents in the branches emanating from the receiving end (of that branch). Therefore, in Figure 2.1, the three-phase current in the branch q between the buses m and n at iteration k is given by,

$$\begin{bmatrix}\bar{J}_{qa}^{(k)}\\ \bar{J}_{qb}^{(k)}\\ \bar{J}_{qc}^{(k)}\end{bmatrix} = \begin{bmatrix}\bar{I}_{na}^{(k)}\\ \bar{I}_{nb}^{(k)}\\ \bar{I}_{nc}^{(k)}\end{bmatrix} + \sum_{l\in M}\begin{bmatrix}\bar{J}_{la}^{(k)}\\ \bar{J}_{lb}^{(k)}\\ \bar{J}_{lc}^{(k)}\end{bmatrix} \quad (2.5)$$

In (2.5), $\bar{J}_{qp}^{(k)}$ denotes the complex branch current in the branch q corresponding to phase p $(p=a,b,c)$ at iteration k while M is the set of feeder sections connected at bus n. By this process, the complex three-phase currents in all the branches are calculated.

Step 6: In this step (known as forward sweep), starting from the root node and moving towards the leaf nodes, the three-phase voltages at every node are updated from the knowledge of the three-phase voltage at the respective previous node. For example, in Figure 2.1, from the value of $\bar{V}_{mp}^{(k)}$, the value of $\bar{V}_{np}^{(k)}$ $(p=a,b,c)$ is calculated as:

$$\begin{bmatrix}\bar{V}_{na}^{(k)}\\ \bar{V}_{nb}^{(k)}\\ \bar{V}_{nc}^{(k)}\end{bmatrix} = \begin{bmatrix}\bar{V}_{ma}^{(k)}\\ \bar{V}_{mb}^{(k)}\\ \bar{V}_{mc}^{(k)}\end{bmatrix} - \bar{\mathbf{Z}}_{\mathbf{mn}}\begin{bmatrix}\bar{J}_{qa}^{(k)}\\ \bar{J}_{qb}^{(k)}\\ \bar{J}_{qc}^{(k)}\end{bmatrix} = \begin{bmatrix}\bar{V}_{ma}^{(k)}\\ \bar{V}_{mb}^{(k)}\\ \bar{V}_{mc}^{(k)}\end{bmatrix} - \begin{bmatrix}\bar{Z}_{mn}^{aa} & \bar{Z}_{mn}^{ab} & \bar{Z}_{mn}^{ac}\\ \bar{Z}_{mn}^{ba} & \bar{Z}_{mn}^{bb} & \bar{Z}_{mn}^{bc}\\ \bar{Z}_{mn}^{ca} & \bar{Z}_{mn}^{cb} & \bar{Z}_{mn}^{cc}\end{bmatrix}\begin{bmatrix}\bar{J}_{qa}^{(k)}\\ \bar{J}_{qb}^{(k)}\\ \bar{J}_{qc}^{(k)}\end{bmatrix} \quad (2.6)$$

In (2.6), $\bar{\mathbf{Z}}_{\mathbf{mn}}$ is the (3×3) impedance matrix of the branch q between buses m and n. By this process, the three-phase voltages at all the nodes are updated.

Table 2.3 System data for the unbalanced system having same topology of Figure 2.2

Line data				Load at bus j					
Bus i	**Bus j**	**Length (ft)**	**Impedance matrix**	P_a **(kW)**	Q_a **(kVAR)**	P_b **(kW)**	Q_b **(kVAR)**	P_c **(kW)**	Q_c **(kVAR)**
1	2	1850	$\bar{\mathbf{Z}}_1$	140	70	140	70	350	175
2	3	960	$\bar{\mathbf{Z}}_2$	160	110	120	90	120	90
3	4	1320	$\bar{\mathbf{Z}}_2$	394	225	418	239	128	86
4	5	360	$\bar{\mathbf{Z}}_1$	170	125	230	132	85	40
3	6	520	$\bar{\mathbf{Z}}_2$	485	190	68	60	290	212

Step 7: Compute $e_i^{(k)} = \max\,(|\bar{V}_{ia}^{(k)} - \bar{V}_{ia}^{(k-1)}|,\ |\bar{V}_{ib}^{(k)} - \bar{V}_{ib}^{(k-1)}|,\ |\bar{V}_{ic}^{(k)} - \bar{V}_{ic}^{(k-1)}|)$ for all $i = 2, \ldots, N$.
Step 8: Compute $e^{(k)} = \max\,(e_2^{(k)}, e_3^{(k)}, \ldots, e_N^{(k)})$.
Step 9: If $e^{(k)} \leq \epsilon$ (tolerance limit), stop and print the solution. Else set $k = k + 1$ and go to step 4.

2.2.2 Illustrative example for unbalanced system

To illustrate the load flow algorithm for unbalanced radial distribution system, a small unbalanced radial distribution system having the same topology as shown in Figure 2.2 is considered. The data of this system are given in Table 2.3. In this system, $V_s = 4.16$ kV. In Table 2.3, P_p and Q_p denote real and reactive power load at phase p (=a, b, c), respectively. Further, the quantities $\bar{\mathbf{Z}}_1$ and $\bar{\mathbf{Z}}_2$ denote the branch impedance matrices per mile (in Ω) and are given by

$$\bar{\mathbf{Z}}_1 = \begin{bmatrix} 0.3465 + j1.0179 & 0.1560 + j0.5017 & 0.1580 + j0.4236 \\ 0.1560 + j0.5017i & 0.3375 + j1.0478 & 0.1535 + 0.3849 \\ 0.1580 + j0.4236i & 0.1535 + j0.3849 & 0.3414 + j1.0348 \end{bmatrix}$$

$$\bar{\mathbf{Z}}_2 = \begin{bmatrix} 0.7526 + j1.1814 & 0.1580 + j0.4236 & 0.1560 + 0.5017i \\ 0.1580 + j0.4236 & 0.7475 + j1.1983 & 0.1535 + j0.3849 \\ 0.1560 + j0.5017 & 0.1535 + j0.3849 & 0.7436 + j1.2112 \end{bmatrix}$$

For computing the load flow, initially all the three-phase voltages are initialised as: $\bar{V}_{ja}^{(0)} = 4.16\angle 0^\circ$, $\bar{V}_{jb}^{(0)} = 4.16\angle{-120^\circ}$ and $\bar{V}_{jc}^{(0)} = 4.16\angle 120^\circ$ for $j = 1, 2, \ldots, 6$. Subsequently, at each iteration k, following three steps are executed:

i. Calculation of three-phase load current vector $\bar{\mathbf{I}}_{\mathbf{jabc}}^{\mathbf{(k)}} = [\bar{I}_{ja}^{(k)}\ \bar{I}_{jb}^{(k)}\ \bar{I}_{jc}^{(k)}]^T$; $j = 2, 3, \ldots, 6$, following equation (2.4).
ii. Backward sweep: In this step, the three-phase feeder current vector in the feeder section between buses m and n ($\bar{\mathbf{I}}_{\mathbf{mnabc}}^{\mathbf{(k)}}$) is calculated. This current vector is

Table 2.4 Load flow result of the small unbalanced radial distribution system

Bus no.	V_a (kV)	θ_a (°)	V_b (kV)	θ_b (°)	V_c (kV)	θ_c (°)
1	4.16	0	4.16	−120.0	4.16	120.0
2	4.085	−0.8742	4.1404	−120.6711	4.1039	119.6447
3	4.0225	−1.2837	4.1101	−120.9792	4.0689	119.7116
4	3.9739	−1.3863	4.0694	−121.4193	4.0636	119.7703
5	3.9709	−1.3882	4.0674	−121.4675	4.0640	119.7657
6	4.0122	−1.4448	4.1126	−120.9583	4.0558	119.7271

represented as $\bar{\mathbf{I}}^{(\mathbf{k})}_{\mathbf{mnabc}} = [\bar{I}^{(k)}_{mna}\ \bar{I}^{(k)}_{mnb}\ \bar{I}^{(k)}_{mnc}]^T$ with $\bar{I}^{(k)}_{mnp}$ denoting the complex feeder current in phase p $(=a,b,c)$ of the feeder section between buses m and n. From Figure 2.2, these currents are updated as: $\bar{\mathbf{I}}^{(\mathbf{k})}_{\mathbf{45abc}} = \bar{\mathbf{I}}^{(\mathbf{k})}_{\mathbf{5abc}}$; $\bar{\mathbf{I}}^{(\mathbf{k})}_{\mathbf{34abc}} = \bar{\mathbf{I}}^{(\mathbf{k})}_{\mathbf{4abc}} + \bar{\mathbf{I}}^{(\mathbf{k})}_{\mathbf{45abc}}$; $\bar{\mathbf{I}}^{(\mathbf{k})}_{\mathbf{36abc}} = \bar{\mathbf{I}}^{(\mathbf{k})}_{\mathbf{6abc}}$; $\bar{\mathbf{I}}^{(\mathbf{k})}_{\mathbf{23abc}} = \bar{\mathbf{I}}^{(\mathbf{k})}_{\mathbf{3abc}} + \bar{\mathbf{I}}^{(\mathbf{k})}_{\mathbf{34abc}} + \bar{\mathbf{I}}^{(\mathbf{k})}_{\mathbf{36abc}}$; $\bar{\mathbf{I}}^{(\mathbf{k})}_{\mathbf{12abc}} = \bar{\mathbf{I}}^{(\mathbf{k})}_{\mathbf{2abc}} + \bar{\mathbf{I}}^{(\mathbf{k})}_{\mathbf{23abc}}$.

iii. Forward sweep: In this vector, the three-phase complex bus voltage vector at each jth bus ($\bar{\mathbf{V}}^{(\mathbf{k})}_{\mathbf{jabc}}$) for $j=2,3,\ldots,6$ is updated, where $\bar{\mathbf{V}}^{(\mathbf{k})}_{\mathbf{jabc}}$ is represented as: $\bar{\mathbf{V}}^{(\mathbf{k})}_{\mathbf{jabc}} = [\bar{V}^{(k)}_{ja}\ \bar{V}^{(k)}_{jb}\ \bar{V}^{(k)}_{jc}]^T$ for $j=1,2,\ldots,6$. Now, as bus 1 is the substation bus, $\bar{\mathbf{V}}^{(\mathbf{k})}_{\mathbf{1abc}} = [\bar{V}^{(0)}_{1a}\ \bar{V}^{(0)}_{1b}\ \bar{V}^{(0)}_{1c}]^T$ for all iterations k. Hence, from Figure 2.2, these voltages at all the other buses are updated as: $\bar{\mathbf{V}}^{(\mathbf{k})}_{\mathbf{2abc}} = \bar{\mathbf{V}}^{(\mathbf{k})}_{\mathbf{1abc}} - \bar{\mathbf{Z}}_1\bar{\mathbf{I}}^{(\mathbf{k})}_{\mathbf{12abc}}$; $\bar{\mathbf{V}}^{(\mathbf{k})}_{\mathbf{3abc}} = \bar{\mathbf{V}}^{(\mathbf{k})}_{\mathbf{2abc}} - \bar{\mathbf{Z}}_2\bar{\mathbf{I}}^{(\mathbf{k})}_{\mathbf{23abc}}$; $\bar{\mathbf{V}}^{(\mathbf{k})}_{\mathbf{4abc}} = \bar{\mathbf{V}}^{(\mathbf{k})}_{\mathbf{3abc}} - \bar{\mathbf{Z}}_2\bar{\mathbf{I}}^{(\mathbf{k})}_{\mathbf{34abc}}$; $\bar{\mathbf{V}}^{(\mathbf{k})}_{\mathbf{5abc}} = \bar{\mathbf{V}}^{(\mathbf{k})}_{\mathbf{4abc}} - \bar{\mathbf{Z}}_1\bar{\mathbf{I}}^{(\mathbf{k})}_{\mathbf{45abc}}$; $\bar{\mathbf{V}}^{(\mathbf{k})}_{\mathbf{6abc}} = \bar{\mathbf{V}}^{(\mathbf{k})}_{\mathbf{3abc}} - \bar{\mathbf{Z}}_2\bar{\mathbf{I}}^{(\mathbf{k})}_{\mathbf{36abc}}$.

After the voltages are updated, the convergence of the algorithm is checked following the procedure described in steps 7–9 of the detailed algorithm given in Section 2.2.1. If the algorithm is not converged, then the iteration count is advanced and steps (i)–(iii) described above are repeated.

The load flow results of the above system for a convergence tolerance of 10^{-12} (in kV) are shown in Table 2.4. The algorithm took 11 iterations for convergence. In Table 2.4, the quantities V_p and θ_p denote the voltage magnitude and angle at phase p $(=a,b,c)$, respectively.

As second example, IEEE-37 bus unbalanced radial distribution system [5] is considered. For the purpose of illustration, some modifications have been made to the original system as: (i) the voltage regulator and the transformer in the original system have been neglected, (ii) the shunt charging susceptances of the feeder sections have been neglected and (iii) the nodes of the original system have been re-numbered. As a result, the system now contains 36 nodes and the detailed data of this modified system are given in Table 2.5. In this table, the branch impedance matrices $\bar{\mathbf{Z}}_1$, $\bar{\mathbf{Z}}_2$, $\bar{\mathbf{Z}}_3$ and $\bar{\mathbf{Z}}_4$ are given by (Ω per mile),

$$\bar{\mathbf{Z}}_1 = \begin{bmatrix} 0.2926+j0.1973 & 0.0673-j0.0368 & 0.0337-j0.0417 \\ 0.0673-j0.0368 & 0.2646+j0.1900 & 0.0673-j0.0368 \\ 0.0337-j0.0417 & 0.0673-j0.0368 & 0.2926+j0.1973 \end{bmatrix}$$

Table 2.5 System data for the modified 37-bus unbalanced radial distribution system

Line data				Load at bus *j*					
Bus *i*	**Bus** *j*	**Length (ft)**	**Impedance matrix**	P_a **(kW)**	Q_a **(kVAR)**	P_b **(kW)**	Q_b **(kVAR)**	P_c **(kW)**	Q_c **(kVAR)**
1	2	1850	$\bar{Z}_1$	140	70	140	70	350	175
2	3	960	$\bar{Z}_2$	0	0	0	0	0	0
3	4	1320	$\bar{Z}_2$	0	0	0	0	0	0
3	5	360	$\bar{Z}_3$	0	0	0	0	85	40
5	6	520	$\bar{Z}_3$	0	0	0	0	0	0
6	7	80	$\bar{Z}_4$	17	8	21	10	0	0
7	8	520	$\bar{Z}_4$	85	40	0	0	0	0
3	9	400	$\bar{Z}_4$	0	0	0	0	0	0
9	10	320	$\bar{Z}_4$	8	4	85	40	0	0
9	11	240	$\bar{Z}_4$	0	0	0	0	85	40
4	12	600	$\bar{Z}_3$	0	0	0	0	85	40
12	13	200	$\bar{Z}_3$	0	0	0	0	0	0
13	14	600	$\bar{Z}_3$	0	0	85	40	0	0
13	15	320	$\bar{Z}_3$	0	0	0	0	0	0
15	16	320	$\bar{Z}_4$	0	0	0	0	42	21
15	17	320	$\bar{Z}_3$	85	40	0	0	0	0
17	18	560	$\bar{Z}_3$	0	0	0	0	42	21
18	19	520	$\bar{Z}_4$	0	0	0	0	0	0
19	20	1280	$\bar{Z}_4$	0	0	42	21	0	0
19	21	200	$\bar{Z}_4$	0	0	0	0	85	40
18	22	640	$\bar{Z}_3$	140	70	0	0	0	0
22	23	400	$\bar{Z}_3$	126	62	0	0	0	0
23	24	400	$\bar{Z}_3$	0	0	0	0	0	0
24	25	400	$\bar{Z}_3$	0	0	0	0	42	21
24	26	200	$\bar{Z}_4$	0	0	0	0	85	40
4	27	240	$\bar{Z}_4$	0	0	0	0	42	21
27	28	280	$\bar{Z}_3$	42	21	0	0	0	0
28	29	200	$\bar{Z}_4$	42	21	42	21	42	21
28	30	280	$\bar{Z}_4$	42	21	0	0	0	0
6	31	800	$\bar{Z}_3$	0	0	0	0	85	40
31	32	600	$\bar{Z}_3$	0	0	0	0	0	0
32	33	280	$\bar{Z}_4$	0	0	42	21	0	0
31	34	920	$\bar{Z}_4$	0	0	0	0	0	0
34	35	120	$\bar{Z}_4$	0	0	140	70	21	10
34	36	760	$\bar{Z}_4$	0	0	42	21	0	0

$$\bar{\mathbf{Z}}_2 = \begin{bmatrix} 0.4751 + j0.2973 & 0.1629 - j0.0326 & 0.1234 - j0.0607 \\ 0.1629 - j0.0326 & 0.4488 + j0.2678 & 0.1629 - j0.0326 \\ 0.1234 - j0.0607 & 0.1629 - j0.0326 & 0.4751 + j0.2973 \end{bmatrix}$$

Table 2.6 Load flow result of the modified 37-bus unbalanced radial distribution system

Bus no.	V_a (kV)	θ_a (°)	V_b (kV)	θ_b (°)	V_c (kV)	θ_c (°)
1	4.8	0	4.8	−120.0	4.8	120.0
2	4.7795	−0.0686	4.7881	−120.0756	4.7701	119.8923
3	4.7671	−0.1179	4.7813	−120.1117	4.7555	119.8320
4	4.7515	−0.1711	4.7798	−120.1514	4.7439	119.7984
5	4.7665	−0.1142	4.7769	−120.1038	4.7525	119.8152
6	4.7652	−0.0984	4.7699	−120.1028	4.7510	119.7921
7	4.7644	−0.0971	4.7699	−120.1047	4.7511	119.7936
8	4.7601	−0.0926	4.7705	−120.1163	4.7516	119.8043
9	4.7676	−0.1167	4.7785	−120.0991	4.7526	119.8273
10	4.7677	−0.1089	4.7759	−120.0965	4.7530	119.8208
11	4.7678	−0.1218	4.7787	−120.0935	4.7506	119.8294
12	4.7406	−0.2112	4.7801	−120.1502	4.7326	119.8171
13	4.7369	−0.2206	4.7800	−120.1539	4.7300	119.8239
14	4.7375	−0.2084	4.7767	−120.1543	4.7306	119.8112
15	4.7305	−0.2421	4.7815	−120.1595	4.7255	119.8415
16	4.7306	−0.2456	4.7817	−120.1558	4.7241	119.8432
17	4.7240	−0.2604	4.7829	−120.1683	4.7218	119.8593
18	4.7158	−0.2912	4.7848	−120.1719	4.7148	119.8806
19	4.7165	−0.2964	4.7832	−120.1564	4.7107	119.8792
20	4.7172	−0.2814	4.7779	−120.1484	4.7115	119.8646
21	4.7167	−0.3007	4.7834	−120.1516	4.7091	119.8809
22	4.7053	−0.3132	4.7877	−120.1957	4.7118	119.9138
23	4.7024	−0.3264	4.7889	−120.1963	4.7092	119.9231
24	4.7028	−0.3387	4.7895	−120.1841	4.7059	119.9218
25	4.7030	−0.3428	4.7897	−120.1801	4.7048	119.9216
26	4.7030	−0.3430	4.7897	−120.1794	4.7042	119.9236
27	4.7489	−0.1696	4.7794	−120.1525	4.7424	119.8056
28	4.7468	−0.1701	4.7792	−120.1585	4.7422	119.8097
29	4.7462	−0.1688	4.7786	−120.1572	4.7416	119.8106
30	4.7457	−0.1685	4.7794	−120.1617	4.7424	119.8126
31	4.7682	−0.0756	4.7594	−120.0811	4.7474	119.7440
32	4.7686	−0.0696	4.7578	−120.0808	4.7477	119.7376
33	4.7687	−0.0664	4.7566	−120.0791	4.7479	119.7344
34	4.7706	−0.0342	4.7431	−120.0505	4.7478	119.7008
35	4.7708	−0.0302	4.7414	−120.0472	4.7478	119.6965
36	4.7710	−0.0254	4.7399	−120.0457	4.7483	119.6922

$$\bar{\mathbf{Z}}_3 = \begin{bmatrix} 1.2936+j0.6713 & 0.4871+j0.2111 & 0.4585+j0.1521 \\ 0.4871+j0.2111 & 1.3022+j0.6326 & 0.4871+j0.2111 \\ 0.4585+j0.1521 & 0.4871+j0.2111 & 1.2936+j0.6713 \end{bmatrix}$$

$$\bar{\mathbf{Z}}_4 = \begin{bmatrix} 2.0952+j0.7758 & 0.5204+j0.2738 & 0.4926+j0.2123 \\ 0.5204+j0.2738 & 2.1068+j0.7398 & 0.5204+j0.2738 \\ 0.4926+j0.2123 & 0.5204+j0.2738 & 2.0952+j0.7758 \end{bmatrix}$$

In this system, bus 1 is the substation bus (root node) with a voltage of 4.8 kV (L-N). By following the steps of the algorithm, the load flow solution of this system has been computed. In this case, the algorithm converged in eight iterations for a tolerance of 10^{-12} (in kV). The voltage magnitudes and angles (in °, denoted as θ_{lf}) at all the buses corresponding to all three phases obtained after the solution are shown in Table 2.6.

The load flow solution of the modified 37-bus system has also been computed using PSCAD/EMTDC. The voltage magnitudes obtained by the backward/forward algorithm and PSCAD simulation are almost identical to each other. In fact, the maximum absolute percentage deviation of the voltage magnitudes (with respect to the results obtained from PSCAD) for phase a, phase b and phase c are 0.0031%, 0.0030% and 0.0031%, respectively. Hence, the results obtained from PSCAD are not plotted along with the results obtained by the backward/forward algorithm as both these curves would overlap each other.

2.3 Load flow analysis of balanced weakly meshed distribution system

For solving the power flow problem of a balanced weakly meshed distribution system, different approaches have been developed in the literature such as breakpoint-based techniques [1, 6], direct approach-based method [7], Z_{bus} matrix-based technique [8], etc. Among all these methods, the method developed in Reference 7 is quite straightforward to apply as it depends on direct manipulation of two constant matrices. For any topology of the distribution system (radial or weakly meshed), these two matrices can be easily constructed following two simple algorithms. Therefore, this method is described in detail below. Initially, construction of these two matrices for a radial distribution system is described and subsequently, modification of these two matrices in the presence of loops in the network is discussed.

Let us consider Figure 2.2 again. In Section 2.1.2, while describing the backward sweep, the relations between the branch currents and the node (load) currents have been shown. Working again with those relations, we get (for brevity, the superscript k is omitted),

$$\left.\begin{aligned}
\bar{I}_{45} &= \bar{I}_5 \\
\bar{I}_{34} &= \bar{I}_4 + \bar{I}_{45} = \bar{I}_4 + \bar{I}_5 \\
\bar{I}_{36} &= \bar{I}_6 \\
\bar{I}_{23} &= \bar{I}_3 + \bar{I}_{34} + \bar{I}_{36} = \bar{I}_3 + \bar{I}_4 + \bar{I}_5 + \bar{I}_6 \\
\bar{I}_{12} &= \bar{I}_2 + \bar{I}_{23} = \bar{I}_2 + \bar{I}_3 + \bar{I}_4 + \bar{I}_5 + \bar{I}_6
\end{aligned}\right\} \tag{2.7}$$

Writing the above equations in a matrix form we get,

$$\begin{bmatrix} \bar{I}_{12} \\ \bar{I}_{23} \\ \bar{I}_{34} \\ \bar{I}_{36} \\ \bar{I}_{45} \end{bmatrix} = \begin{bmatrix} 1 & 1 & 1 & 1 & 1 \\ 0 & 1 & 1 & 1 & 1 \\ 0 & 0 & 1 & 1 & 0 \\ 0 & 0 & 0 & 0 & 1 \\ 0 & 0 & 0 & 1 & 0 \end{bmatrix} \begin{bmatrix} \bar{I}_2 \\ \bar{I}_3 \\ \bar{I}_4 \\ \bar{I}_5 \\ \bar{I}_6 \end{bmatrix} \tag{2.8}$$

In (2.8), the matrix having entries 0 and 1 represents the branch currents in terms of the node currents. This matrix is termed as bus injection to branch current ([**BIBC**]) matrix in Reference 7. It is to be noted that this matrix is a constant matrix. Equation (2.8) can be written in compact form as:

$$\bar{\mathbf{I}}_{\mathbf{branch}} = [\mathbf{BIBC}]\bar{\mathbf{I}}_{\mathbf{node}} \tag{2.9}$$

where $\bar{\mathbf{I}}_{\mathbf{branch}} = \left[\bar{I}_{12}\ \bar{I}_{23}\ \bar{I}_{34}\ \bar{I}_{36}\ \bar{I}_{45}\right]^T$ and $\bar{\mathbf{I}}_{\mathbf{node}} = \left[\bar{I}_2\ \bar{I}_3\ \bar{I}_4\ \bar{I}_5\ \bar{I}_6\right]^T$.

For a general n bus, m branch radial distribution system ($m = n - 1$), the algorithm for constructing the [**BIBC**] matrix given in Reference 7 is as follows:

1. The size of the [**BIBC**] matrix is $m \times m$.
2. If ℓth branch section is connected between bus i and bus j, copy the column corresponding to bus i of the [**BIBC**] matrix to the column of jth bus. Also, fill a +1 to the element corresponding to ℓth row and jth column.
3. Repeat step 2 for all branches in the radial distribution system.

Again, in Section 2.1.2, while describing the forward sweep, the relationship between the branch currents and the bus voltages is shown. Working with those relations further, we have (for brevity, the superscript k is omitted again),

$$\left.\begin{aligned}
\bar{V}_2 &= \bar{V}_1 - \bar{Z}_{12}\bar{I}_{12} \\
\bar{V}_3 &= \bar{V}_2 - \bar{Z}_{23}\bar{I}_{23} = \bar{V}_1 - \bar{Z}_{12}\bar{I}_{12} - \bar{Z}_{23}\bar{I}_{23} \\
\bar{V}_4 &= \bar{V}_3 - \bar{Z}_{34}\bar{I}_{34} = \bar{V}_1 - \bar{Z}_{12}\bar{I}_{12} - \bar{Z}_{23}\bar{I}_{23} - \bar{Z}_{34}\bar{I}_{34} \\
\bar{V}_5 &= \bar{V}_4 - \bar{Z}_{45}\bar{I}_{45} = \bar{V}_1 - \bar{Z}_{12}\bar{I}_{12} - \bar{Z}_{23}\bar{I}_{23} - \bar{Z}_{34}\bar{I}_{34} - \bar{Z}_{45}\bar{I}_{45} \\
\bar{V}_6 &= \bar{V}_3 - \bar{Z}_{36}\bar{I}_{36} = \bar{V}_1 - \bar{Z}_{12}\bar{I}_{12} - \bar{Z}_{23}\bar{I}_{23} - \bar{Z}_{36}\bar{I}_{36}
\end{aligned}\right\} \tag{2.10}$$

Again, writing the above equations in a matrix form we get,

$$\begin{bmatrix} \bar{V}_1 \\ \bar{V}_1 \\ \bar{V}_1 \\ \bar{V}_1 \\ \bar{V}_1 \end{bmatrix} - \begin{bmatrix} \bar{V}_2 \\ \bar{V}_3 \\ \bar{V}_4 \\ \bar{V}_5 \\ \bar{V}_6 \end{bmatrix} = \begin{bmatrix} \bar{Z}_{12} & 0 & 0 & 0 & 0 \\ \bar{Z}_{12} & \bar{Z}_{23} & 0 & 0 & 0 \\ \bar{Z}_{12} & \bar{Z}_{23} & \bar{Z}_{34} & 0 & 0 \\ \bar{Z}_{12} & \bar{Z}_{23} & \bar{Z}_{34} & 0 & \bar{Z}_{45} \\ \bar{Z}_{12} & \bar{Z}_{23} & 0 & \bar{Z}_{36} & 0 \end{bmatrix} \begin{bmatrix} \bar{I}_{12} \\ \bar{I}_{23} \\ \bar{I}_{34} \\ \bar{I}_{36} \\ \bar{I}_{45} \end{bmatrix} \tag{2.11}$$

In (2.11), the matrix having entries 0 and branch impedances represents the voltage drops in terms of the branch currents. This matrix is termed as branch current to bus voltage ([**BCBV**]) matrix in Reference 7. It is to be noted that this matrix is also a constant matrix. Equation (2.11) can be written in compact form as:

$$\mathbf{\Delta\bar{V}} = [\mathbf{BCBV}]\bar{\mathbf{I}}_{\mathbf{branch}} \tag{2.12}$$

where $\mathbf{\Delta\bar{V}} = \left[\bar{V}_1 - \bar{V}_2 \quad \bar{V}_1 - \bar{V}_3 \quad \bar{V}_1 - \bar{V}_4 \quad \bar{V}_1 - \bar{V}_5 \quad \bar{V}_1 - \bar{V}_6\right]^T$.

Again, for a general n bus, m branch radial distribution system ($m = n - 1$), the algorithm for constructing the [**BCBV**] matrix given in Reference 7 is as follows:

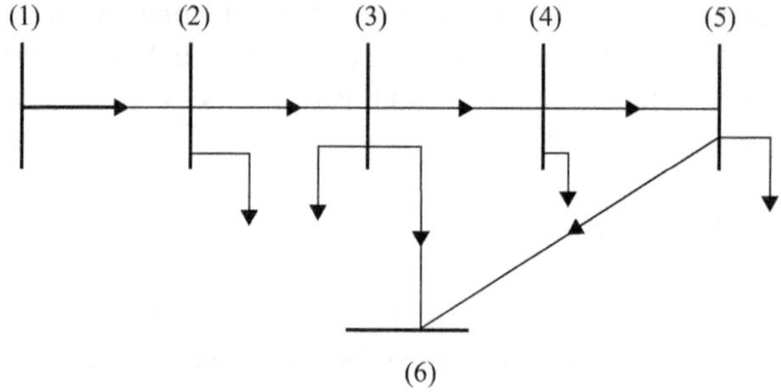

Figure 2.4 A small weakly meshed distribution system

1. The size of the **[BCBV]** matrix is $m \times m$.
2. If ℓth branch section is connected between bus i and bus j, copy the row corresponding to bus i of the **[BCBV]** matrix to the row corresponding to jth bus. Also, fill the line impedance $\bar{Z}_{ij}$ to the element corresponding to jth row and ℓth column.
3. Repeat step 2 for all branches in the radial distribution system.

Now let us assume that in Figure 2.2, a branch between bus 5 and bus 6 exists thereby making the system a weakly meshed system. The diagram of the system is shown in Figure 2.4. Let the current flowing in this branch is $\bar{I}_{56}$ (i.e. flowing from bus 5 to bus 6). The directions of currents flowing in all the other branches and loads are also shown in this figure.

For the system shown in Figure 2.4, the relations between the branch currents and the load currents can be written as:

$$\left.\begin{aligned}
\bar{I}_{45} &= \bar{I}_5 + \bar{I}_{56} \\
\bar{I}_{34} &= \bar{I}_4 + \bar{I}_{45} = \bar{I}_4 + \bar{I}_5 + \bar{I}_{56} \\
\bar{I}_{36} &= \bar{I}_6 - \bar{I}_{56} \\
\bar{I}_{23} &= \bar{I}_3 + \bar{I}_{34} + \bar{I}_{36} = \bar{I}_3 + \bar{I}_4 + \bar{I}_5 + \bar{I}_6 \\
\bar{I}_{12} &= \bar{I}_2 + \bar{I}_{23} = \bar{I}_2 + \bar{I}_3 + \bar{I}_4 + \bar{I}_5 + \bar{I}_6
\end{aligned}\right\} \tag{2.13}$$

The above equations can be written in a matrix form as:

$$\begin{bmatrix} \bar{I}_{12} \\ \bar{I}_{23} \\ \bar{I}_{34} \\ \bar{I}_{36} \\ \bar{I}_{45} \end{bmatrix} = \begin{bmatrix} 1 & 1 & 1 & 1 & 1 \\ 0 & 1 & 1 & 1 & 1 \\ 0 & 0 & 1 & 1 & 0 \\ 0 & 0 & 0 & 0 & 1 \\ 0 & 0 & 0 & 1 & 0 \end{bmatrix} \begin{bmatrix} \bar{I}_2 \\ \bar{I}_3 \\ \bar{I}_4 \\ \bar{I}_5 + \bar{I}_{56} \\ \bar{I}_6 - \bar{I}_{56} \end{bmatrix} = \begin{bmatrix} 1 & 1 & 1 & 1 & 1 \\ 0 & 1 & 1 & 1 & 1 \\ 0 & 0 & 1 & 1 & 0 \\ 0 & 0 & 0 & 0 & 1 \\ 0 & 0 & 0 & 1 & 0 \end{bmatrix} \begin{bmatrix} \bar{I}_2 \\ \bar{I}_3 \\ \bar{I}_4 \\ \bar{I}_5 \\ \bar{I}_6 \end{bmatrix} + \begin{bmatrix} 1 & 1 \\ 1 & 1 \\ 1 & 0 \\ 0 & 1 \\ 1 & 0 \end{bmatrix} \begin{bmatrix} \bar{I}_{56} \\ -\bar{I}_{56} \end{bmatrix} \tag{2.14}$$

Equation (2.14) can be written in the following way to get the modified [**BIBC**] matrix:

$$\begin{bmatrix} \bar{I}_{12} \\ \bar{I}_{23} \\ \bar{I}_{34} \\ \bar{I}_{36} \\ \bar{I}_{45} \\ \bar{I}_{56} \end{bmatrix} = \begin{bmatrix} 1 & 1 & 1 & 1 & 1 & 0 \\ 0 & 1 & 1 & 1 & 1 & 0 \\ 0 & 0 & 1 & 1 & 0 & 1 \\ 0 & 0 & 0 & 0 & 1 & -1 \\ 0 & 0 & 0 & 1 & 0 & 1 \\ 0 & 0 & 0 & 0 & 0 & 1 \end{bmatrix} \begin{bmatrix} \bar{I}_2 \\ \bar{I}_3 \\ \bar{I}_4 \\ \bar{I}_5 \\ \bar{I}_6 \\ \bar{I}_{56} \end{bmatrix} \tag{2.15}$$

Now, in Figure 2.4, branch 5–6 creates a loop. Therefore, let us denote the current $\bar{I}_{56}$ as $\bar{I}_{\text{loop}}$. With this new notation, (2.15) can be re-written as:

$$\begin{bmatrix} \bar{\mathbf{I}}_{\mathbf{branch}} \\ \bar{I}_{\text{loop}} \end{bmatrix} = [\mathbf{BIBC}] \begin{bmatrix} \bar{\mathbf{I}}_{\mathbf{node}} \\ \bar{I}_{\text{loop}} \end{bmatrix} \tag{2.16}$$

In general, in a general n bus, m branch distribution system ($m \geq n$), the number of loops is $l = m - n + 1$. In this general case, the quantity $\bar{I}_{\text{loop}}$ would be replaced by a vector $\bar{\mathbf{I}}_{\mathbf{loop}}$ which will contain the currents in those branches forming l loops in the system. Consequently, the sizes of the vectors $\bar{\mathbf{I}}_{\mathbf{branch}}$, $\bar{\mathbf{I}}_{\mathbf{node}}$ and $\bar{\mathbf{I}}_{\mathbf{loop}}$ would be $(n-1) \times 1$, $(n-1) \times 1$ and $l \times 1$, respectively.

In Reference 7, an algorithm for constructing the [**BIBC**] matrix for a general n bus, m branch distribution system ($m \geq n$) is given as follows:

1. The size of the [**BIBC**] matrix is $m \times m$.
2. If the ℓth branch section is connected between bus i and bus j without forming a loop, follow step 2 of the algorithm for forming [**BIBC**] matrix in a radial distribution system. On the other hand, if the ℓth branch section forms a loop, then copy the column corresponding to ith bus to the ℓth column and subtract the column corresponding to jth bus from this new ℓth column and finally fill a $+1$ to the element corresponding to ℓth row and ℓth column.
3. Repeat step 2 for all branches in the distribution system.

Now, to construct the [**BCBV**] matrix of the system shown in Figure 2.4, the voltage equations of this system are to be written. It is to be noted that all the equations of the equation set (2.10) are still valid for this system. Apart from these equations, another extra equation for the loop of Figure 2.4 can be written as:

$$\bar{Z}_{34}\bar{I}_{34} + \bar{Z}_{45}\bar{I}_{45} + \bar{Z}_{56}\bar{I}_{56} - \bar{Z}_{36}\bar{I}_{36} = 0 \tag{2.17}$$

Writing (2.10) and (2.17) together in a matrix form we get:

$$\begin{bmatrix} \bar{V}_1 \\ \bar{V}_1 \\ \bar{V}_1 \\ \bar{V}_1 \\ \bar{V}_1 \\ 0 \end{bmatrix} - \begin{bmatrix} \bar{V}_2 \\ \bar{V}_3 \\ \bar{V}_4 \\ \bar{V}_5 \\ \bar{V}_6 \\ 0 \end{bmatrix} = \begin{bmatrix} \bar{Z}_{12} & 0 & 0 & 0 & 0 & 0 \\ \bar{Z}_{12} & \bar{Z}_{23} & 0 & 0 & 0 & 0 \\ \bar{Z}_{12} & \bar{Z}_{23} & \bar{Z}_{34} & 0 & 0 & 0 \\ \bar{Z}_{12} & \bar{Z}_{23} & \bar{Z}_{34} & 0 & \bar{Z}_{45} & 0 \\ \bar{Z}_{12} & \bar{Z}_{23} & 0 & \bar{Z}_{36} & 0 & 0 \\ 0 & 0 & \bar{Z}_{34} & -\bar{Z}_{36} & \bar{Z}_{45} & \bar{Z}_{56} \end{bmatrix} \begin{bmatrix} \bar{I}_{12} \\ \bar{I}_{23} \\ \bar{I}_{34} \\ \bar{I}_{36} \\ \bar{I}_{45} \\ \bar{I}_{56} \end{bmatrix} \tag{2.18}$$

The above equation can be re-written as:

$$\begin{bmatrix} \Delta\bar{\mathbf{V}} \\ 0 \end{bmatrix} = [\mathbf{BCBV}] \begin{bmatrix} \bar{\mathbf{I}}_{\mathbf{branch}} \\ \bar{I}_{\text{loop}} \end{bmatrix} \tag{2.19}$$

It is to be noted that in a general n bus, m branch distribution system ($m \geq n$), the single entry 0 in (2.19) would be replaced by a null vector whose dimension would be $l \times 1$. In Reference 7, an algorithm for constructing the [**BCBV**] matrix for a general n bus, m branch distribution system ($m \geq n$) is given, which is described as follows:

1. The size of the [**BCBV**] matrix is $m \times m$.
2. If the ℓth branch section is connected between bus i and bus j without forming a loop, follow step 2 of the algorithm for forming [**BCBV**] matrix in a radial distribution system. On the other hand, if the ℓth branch section forms a loop, add a new row to the original [**BCBV**] matrix corresponding to the KVL equation in the loop.
3. Repeat step 2 for all branches in the distribution system.

Substituting (2.16) into (2.19) one gets:

$$\begin{bmatrix} \Delta\bar{\mathbf{V}} \\ 0 \end{bmatrix} = [\mathbf{BCBV}][\mathbf{BIBC}] \begin{bmatrix} \bar{\mathbf{I}}_{\mathbf{node}} \\ \bar{I}_{\text{loop}} \end{bmatrix} = \begin{bmatrix} \mathbf{A} & \mathbf{B}^T \\ \mathbf{B} & \mathbf{C} \end{bmatrix} \begin{bmatrix} \bar{\mathbf{I}}_{\mathbf{node}} \\ \bar{I}_{\text{loop}} \end{bmatrix} \tag{2.20}$$

In (2.20), the dimensions of the matrices **A**, **B** and **C** are (5×5), (1×5) and (1×1), respectively. These matrices are given by:

$$\mathbf{A} = \begin{bmatrix} \bar{Z}_{12} & \bar{Z}_{12} & \bar{Z}_{12} & \bar{Z}_{12} & \bar{Z}_{12} \\ \bar{Z}_{12} & \bar{Z}_{12}+\bar{Z}_{23} & \bar{Z}_{12}+\bar{Z}_{23} & \bar{Z}_{12}+\bar{Z}_{23} & \bar{Z}_{12}+\bar{Z}_{23} \\ \bar{Z}_{12} & \bar{Z}_{12}+\bar{Z}_{23} & \bar{Z}_{12}+\bar{Z}_{23}+\bar{Z}_{34} & \bar{Z}_{12}+\bar{Z}_{23}+\bar{Z}_{34} & \bar{Z}_{12}+\bar{Z}_{23} \\ \bar{Z}_{12} & \bar{Z}_{12}+\bar{Z}_{23} & \bar{Z}_{12}+\bar{Z}_{23}+\bar{Z}_{34} & \bar{Z}_{12}+\bar{Z}_{23}+\bar{Z}_{34}+\bar{Z}_{45} & \bar{Z}_{12}+\bar{Z}_{23} \\ \bar{Z}_{12} & \bar{Z}_{12}+\bar{Z}_{23} & \bar{Z}_{12}+\bar{Z}_{23} & \bar{Z}_{12}+\bar{Z}_{23} & \bar{Z}_{12}+\bar{Z}_{23}+\bar{Z}_{36} \end{bmatrix} \tag{2.21}$$

$$\mathbf{B} = \begin{bmatrix} 0 & 0 & \bar{Z}_{34} & \bar{Z}_{34}+\bar{Z}_{45} & -\bar{Z}_{36} \end{bmatrix}; \qquad \mathbf{C} = \left(\bar{Z}_{34}+\bar{Z}_{36}+\bar{Z}_{45}+\bar{Z}_{56}\right) \tag{2.22}$$

For a general n bus, m branch distribution system ($m \geq n$) having $l(=m-n+1)$ loops, the sizes of the matrices **A**, **B** and **C** are $(n-1) \times (n-1)$, $l \times (n-1)$ and $l \times l$, respectively. Application of Cron reduction to (2.20) leads to [7],

$$\Delta\bar{\mathbf{V}} = \left[\mathbf{A} - \mathbf{B}^T\mathbf{C}^{-1}\mathbf{B}\right]\bar{\mathbf{I}}_{\mathbf{node}} = [\mathbf{DLF}]\,\bar{\mathbf{I}}_{\mathbf{node}} \tag{2.23}$$

It is to be noted that the matrix **[DLF]** is a constant matrix, which is based only on the feeder parameters. Therefore, this matrix needs to be calculated only once. Based on (2.23), the detailed algorithm for computing the load flow solution of a n bus, m branch ($m \geq n$) weakly meshed distribution system is given in Reference 7 as follows.

2.3.1 Detailed algorithm

Step 1: Initialise all the node voltages to $V_s\angle 0°$ (i.e. $\bar{V}_j^{(0)} = V_s\angle 0°$ for $j = 2, 3, \ldots, n$).
Step 2: Initialise iteration count $k = 0$.
Step 3: Update iteration count $k = k + 1$.
Step 4: At iteration k, the load current $\bar{I}_j^{(k)}$ at node j is calculated as in (2.1).
Step 5: At iteration k, form the vector $\bar{\mathbf{I}}_{\mathbf{node}}^{\mathbf{(k)}}$ as $\bar{\mathbf{I}}_{\mathbf{node}}^{\mathbf{(k)}} = \left[\bar{I}_2^{(k)}\ \bar{I}_3^{(k)}\ \cdots\ \bar{I}_n^{(k)}\right]^T$.
Step 6: Calculate $\mathbf{\Delta\bar{V}}^{\mathbf{(k)}} = [\mathbf{DLF}]\bar{\mathbf{I}}_{\mathbf{node}}^{\mathbf{(k)}}$, where $\mathbf{\Delta\bar{V}}^{\mathbf{(k)}} = \left[\Delta\bar{V}_2^{(k)}\ \Delta\bar{V}_3^{(k)}\ \cdots\ \Delta\bar{V}_n^{(k)}\right]^T$.
Step 7: Update $\bar{V}_j^{(k)} = \bar{V}_j^{(k-1)} + \Delta\bar{V}_j^{(k)}$ for $j = 2, 3, \ldots, n$.
Step 8: Compute $e_i^{(k)} = |\bar{V}_i^{(k)} - \bar{V}_i^{(k-1)}|$ for all $i = 2, \ldots, n$.
Step 9: Compute $e^{(k)} = \max(e_2^{(k)}, e_3^{(k)}, \ldots, e_n^{(k)})$.
Step 10: If $e^{(k)} \leq \epsilon$ (tolerance limit), stop and print the solution. Else set $k = k + 1$ and go to step 3.

2.3.2 Examples

As the first example, the system shown in Figure 2.4 is first considered. For this system, the load data are same as given in Table 2.1. Further, the impedances of the feeder sections 1-2, 2-3, 3-4, 4-5 and 5-6 are also same as given in Table 2.1. The impedance of the feeder section 5-6 is $0.896 + j0.155\ \Omega$. The **BIBC** and **BCBV** matrices for this system are given in (2.15) and (2.18), respectively. The load flow results obtained by the algorithm described in Section 2.3.1 are shown in Table 2.7. In this table, the symbols V_{lf} and θ_{lf} denote the voltage magnitude and angle, respectively, obtained by the load flow algorithm. Further, for validation purpose, the system shown in Figure 2.4 has also been simulated using PSCAD/EMTDC. The voltage magnitudes obtained by PSCAD/EMTDC (denoted by V_{sim}) are shown in Table 2.7. From these results, it is observed that the voltage magnitudes obtained by the load flow algorithm and time domain simulation are almost identical to each other and thus, the algorithm described in Section 2.3.1 is quite accurate.

As a second example, the 31-bus distribution system is again considered. The feeder parameter and load data of the base (radial) system are given in Table 2.2. However, in this base system, two extra feeder sections are now considered between (a) buses 11 and 20 and (b) buses 16 and 29. The impedances of these two feeder sections are $(0.896 + j0.155)\ \Omega$ and $(1.374 + j0.774)\ \Omega$, respectively. Therefore, with the addition of these two feeder sections, the 31-bus system has now become a weakly meshed system having two loops. The load flow solution of this system has been computed by the algorithm and is shown in Table 2.8. Further, this system has also been simulated

Table 2.7 Load flow result of 6-bus weakly meshed distribution system

Bus no.	V_{lf} (kV)	θ_{lf} (°)	V_{sim} (kV)	Bus no.	V_{lf} (kV)	θ_{lf} (°)	V_{sim} (kV)
1	11.0	0	11.0	4	10.4257	−1.4739	10.4256
2	10.8654	0.0707	10.8654	5	10.4130	−1.4505	10.4129
3	10.6218	−0.884	10.6218	6	10.4233	−1.4203	10.4232

Table 2.8 Load flow result of 31-bus weakly meshed distribution system

Bus no.	V_{lf} (kV)	θ_{lf} (°)	V_{sim} (kV)	Bus no.	V_{lf} (kV)	θ_{lf} (°)	V_{sim} (kV)
1	23	0	23	17	21.8431	0.088	21.8429
2	22.3422	0.2595	22.342	18	21.8063	0.0699	21.8072
3	22.2364	0.342	22.2373	19	21.5625	−0.2216	21.5617
4	22.0294	0.0983	22.0297	20	21.4728	−0.2988	21.4717
5	21.8914	−0.022	21.8922	21	21.4337	−0.3183	21.434
6	21.8017	−0.0915	21.8025	22	21.6637	−0.1323	21.6633
7	21.6729	−0.1206	21.674	23	21.896	−0.0712	21.8957
8	21.7235	−0.0886	21.7233	24	21.7764	−0.2244	21.7759
9	21.7718	−0.0567	21.7726	25	21.6476	−0.3627	21.648
10	21.6039	−0.168	21.6042	26	21.5211	−0.5027	21.5204
11	21.4498	−0.2736	21.4502	27	21.4498	−0.5798	21.4506
12	21.2566	−0.3731	21.2556	28	21.39	−0.611	21.3901
13	21.0864	−0.4607	21.0868	29	22.2502	0.3199	22.2498
14	20.9691	−0.5226	20.969	30	22.1191	0.255	22.1195
15	20.9116	−0.5527	20.9121	31	22.0547	0.2223	22.0542
16	21.919	0.1268	21.9196				

using PSCAD/EMTDC. The voltage magnitudes obtained by PSCAD/EMTDC are also shown in this table. It can be seen that the voltage magnitudes obtained by these two methods are very close to each other (in fact, the maximum percentage difference between these two sets of results is only 0.005%) thereby again establishing the accuracy of the method.

2.4 Conclusion

In this chapter, load flow techniques for balanced radial distribution system, unbalanced radial distribution system and balanced weakly meshed distribution system are discussed along with suitable examples. These techniques are used to ensure that different tasks in an automated distribution system are accomplished without violating any system constraint.

Acknowledgement

The author would like to thank Mr. Akhilesh Mathur, research scholar in the Department of Electrical Engineering, IIT Roorkee, for carrying out the PSCAD simulation studies.

References

[1] D. Shirmohammadi, H. W. Long, A. Semlyen and G. X. Luo, "A compensation based power flow method for weakly meshed distribution and transmission networks", *IEEE Transactions on Power System*, vol. 3, no. 2, pp. 753–762, May 1988.

[2] PSCAD/EMTDC, Version 4.2, Manitoba HVDC Research Center, Winnipeg, Manitoba, Canada.

[3] S. K. Goswami and S. K. Basu, "Direct solution of distribution system", *IEE Proceeding – Part C, Generation, Transmission and Distribution*, vol. 138, no. 1, pp. 78–88, January 1991.

[4] C. S. Cheng and D. Shirmohammadi, "A three-phase power flow method for real time distribution system analysis", *IEEE Transactions on Power System*, vol. 10, no. 2, pp. 671–679, May 1995.

[5] IEEE Power and Energy Society, "Distribution test feeders". Available from http://ewh.ieee.org/soc/pes/dsacom/testfeeders/.

[6] G. X. Luo and A. Semlyen, "Efficient load flow for large weakly meshed networks", *IEEE Transactions on Power System*, vol. 5, no. 4, pp. 1309–1316, November 1990.

[7] J. H. Teng, "A direct approach for distribution system load flow solutions", *IEEE Transactions on Power Delivery*, vol. 18, no. 3, pp. 882–887, July 2003.

[8] T.-H. Chen, M.-S. Chen, K.-J. Hwang, P. Kotas and E. A. Chebli, "Distribution system power flow analysis – a rigid approach", *IEEE Transactions on Power Delivery*, vol. 6, no. 3, pp. 1146–1152, July 1991.

Chapter 3
Short-circuit analysis

Vinay Pant[1]

3.1 Introduction

Under normal operating conditions, the currents flowing through the components of a power system are usually well within the specified limits. However, when a fault occurs in a system, currents far in excess of normal current values flow through the elements of the network. These large currents, if not interrupted in time or limited in value, can cause serious damage to the equipment through which they flow. The faults in a power system can also cause interruptions in power supply to the consumers and affect the power quality. The severity of a fault depends on the type of fault, location of fault, fault impedance, system impedance and voltage level. The process of estimating the currents and voltages in different parts of a power system under abnormal conditions is called fault analysis or short-circuit analysis [1]. The information obtained from the short-circuit analysis helps in designing a properly coordinated protective system and selection of appropriately rated equipments for the system. This analysis can also be helpful in estimating the size of current limiting reactors or fault current limiters required for limiting the fault current to a safe value.

The faults occurring in a power system can be categorized as balanced or unbalanced faults. In the balanced or symmetrical faults, all the three phases are equally affected and the system remains balanced, e.g. three-phase fault (LLL or LLLG faults). In the unbalanced or unsymmetrical faults, the system no longer remains balanced and the three phases are affected differently. The examples of this type of fault are: (i) line to ground (LG) fault, (ii) line to line (LL) fault and (iii) double line to ground (LLG) fault. The percentage of occurrences of various types of faults is given in Table 3.1. The category *others* includes LLL, LLLG and open conductor faults. The most commonly occurring fault in power systems is the LG fault while the most severe fault (in terms of fault current) is the LLLG fault.

Short-circuit analysis of transmission and distributions systems has usually been carried out using sequence component-based methods [1, 3, 4]. For transmission systems, which are usually transposed and hence balanced in nature, these methods

[1]Department of Electrical Engineering, Indian Institute of Technology Roorkee, Roorkee, Uttarakhand, India

Table 3.1 Frequency of occurrence of short-circuit faults [2]

No.	Fault type	Percentage of occurrence
1	LG	70
2	LL	15
3	LLG	4
4	Others	11

give accurate results. Distribution systems, however, are inherently unbalanced as they also have single phase, two phase and unbalanced loads apart from the usual three-phase balanced loads. Moreover, the distribution feeders are not transposed. As a result, distribution systems do not fulfil the fundamental requirement of a balanced system for the application of sequence component-based methods. Hence, these methods do not give accurate results for the short-circuit analysis of unbalanced distribution systems. A study was carried out in 5 to ascertain the error in the short circuit analysis of distribution systems using symmetrical component-based method and it was observed that the error in the results obtained by this method is not negligible and can significantly affect the coordination of protective devices in the system. Further, it was also established that the error in fault current estimation is a function of the degree of unbalance present in the system. The error in applying symmetrical component-based short-circuit analysis methods to distribution systems lead to the development of various phase variable-based short-circuit analysis methods [6–11]. One of the recent phase variable-based distribution system short-circuit analysis methods is based on bus injection to branch current (BIBC) and branch current to bus voltage (BCBV) matrices [9–11] and this method is described in detail in this chapter.

3.2 Short-circuit analysis of unbalanced radial distribution system

In Chapter 2, the [**BIBC**] and [**BCBV**] matrices have been described for a single-phase distribution system. These matrices can be easily extended to three-phase distribution systems. The (1×1) element of [**BIBC**] matrix for a single-phase element needs to be replaced by a (3×3) identity matrix $[I]$ for a three-phase element. For example, the *ij*th element of the [**BIBC**] matrix is equal to 1 for single-phase case, while for the three-phase case, the (3×3) matrix $[\mathbf{BIBC}_{ij}^{abc}]$ is given by

$$\left[\mathbf{BIBC}_{ij}^{abc}\right] = \begin{array}{c} \\ a \\ b \\ c \end{array}\begin{array}{c} \begin{array}{ccc} a & b & c \end{array} \\ \begin{bmatrix} 1 & 0 & 0 \\ 0 & 1 & 0 \\ 0 & 0 & 1 \end{bmatrix} \end{array}$$

Similarly, the impedance $\bar{Z}_{ij}$ of a single-phase line element connected between ith and jth node is a single element, while for a three-phase element it is a (3×3) matrix as given below:

$$\left[\bar{Z}_{\mathbf{ij}}^{\mathbf{abc}}\right] = \begin{matrix} & a & b & c \\ a & \bar{Z}_{ij}^{aa} & \bar{Z}_{ij}^{ab} & \bar{Z}_{ij}^{ac} \\ b & \bar{Z}_{ij}^{ba} & \bar{Z}_{ij}^{bb} & \bar{Z}_{ij}^{bc} \\ c & \bar{Z}_{ij}^{ca} & \bar{Z}_{ij}^{cb} & \bar{Z}_{ij}^{cc} \end{matrix}$$

Further, the three-phase voltages and currents of an individual bus or branch in the circuit are represented by (3×1) arrays. For two-phase elements, the corresponding entries of **[BIBC]** and **[BCBV]** matrices are (2×2) matrices and the branch currents and bus voltages are (2×1) arrays. The method for the short-circuit analysis of a distribution system for different types of faults is explained next. In this method, the bus active and reactive loads have not been considered while calculating the short-circuit currents and voltages in the network.

3.2.1 Single line to ground fault

For an LG fault through a fault impedance $\bar{Z}_f$ on phase a of ith bus (shown in Figure 3.1a), the boundary conditions can be expressed as [9]

$$\bar{I}_i^a = \bar{I}_{i,f}^a, \quad \bar{I}_i^b = 0, \quad \bar{I}_i^c = 0 \tag{3.1}$$

$$\bar{V}_{i,f}^a = \bar{Z}_f \bar{I}_{i,f}^a \tag{3.2}$$

where $\bar{I}_{i,f}^a$ is the fault current of phase a and $\bar{I}_i^a$, $\bar{I}_i^b$ and $\bar{I}_i^c$ are the bus injection currents of phases a, b and c, of ith bus respectively. $\bar{V}_{i,f}^a$ is the post-fault voltage of phase a of ith bus.

The branch current vector under fault conditions can be expressed as

$$\left[\bar{\mathbf{I}}_{\mathbf{br,f}}\right] = \left[\mathbf{BIBC}\right]\left[0 \quad 0 \quad 0 \quad \cdots \quad \bar{I}_{i,f}^a \quad 0 \quad 0 \quad \cdots \quad 0 \quad 0\right]^T \tag{3.3}$$

Equation (3.3) can be simplified as

$$\left[\bar{\mathbf{I}}_{\mathbf{br,f}}\right] = \left[\mathbf{BIBC}_{\mathbf{i}}^{\mathbf{a}}\right]\bar{I}_{i,f}^a \tag{3.4}$$

where $[\mathbf{BIBC}_{\mathbf{i}}^{\mathbf{a}}]$ is the column vector of **[BIBC]** matrix corresponding to phase a of the faulted bus i. The voltage drop in phase a from the substation bus to the fault point can be written as

$$\Delta\bar{V}_{i,f}^a = \bar{V}_s^a - \bar{V}_{i,f}^a \tag{3.5}$$

where $\bar{V}_s^a$ is the voltage of phase a of the substation bus.

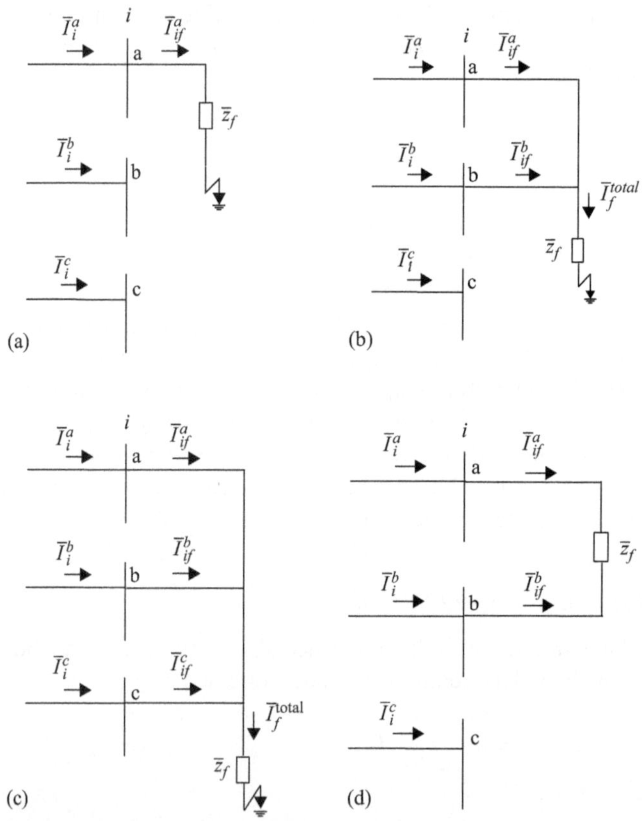

Figure 3.1 Unsymmetrical short-circuit faults: (a) LG fault; (b) LLG fault; (c) LLLG fault; and (d) LL fault

The voltage drop $\Delta\bar{V}^a_{i,f}$ can be written as [10]

$$\Delta\bar{V}^a_{i,f} = [\mathbf{BCBV^a_i}][\mathbf{BIBC^a_i}]\bar{I}^a_{i,f} \tag{3.6}$$

where $[\mathbf{BCBV^a_i}]$ is the row vector of [**BCBV**] matrix corresponding to phase *a* of *i*th bus.

From (3.2), (3.5) and (3.6), we get

$$\bar{V}^a_s - \bar{Z}_f\bar{I}^a_{i,f} = [\mathbf{BCBV^a_i}][\mathbf{BIBC^a_i}]\bar{I}^a_{i,f} \tag{3.7}$$

Now, from (3.7), the fault current $\bar{I}^a_{i,f}$ can be calculated as

$$\begin{aligned}\bar{I}^a_{i,f} &= \left([\mathbf{BCBV^a_i}][\mathbf{BIBC^a_i}] + \bar{Z}_f\right)^{-1}\bar{V}^a_s \\ &= \left(\left[\bar{Z}^{\mathrm{LG}}_{\mathrm{sc}}\right]^{-1}\bar{V}^a_s\right)\end{aligned} \tag{3.8}$$

where $\left[\bar{Z}_{\mathrm{sc}}^{\mathrm{LG}}\right]$ is the short-circuit impedance matrix for an LG fault and it given as

$$\left[\bar{Z}_{\mathrm{sc}}^{\mathrm{LG}}\right] = \left[\mathbf{BCBV_i^a}\right]\left[\mathbf{BIBC_i^a}\right] + \bar{Z}_f \tag{3.9}$$

For a single line to ground fault, the short-circuit impedance matrix is (1×1) in size. After calculating the fault current from (3.8), the branch currents under fault conditions can be calculated from (3.4). The post-fault bus voltage of *j*th bus $[\bar{\mathbf{V}}_{\mathbf{j,f}}]$ can be calculated as [10]

$$\left[\bar{\mathbf{V}}_{\mathbf{j,f}}\right] = \left[\bar{\mathbf{V}}_{\mathbf{s}}\right] - \left[\mathbf{BCBV_j}\right]\left[\mathbf{I_{br,f}}\right] \qquad \forall j = 1,\ldots,n, \quad j \neq i \tag{3.10}$$

where $[\bar{\mathbf{V}}_{\mathbf{s}}]$ is the substation bus voltage vector, $[\mathbf{BCBV_j}]$ is the portion of $[\mathbf{BCBV}]$ matrix corresponding to *j*th bus and *n* is the number of buses in the system. The sizes of $[\bar{\mathbf{V}}_{\mathbf{s}}]$ and $[\mathbf{BCBV_j}]$ depend on the number of phases at *j*th bus. The numbers of rows for these two matrices can be one or two or three depending on whether the *j*th bus is a single-phase bus or a two-phase bus or a three-phase bus.

3.2.2 *Double line to ground fault*

For an LLG fault through a fault impedance $\bar{Z}_f$ on phases *a* and *b* of *i*th bus (shown in Figure 3.1b), the boundary conditions can be expressed as

$$\bar{I}_i^a = \bar{I}_{i,f}^a, \quad \bar{I}_i^b = \bar{I}_{i,f}^b, \quad \bar{I}_i^c = 0 \tag{3.11}$$

$$\bar{V}_{i,f}^a = \bar{V}_{i,f}^b = \bar{Z}_f\left(\bar{I}_{i,f}^a + \bar{I}_{i,f}^b\right) \tag{3.12}$$

where $\bar{I}_{i,f}^a, \bar{I}_{i,f}^b$ and $\bar{V}_{i,f}^a, \bar{V}_{i,f}^b$ are the fault currents and the post-fault voltages of phases *a* and *b* of *i*th bus, respectively.

The branch current vector under fault conditions can be expressed as

$$\left[\bar{\mathbf{I}}_{\mathbf{br,f}}\right] = \left[\mathbf{BIBC}\right]\left[0 \quad 0 \quad 0 \quad \cdots \quad \bar{I}_{i,f}^a \quad \bar{I}_{i,f}^b \quad 0 \quad \cdots \quad 0 \quad 0\right]^T \tag{3.13}$$

Equation (3.13) can be rewritten as

$$\left[\bar{\mathbf{I}}_{\mathbf{br,f}}\right] = \left[\mathbf{BIBC_i^a} \quad \mathbf{BIBC_i^b}\right]\begin{bmatrix}\bar{I}_{i,f}^a \\ \bar{I}_{i,f}^b\end{bmatrix} \tag{3.14}$$

In (3.14), $[\mathbf{BIBC_i^a}]$ and $[\mathbf{BIBC_i^b}]$ are the column vectors of $[\mathbf{BIBC}]$ matrix corresponding to phases *a* and *b* of the faulted bus *i*. The voltage drops in phases *a* and *b* from the substation bus to the fault point can be expressed as

$$\Delta\bar{V}_{i,f}^a = \bar{V}_s^a - \bar{V}_{i,f}^a \tag{3.15a}$$

$$\Delta\bar{V}_{i,f}^b = \bar{V}_s^b - \bar{V}_{i,f}^b \tag{3.15b}$$

where $\bar{V}_s^a$ and $\bar{V}_s^b$ are the voltages of phases *a* and *b* of the substation bus.

The voltage drops $\Delta\bar{V}^a_{i,f}$ and $\Delta\bar{V}^b_{i,f}$ can be written as [9]

$$\begin{bmatrix} \Delta\bar{V}^a_{i,f} \\ \Delta\bar{V}^b_{i,f} \end{bmatrix} = \begin{bmatrix} \mathbf{BCBV^a_i} \\ \mathbf{BCBV^b_i} \end{bmatrix} \begin{bmatrix} \mathbf{BIBC^a_i} & \mathbf{BIBC^b_i} \end{bmatrix} \begin{bmatrix} \bar{I}^a_{i,f} \\ \bar{I}^b_{i,f} \end{bmatrix} \tag{3.16}$$

where $[\mathbf{BCBV^a_i}]$ and $[\mathbf{BCBV^b_i}]$ are the row vectors of $[\mathbf{BCBV}]$ matrix, corresponding to phases a and b of ith bus.

From (3.12), (3.15) and (3.16), we get

$$\begin{bmatrix} \bar{V}^a_s - \bar{Z}_f\left(\bar{I}^a_{i,f} + \bar{I}^b_{i,f}\right) \\ \bar{V}^b_s - \bar{Z}_f\left(\bar{I}^a_{i,f} + \bar{I}^b_{i,f}\right) \end{bmatrix} = \begin{bmatrix} \mathbf{BCBV^a_i} \\ \mathbf{BCBV^b_i} \end{bmatrix} \begin{bmatrix} \mathbf{BIBC^a_i} & \mathbf{BIBC^b_i} \end{bmatrix} \begin{bmatrix} \bar{I}^a_{i,f} \\ \bar{I}^b_{i,f} \end{bmatrix} \tag{3.17}$$

Equation (3.17) can be rearranged as

$$\begin{bmatrix} \bar{V}^a_s \\ \bar{V}^b_s \end{bmatrix} = \left(\begin{bmatrix} \mathbf{BCBV^a_i} \\ \mathbf{BCBV^b_i} \end{bmatrix} \begin{bmatrix} \mathbf{BIBC^a_i} & \mathbf{BIBC^b_i} \end{bmatrix} + \begin{bmatrix} \bar{Z}_f & \bar{Z}_f \\ \bar{Z}_f & \bar{Z}_f \end{bmatrix} \right) \begin{bmatrix} \bar{I}^a_{i,f} \\ \bar{I}^b_{i,f} \end{bmatrix} \tag{3.18}$$

Now the fault currents of phases a and b can be calculated as

$$\begin{bmatrix} \bar{I}^a_{i,f} \\ \bar{I}^b_{i,f} \end{bmatrix} = \left(\begin{bmatrix} \mathbf{BCBV^a_i} \\ \mathbf{BCBV^b_i} \end{bmatrix} \begin{bmatrix} \mathbf{BIBC^a_i} & \mathbf{BIBC^b_i} \end{bmatrix} + \begin{bmatrix} \bar{Z}_f & \bar{Z}_f \\ \bar{Z}_f & \bar{Z}_f \end{bmatrix} \right)^{-1} \begin{bmatrix} \bar{V}^a_s \\ \bar{V}^b_s \end{bmatrix}$$

$$\begin{bmatrix} \bar{I}^a_{i,f} \\ \bar{I}^b_{i,f} \end{bmatrix} = \left[\bar{\mathbf{Z}}^{\mathbf{LLG}}_{\mathbf{sc}}\right]^{-1} \begin{bmatrix} \bar{V}^a_s \\ \bar{V}^b_s \end{bmatrix} \tag{3.19}$$

where

$$\left[\bar{\mathbf{Z}}^{\mathbf{LLG}}_{\mathbf{sc}}\right] = \begin{bmatrix} \mathbf{BCBV^a_i} \\ \mathbf{BCBV^b_i} \end{bmatrix} \begin{bmatrix} \mathbf{BIBC^a_i} & \mathbf{BIBC^b_i} \end{bmatrix} \begin{bmatrix} \bar{Z}_f & \bar{Z}_f \\ \bar{Z}_f & \bar{Z}_f \end{bmatrix} \tag{3.20}$$

where $[\bar{\mathbf{Z}}^{\mathbf{LLG}}_{\mathbf{sc}}]$ is the short-circuit impedance matrix for a double line to ground fault and is (2×2) in size.

Once the fault current is calculated from (3.19), the branch currents can be calculated using (3.14) and the post-fault bus voltage of jth bus $[\bar{\mathbf{V}}_{\mathbf{j,f}}]$ can be estimated using (3.10).

3.2.3 Triple line to ground fault

For an LLLG fault through a fault impedance $\bar{Z}_f$ on phases a, b and c of ith bus (shown in Figure 3.1c), the boundary conditions can be expressed as

$$\bar{I}^a_i = \bar{I}^a_{i,f}, \quad \bar{I}^b_i = \bar{I}^b_{i,f}, \quad \bar{I}^c_i = \bar{I}^c_{i,f} \tag{3.21}$$

$$\bar{V}^a_{i,f} = \bar{V}^b_{i,f} = \bar{V}^c_{i,f} = \bar{Z}_f\left(\bar{I}^a_{i,f} + \bar{I}^b_{i,f} + \bar{I}^c_{i,f}\right) \tag{3.22}$$

where $\bar{I}^a_{i,f}$, $\bar{I}^b_{i,f}$, $\bar{I}^c_{i,f}$ and $\bar{V}^a_{i,f}$, $\bar{V}^b_{i,f}$, $\bar{V}^c_{i,f}$ are the fault currents and the post-fault voltages of phases *a*, *b* and *c* of *i*th bus, respectively.

The branch current vector under fault conditions can be expressed as

$$\left[\bar{\mathbf{I}}_{\mathbf{br,f}}\right] = \left[\mathbf{BIBC}\right]\left[0 \quad 0 \quad 0 \quad \cdots \quad \bar{I}^a_{i,f} \quad \bar{I}^b_{i,f} \quad \bar{I}^c_{i,f} \quad \cdots \quad 0 \quad 0\right]^T \tag{3.23}$$

Equation (3.23) can be rewritten as

$$\left[\bar{\mathbf{I}}_{\mathbf{br,f}}\right] = \left[\mathbf{BIBC^a_i} \quad \mathbf{BIBC^b_i} \quad \mathbf{BIBC^c_i}\right]\begin{bmatrix} \bar{I}^a_{i,f} \\ \bar{I}^b_{i,f} \\ \bar{I}^c_{i,f} \end{bmatrix} \tag{3.24}$$

In (3.24), $[\mathbf{BIBC^a_i}]$, $[\mathbf{BIBC^b_i}]$ and $[\mathbf{BIBC^c_i}]$ are the column vectors of $[\mathbf{BIBC}]$ matrix corresponding to phases *a*, *b* and *c* of the faulted bus *i*, respectively. The voltage drops in phases *a*, *b* and *c* from the substation bus to the fault point can be expressed as

$$\Delta\bar{V}^a_{i,f} = \bar{V}^a_s - \bar{V}^a_{i,f} \tag{3.25a}$$

$$\Delta\bar{V}^b_{i,f} = \bar{V}^b_s - \bar{V}^b_{i,f} \tag{3.25b}$$

$$\Delta\bar{V}^c_{i,f} = \bar{V}^c_s - \bar{V}^c_{i,f} \tag{3.25c}$$

where $\bar{V}^a_s$, $\bar{V}^b_s$ and $\bar{V}^c_s$ are the voltages of phases *a*, *b* and *c* of the substation bus.

The voltage drops $\Delta\bar{V}^a_{i,f}$, $\Delta\bar{V}^b_{i,f}$ and $\Delta\bar{V}^c_{i,f}$ can be written as [10]

$$\begin{bmatrix} \Delta\bar{V}^a_{i,f} \\ \Delta\bar{V}^b_{i,f} \\ \Delta\bar{V}^c_{i,f} \end{bmatrix} = \begin{bmatrix} \mathbf{BCBV^a_i} \\ \mathbf{BCBV^b_i} \\ \mathbf{BCBV^c_i} \end{bmatrix}\left[\mathbf{BIBC^a_i} \quad \mathbf{BIBC^b_i} \quad \mathbf{BIBC^c_i}\right]\begin{bmatrix} \bar{I}^a_{i,f} \\ \bar{I}^b_{i,f} \\ \bar{I}^c_{i,f} \end{bmatrix} \tag{3.26}$$

where $[\mathbf{BCBV^a_i}]$, $[\mathbf{BCBV^b_i}]$ and $[\mathbf{BCBV^c_i}]$ are the row vectors of $[\mathbf{BCBV}]$ matrix, corresponding to phases *a*, *b* and *c* of *i*th bus.

From (3.22), (3.25) and (3.26), we get

$$\begin{bmatrix} \bar{V}^a_s - \bar{Z}_f\left(\bar{I}^a_{i,f} + \bar{I}^b_{i,f} + \bar{I}^c_{i,f}\right) \\ \bar{V}^b_s - \bar{Z}_f\left(\bar{I}^a_{i,f} + \bar{I}^b_{i,f} + \bar{I}^c_{i,f}\right) \\ \bar{V}^c_s - \bar{Z}_f\left(\bar{I}^a_{i,f} + \bar{I}^b_{i,f} + \bar{I}^c_{i,f}\right) \end{bmatrix} = \begin{bmatrix} \mathbf{BCBV^a_i} \\ \mathbf{BCBV^b_i} \\ \mathbf{BCBV^c_i} \end{bmatrix}\left[\mathbf{BIBC^a_i} \quad \mathbf{BIBC^b_i} \quad \mathbf{BIBC^c_i}\right]\begin{bmatrix} \bar{I}^a_{i,f} \\ \bar{I}^b_{i,f} \\ \bar{I}^c_{i,f} \end{bmatrix} \tag{3.27}$$

Equation (3.27) can be written as

$$\begin{bmatrix} \bar{V}_s^a \\ \bar{V}_s^b \\ \bar{V}_s^c \end{bmatrix} = \left(\begin{bmatrix} \mathbf{BCBV_i^a} \\ \mathbf{BCBV_i^b} \\ \mathbf{BCBV_i^c} \end{bmatrix} \begin{bmatrix} \mathbf{BIBC_i^a} & \mathbf{BIBC_i^b} & \mathbf{BIBC_i^c} \end{bmatrix} + \begin{bmatrix} \bar{Z}_f & \bar{Z}_f & \bar{Z}_f \\ \bar{Z}_f & \bar{Z}_f & \bar{Z}_f \\ \bar{Z}_f & \bar{Z}_f & \bar{Z}_f \end{bmatrix} \right) \begin{bmatrix} \bar{I}_{i,f}^a \\ \bar{I}_{i,f}^b \\ \bar{I}_{i,f}^c \end{bmatrix} \tag{3.28}$$

Now the fault currents of phases *a*, *b* and *c* can be calculated as

$$\begin{bmatrix} \bar{I}_{i,f}^a \\ \bar{I}_{i,f}^b \\ \bar{I}_{i,f}^c \end{bmatrix} = \left[\bar{\mathbf{Z}}_{\mathbf{sc}}^{\mathbf{LLLG}} \right]^{-1} \begin{bmatrix} \bar{V}_s^a \\ \bar{V}_s^b \\ \bar{V}_s^c \end{bmatrix} \tag{3.29}$$

where

$$\left[\bar{\mathbf{Z}}_{\mathbf{sc}}^{\mathbf{LLLG}} \right] = \begin{bmatrix} \mathbf{BCBV_i^a} \\ \mathbf{BCBV_i^b} \\ \mathbf{BCBV_i^c} \end{bmatrix} \begin{bmatrix} \mathbf{BIBC_i^a} & \mathbf{BIBC_i^b} & \mathbf{BIBC_i^c} \end{bmatrix} + \begin{bmatrix} \bar{Z}_f & \bar{Z}_f & \bar{Z}_f \\ \bar{Z}_f & \bar{Z}_f & \bar{Z}_f \\ \bar{Z}_f & \bar{Z}_f & \bar{Z}_f \end{bmatrix} \tag{3.30}$$

where $[\bar{\mathbf{Z}}_{\mathbf{sc}}^{\mathbf{LLLG}}]$ is the short-circuit impedance matrix for a triple line to ground fault and is (3 × 3) in size. Once the fault current is calculated from (3.29), the branch currents can be calculated using (3.24) and the post-fault bus voltage $[\bar{\mathbf{V}}_{\mathbf{j,f}}]$ of *j*th bus can be found out from (3.10).

3.2.4 Line to line fault

When an LL fault occurs through a fault impedance $\bar{Z}_f$ on phases *a* and *b* of *i*th bus (shown in Figure 3.1d), the boundary conditions can be expressed as

$$\bar{I}_i^a = \bar{I}_{i,f}^a, \quad \bar{I}_i^b = -\bar{I}_{i,f}^a, \quad \bar{I}_i^c = 0 \tag{3.31}$$

$$\bar{V}_{i,f}^a - \bar{V}_{i,f}^b = \bar{Z}_f \bar{I}_{i,f}^a \tag{3.32}$$

Equation (3.32) can be written in terms of voltage drops in phases *a* and *b* ($\Delta\bar{V}_{i,f}^a$ and $\Delta\bar{V}_{i,f}^b$) from the substation bus to the fault point as

$$\begin{aligned} \left(\bar{V}_s^a - \bar{V}_{i,f}^a\right) - \left(\bar{V}_s^b - \bar{V}_{i,f}^b\right) &= \left(\bar{V}_s^a - \bar{V}_s^b\right) - \bar{Z}_f \bar{I}_{i,f}^a \\ \Delta\bar{V}_{i,f}^a - \Delta\bar{V}_{i,f}^b &= \left(\bar{V}_s^a - \bar{V}_s^b\right) - \bar{Z}_f \bar{I}_{i,f}^a \end{aligned} \tag{3.33}$$

The branch current vector under LL fault conditions can be expressed as

$$\begin{aligned}
\left[\bar{\mathbf{I}}_{\mathbf{br,f}}\right] &= \left[\mathbf{BIBC}\right]\left[0 \quad 0 \quad 0 \quad \cdots \quad \bar{I}^{a}_{i,f} \quad -\bar{I}^{a}_{i,f} \quad 0 \quad \cdots \quad 0 \quad 0\right]^{T} \\
\left[\bar{\mathbf{I}}_{\mathbf{br,f}}\right] &= \left[\mathbf{BIBC^{a}_{i}} \quad \mathbf{BIBC^{b}_{i}}\right]\begin{bmatrix} \bar{I}^{a}_{i,f} \\ -\bar{I}^{a}_{i,f} \end{bmatrix}
\end{aligned} \tag{3.34}$$

The voltage drops $\Delta\bar{V}^{a}_{i,f}$ and $\Delta\bar{V}^{b}_{i,f}$ can be expressed in terms of the elements of [**BIBC**], [**BCBV**] and $[\bar{\mathbf{I}}_{\mathbf{br,f}}]$ as

$$\begin{bmatrix} \Delta\bar{V}^{a}_{i,f} \\ \Delta\bar{V}^{b}_{i,f} \end{bmatrix} = \begin{bmatrix} \mathbf{BCBV^{a}_{i}} \\ \mathbf{BCBV^{b}_{i}} \end{bmatrix}\left[\mathbf{BIBC^{a}_{i}} \quad \mathbf{BIBC^{b}_{i}}\right]\begin{bmatrix} \bar{I}^{a}_{i,f} \\ -\bar{I}^{a}_{i,f} \end{bmatrix} \tag{3.35}$$

Equation (3.35) can be expanded as

$$\begin{aligned}
\begin{bmatrix} \Delta\bar{V}^{a}_{i,f} \\ \Delta\bar{V}^{b}_{i,f} \end{bmatrix} &= \begin{bmatrix} [\mathbf{BCBV^{a}_{i}}][\mathbf{BIBC^{a}_{i}}] & [\mathbf{BCBV^{a}_{i}}][\mathbf{BIBC^{b}_{i}}] \\ [\mathbf{BCBV^{b}_{i}}][\mathbf{BIBC^{a}_{i}}] & [\mathbf{BCBV^{b}_{i}}][\mathbf{BIBC^{b}_{i}}] \end{bmatrix}\begin{bmatrix} \bar{I}^{a}_{i,f} \\ -\bar{I}^{a}_{i,f} \end{bmatrix} \\
\begin{bmatrix} \Delta\bar{V}^{a}_{i,f} \\ \Delta\bar{V}^{b}_{i,f} \end{bmatrix} &= \begin{bmatrix} \bar{Z}^{aa}_{i} & \bar{Z}^{ab}_{i} \\ \bar{Z}^{ba}_{i} & \bar{Z}^{bb}_{i} \end{bmatrix}\begin{bmatrix} \bar{I}^{a}_{i,f} \\ -\bar{I}^{a}_{i,f} \end{bmatrix}
\end{aligned} \tag{3.36}$$

where

$$\begin{aligned}
\bar{Z}^{aa}_{i} &= \left[\mathbf{BCBV^{a}_{i}}\right]\left[\mathbf{BIBC^{a}_{i}}\right]; \quad \bar{Z}^{ab}_{i} = \left[\mathbf{BCBV^{a}_{i}}\right]\left[\mathbf{BIBC^{b}_{i}}\right] \\
\bar{Z}^{ba}_{i} &= \left[\mathbf{BCBV^{b}_{i}}\right]\left[\mathbf{BIBC^{a}_{i}}\right]; \quad \bar{Z}^{bb}_{i} = \left[\mathbf{BCBV^{b}_{i}}\right]\left[\mathbf{BIBC^{b}_{i}}\right]
\end{aligned} \tag{3.37}$$

From the above equation, $(\Delta\bar{V}^{a}_{i,f} - \Delta\bar{V}^{b}_{i,f})$ can be calculated as

$$\left(\Delta\bar{V}^{a}_{i,f} - \Delta\bar{V}^{b}_{i,f}\right) = \left(\bar{Z}^{aa}_{i} - 2\bar{Z}^{ab}_{i} + \bar{Z}^{bb}_{i}\right)\bar{I}^{a}_{i,f} \tag{3.38}$$

From (3.33) and (3.38), the fault current $\bar{I}^{a}_{i,f}$ can be calculated as

$$\begin{aligned}
\bar{I}^{a}_{i,f} &= \left[\bar{Z}^{aa}_{i} - 2\bar{Z}^{ab}_{i} + \bar{Z}^{bb}_{i} + \bar{Z}_{f}\right]^{-1}\left(\bar{V}^{a}_{s} - \bar{V}^{b}_{s}\right) \\
\bar{I}^{a}_{i,f} &= \left[\bar{Z}^{\mathrm{LL}}_{\mathrm{sc}}\right]^{-1}\left(\bar{V}^{a}_{s} - \bar{V}^{b}_{s}\right)
\end{aligned} \tag{3.39}$$

where

$$\left[\bar{Z}^{\mathrm{LL}}_{\mathrm{sc}}\right] = \left[\bar{Z}^{aa}_{i} - 2\bar{Z}^{ab}_{i} + \bar{Z}^{bb}_{i} + \bar{Z}_{f}\right] \tag{3.40}$$

where $[\bar{Z}^{\mathrm{LL}}_{\mathrm{sc}}]$ is the short-circuit impedance matrix for a line to line fault and is (1×1) in size. After calculating the fault current from (3.40), the branch currents and post-fault bus voltages can be found out using (3.35) and (3.10), respectively.

3.3 Short-circuit analysis of unbalanced weakly meshed distribution system

3.3.1 Single line to ground fault

For an LG fault in a weakly meshed distribution system through a fault impedance $\bar{Z}_f$ on phase a of ith bus (shown in Figure 3.1a), the boundary conditions are identical to those given in (3.1) and (3.2).

Equation (3.4) can be extended for a weakly meshed distribution system as follows:

$$\begin{bmatrix} \bar{\mathbf{I}}^{\mathbf{rad}}_{\mathbf{br,f}} \\ \cdots \\ \bar{\mathbf{I}}^{\mathbf{loop}}_{\mathbf{br,f}} \end{bmatrix} = \begin{bmatrix} \mathbf{BIBC}^{\mathbf{rad,a}}_{\mathbf{i}} & \vdots & \mathbf{BIBC}^{\mathbf{loop}} \end{bmatrix} \begin{bmatrix} \bar{I}^{a}_{i,f} \\ \cdots \\ \bar{\mathbf{I}}^{\mathbf{loop}}_{\mathbf{br,f}} \end{bmatrix} \tag{3.41}$$

where $\bar{\mathbf{I}}^{\mathbf{rad}}_{\mathbf{br,f}}$ and $\bar{\mathbf{I}}^{\mathbf{loop}}_{\mathbf{br,f}}$ are the branch currents of the radial and meshed branches of the distribution network under fault conditions, respectively. $[\mathbf{BIBC}^{\mathbf{rad,a}}_{\mathbf{i}}]$ is the column vector of $[\mathbf{BIBC}]$ matrix for the radial part of the distribution network corresponding to phase a of the faulted bus i and $[\mathbf{BIBC}^{\mathbf{loop}}]$ matrix is the $[\mathbf{BIBC}]$ matrix corresponding to the meshed part of the distribution system.

The voltage drops for radial and meshed parts of the network can be written as [11]

$$\begin{bmatrix} \Delta\bar{V}^{a}_{i,f} \\ \cdots \\ \mathbf{0} \end{bmatrix} = \begin{bmatrix} \mathbf{BCBV}^{\mathbf{rad,a}}_{\mathbf{i}} \\ \cdots \\ \mathbf{BCBV}^{\mathbf{loop}} \end{bmatrix} \begin{bmatrix} \mathbf{BIBC}^{\mathbf{rad,a}}_{\mathbf{i}} & | & \mathbf{BIBC}^{\mathbf{loop}} \end{bmatrix} \begin{bmatrix} \bar{I}^{a}_{i,f} \\ \cdots \\ \bar{\mathbf{I}}^{\mathbf{loop}}_{\mathbf{br,f}} \end{bmatrix} \tag{3.42}$$

where $[\mathbf{BCBV}^{\mathbf{rad,a}}_{\mathbf{i}}]$ is the row vector of $[\mathbf{BCBV}]$ matrix for the radial part of the distribution network corresponding to phase a of ith bus and $[\mathbf{BCBV}^{\mathbf{loop}}]$ represents the $[\mathbf{BCBV}]$ matrix of the meshed part of the network.

From (3.2), (3.5) and (3.42), we get

$$\begin{bmatrix} \bar{V}^{a}_{s} - \bar{Z}_f\bar{I}^{a}_{i,f} \\ \cdots \\ \mathbf{0} \end{bmatrix} = \begin{bmatrix} \left[\mathbf{BCBV}^{\mathbf{rad,a}}_{\mathbf{i}}\right]\left[\mathbf{BIBC}^{\mathbf{rad,a}}_{\mathbf{i}}\right] & \left[\mathbf{BCBV}^{\mathbf{rad,a}}_{\mathbf{i}}\right]\left[\mathbf{BIBC}^{\mathbf{loop}}\right] \\ \cdots & \cdots \\ \left[\mathbf{BCBV}^{\mathbf{loop}}\right]\left[\mathbf{BIBC}^{\mathbf{rad,a}}_{\mathbf{i}}\right] & \left[\mathbf{BCBV}^{\mathbf{loop}}\right]\left[\mathbf{BIBC}^{\mathbf{loop}}\right] \end{bmatrix} \begin{bmatrix} \bar{I}^{a}_{i,f} \\ \cdots \\ \bar{\mathbf{I}}^{\mathbf{loop}}_{\mathbf{br,f}} \end{bmatrix}$$

$$\begin{bmatrix} \bar{V}^{a}_{s} - \bar{Z}_f\bar{I}^{a}_{i,f} \\ \cdots \\ \mathbf{0} \end{bmatrix} = \begin{bmatrix} \bar{Z}^{rr}_{i} & [\bar{Z}^{rl}_{i}] \\ \cdots & \cdots \\ [\bar{Z}^{lr}_{i}] & [\bar{Z}^{ll}_{i}] \end{bmatrix} \begin{bmatrix} \bar{I}^{a}_{i,f} \\ \cdots \\ \bar{\mathbf{I}}^{\mathbf{loop}}_{\mathbf{br,f}} \end{bmatrix} \tag{3.43}$$

For a network with ℓ number of three-phase loops, $\bar{Z}^{rr}_{i} = [\mathbf{BCBV}^{\mathbf{rad,a}}_{\mathbf{i}}][\mathbf{BIBC}^{\mathbf{rad,a}}_{\mathbf{i}}]$ is (1×1) in size, $[\bar{Z}^{rl}_{i}] = [\mathbf{BCBV}^{\mathbf{rad,a}}_{\mathbf{i}}][\mathbf{BIBC}^{\mathbf{loop}}]$ is $(1 \times 3\ell)$ in size,

$[\bar{Z}_i^{lr}] = [\mathbf{BCBV^{loop}}][\mathbf{BIBC_i^{rad,a}}]$ is $(3\ell \times 1)$ in size and $[\bar{Z}_i^{ll}] = [\mathbf{BCBV^{loop}}][\mathbf{BIBC^{loop}}]$ is $(3\ell \times 3\ell)$ in size.

Now, from (3.43), the fault current $\bar{I}_{i,f}^a$ and $[\bar{\mathbf{I}}_{\mathbf{br,f}}^{\mathbf{loop}}]$ can be calculated as

$$\begin{bmatrix} \bar{I}_{i,f}^a \\ \cdots \\ \bar{\mathbf{I}}_{\mathbf{br,f}}^{\mathbf{loop}} \end{bmatrix} = \begin{bmatrix} \bar{Z}_i^{rr} + \bar{Z}_f & [\bar{Z}_i^{rl}] \\ \cdots & \cdots \\ [\bar{Z}_i^{lr}] & [\bar{Z}_i^{ll}] \end{bmatrix}^{-1} \begin{bmatrix} \bar{V}_s^a \\ \cdots \\ \mathbf{0} \end{bmatrix}$$

$$= \left[\bar{\mathbf{Z}}_{\mathbf{sc}}^{\mathbf{LGm}}\right]^{-1} \begin{bmatrix} \bar{V}_s^a \\ \cdots \\ \mathbf{0} \end{bmatrix} \tag{3.44}$$

The size of the short-circuit impedance matrix for an LG fault $\left[\bar{\mathbf{Z}}_{\mathbf{sc}}^{\mathbf{LGm}}\right]$ is $(3\ell + 1) \times (3\ell + 1)$.

After calculating the fault current and the branch currents for the meshed part of the network from (3.44), the branch currents and post-fault bus voltages can be calculated using (3.41). The post-fault voltage of $[\bar{\mathbf{V}}_{\mathbf{j,f}}]$ of jth bus can be determined using the following relation [11]:

$$\left[\bar{\mathbf{V}}_{\mathbf{j,f}}\right] = \left[\bar{\mathbf{V}}_{\mathbf{s}}\right] - \left[\mathbf{BCBV_j^{rad}}\right]\left[\mathbf{I_{br,f}^{rad}}\right] \qquad \forall j = 1, \ldots, n, \quad j \neq i \tag{3.45}$$

where $[\bar{\mathbf{V}}_{\mathbf{s}}]$ is the substation bus voltage vector, $[\mathbf{BCBV_j^{rad}}]$ is the row vector of $[\mathbf{BCBV^{rad}}]$ matrix corresponding to jth bus and n is the number of buses in the system.

3.3.2 *Double line to ground fault*

For an LLG fault through a fault impedance $\bar{Z}_f$ on phases a and b of ith bus (shown in Figure 3.1b), the boundary conditions are identical to those described in (3.11) and (3.12).

Equation (3.14) can be modified for a weakly meshed distribution system as follows:

$$\begin{bmatrix} \bar{\mathbf{I}}_{\mathbf{br,f}}^{\mathbf{rad}} \\ \cdots \\ \bar{\mathbf{I}}_{\mathbf{br,f}}^{\mathbf{loop}} \end{bmatrix} = \left[\mathbf{BIBC_i^{rad,a}} \quad \mathbf{BIBC_i^{rad,b}} \ \vdots \ \mathbf{BIBC^{loop}}\right] \begin{bmatrix} \bar{I}_{i,f}^a \\ \bar{I}_{i,f}^b \\ \cdots \\ \bar{\mathbf{I}}_{\mathbf{br,f}}^{\mathbf{loop}} \end{bmatrix} \tag{3.46}$$

where $[\mathbf{BIBC_i^{rad,a}}]$ and $[\mathbf{BIBC_i^{rad,b}}]$ are the column vectors of $[\mathbf{BIBC}]$ matrix for the radial part of the distribution network corresponding to phases a and b, respectively, of the faulted bus i.

The voltage drops for radial and meshed parts of the network can be written as [11]

$$\begin{bmatrix} \Delta\bar{V}_{i,f}^{a} \\ \Delta\bar{V}_{i,f}^{b} \\ \cdots \\ \mathbf{0} \end{bmatrix} = \begin{bmatrix} \mathbf{BCBV_i^{rad,a}} \\ \mathbf{BCBV_i^{rad,b}} \\ \cdots \\ \mathbf{BCBV^{loop}} \end{bmatrix} \begin{bmatrix} \mathbf{BIBC_i^{rad,a}} & \mathbf{BIBC_i^{rad,b}} & \vdots & \mathbf{BIBC^{loop}} \end{bmatrix} \begin{bmatrix} \bar{I}_{i,f}^{a} \\ \bar{I}_{i,f}^{b} \\ \cdots \\ \bar{\mathbf{I}}_{\mathbf{br,f}}^{\mathbf{loop}} \end{bmatrix} \tag{3.47}$$

where $[\mathbf{BCBV_i^{rad,a}}]$ and $[\mathbf{BCBV_i^{rad,b}}]$ are the row vectors of $[\mathbf{BCBV}]$ matrix for the radial part of the distribution network corresponding to phases a and b, respectively, of ith bus.

From (3.12), (3.15) and (3.47), we get

$$\begin{bmatrix} \bar{I}_{i,f}^{a} \\ \bar{I}_{i,f}^{b} \\ \cdots \\ \bar{\mathbf{I}}_{\mathbf{br,f}}^{\mathbf{loop}} \end{bmatrix} = \left[\bar{\mathbf{Z}}_{\mathbf{sc}}^{\mathbf{LLGm}}\right]^{-1} \begin{bmatrix} \bar{V}_{s}^{a} \\ \bar{V}_{s}^{b} \\ \cdots \\ \mathbf{0} \end{bmatrix} \tag{3.48}$$

where

$$\left[\bar{\mathbf{Z}}_{\mathbf{sc}}^{\mathbf{LLGm}}\right] = \begin{bmatrix} \mathbf{BCBV_i^{rad,a}} \\ \mathbf{BCBV_i^{rad,b}} \\ \cdots \\ \mathbf{BCBV^{loop}} \end{bmatrix} \begin{bmatrix} \mathbf{BIBC_i^{rad,a}} & \mathbf{BIBC_i^{rad,b}} & \vdots & \mathbf{BIBC^{loop}} \end{bmatrix} + \begin{bmatrix} \bar{Z}_f & \bar{Z}_f & \vdots & \\ \bar{Z}_f & \bar{Z}_f & \vdots & \\ \cdots & \cdots & \vdots & \cdots \\ & & \vdots & 0 \end{bmatrix}$$

For a distribution network with ℓ number of three-phase loops, the size of the $\left[\bar{\mathbf{Z}}_{\mathbf{sc}}^{\mathbf{LLGm}}\right]$ is $(3\ell + 2) \times (3\ell + 2)$.

After calculating the fault current and the branch currents for the network from (3.48), the branch currents and post-fault bus voltages can be calculated using (3.46) and (3.45), respectively.

3.3.3 Triple line to ground fault

For an LLLG fault through a fault impedance $\bar{Z}_f$ on ith bus (shown in Figure 3.1c), the boundary conditions are identical to those described in (3.21) and (3.22). Equation (3.24) can be modified for a weakly meshed distribution system as follows:

$$\begin{bmatrix} \bar{\mathbf{I}}^{\mathbf{rad}}_{\mathbf{br,f}} \\ \cdots \\ \bar{\mathbf{I}}^{\mathbf{loop}}_{\mathbf{br,f}} \end{bmatrix} = \begin{bmatrix} \mathbf{BIBC}^{\mathbf{rad,a}}_{\mathbf{i}} & \mathbf{BIBC}^{\mathbf{rad,b}}_{\mathbf{i}} & \mathbf{BIBC}^{\mathbf{rad,a}}_{\mathbf{i}} & \vdots & \mathbf{BIBC}^{\mathbf{loop}} \end{bmatrix} \begin{bmatrix} \bar{I}^{a}_{i,f} \\ \bar{I}^{b}_{i,f} \\ \bar{I}^{c}_{i,f} \\ \cdots \\ \bar{\mathbf{I}}^{\mathbf{loop}}_{\mathbf{br,f}} \end{bmatrix} \tag{3.49}$$

where $[\mathbf{BIBC}^{\mathbf{rad,a}}_{\mathbf{i}}]$, $[\mathbf{BIBC}^{\mathbf{rad,b}}_{\mathbf{i}}]$ and $[\mathbf{BIBC}^{\mathbf{rad,c}}_{\mathbf{i}}]$ are the column vectors of $[\mathbf{BIBC}]$ matrix for the radial part of the distribution network corresponding to the three phases of the faulted bus i. The voltage drops for radial and meshed parts of the network can be written as [11]

$$\begin{bmatrix} \Delta\bar{V}^{a}_{i,f} \\ \Delta\bar{V}^{b}_{i,f} \\ \Delta\bar{V}^{c}_{i,f} \\ \cdots \\ \mathbf{0} \end{bmatrix} = \begin{bmatrix} \mathbf{BCBV}^{\mathbf{rad,a}}_{\mathbf{i}} \\ \mathbf{BCBV}^{\mathbf{rad,b}}_{\mathbf{i}} \\ \mathbf{BCBV}^{\mathbf{rad,c}}_{\mathbf{i}} \\ \cdots \\ \mathbf{BCBV}^{\mathbf{loop}} \end{bmatrix} \begin{bmatrix} \mathbf{BIBC}^{\mathbf{rad,a}}_{\mathbf{i}} & \mathbf{BIBC}^{\mathbf{rad,b}}_{\mathbf{i}} & \mathbf{BIBC}^{\mathbf{rad,c}}_{\mathbf{i}} & \vdots & \mathbf{BIBC}^{\mathbf{loop}} \end{bmatrix} \begin{bmatrix} \bar{I}^{a}_{i,f} \\ \bar{I}^{b}_{i,f} \\ \bar{I}^{c}_{i,f} \\ \cdots \\ \bar{\mathbf{I}}^{\mathbf{loop}}_{\mathbf{br,f}} \end{bmatrix} \tag{3.50}$$

where $[\mathbf{BCBV}^{\mathbf{rad,a}}_{\mathbf{i}}]$, $[\mathbf{BCBV}^{\mathbf{rad,b}}_{\mathbf{i}}]$ and $[\mathbf{BCBV}^{\mathbf{rad,c}}_{\mathbf{i}}]$ are the row vectors of $[\mathbf{BCBV}]$ matrix for the radial part of the distribution network corresponding to phases a, b and c, respectively, of ith bus. From (3.22), (3.25) and (3.50), we get

$$\begin{bmatrix} \bar{I}^{a}_{i,f} \\ \bar{I}^{b}_{i,f} \\ \bar{I}^{c}_{i,f} \\ \cdots \\ \bar{\mathbf{I}}^{\mathbf{loop}}_{\mathbf{br,f}} \end{bmatrix} = \left[\bar{\mathbf{Z}}^{\mathbf{LLLGm}}_{\mathbf{sc}}\right]^{-1} \begin{bmatrix} \bar{V}^{a}_{s} \\ \bar{V}^{b}_{s} \\ \bar{V}^{c}_{s} \\ \cdots \\ \mathbf{0} \end{bmatrix} \tag{3.51}$$

For a distribution network with ℓ number of three-phase loops, the size of the $\left[\bar{\mathbf{Z}}^{\mathbf{LLLGm}}_{\mathbf{sc}}\right]$ is $(3\ell + 3) \times (3\ell + 3)$. After calculating the fault current and the branch currents for the meshed part of the network from (3.51), the branch currents and post-fault bus voltages can be calculated using (3.49) and (3.45), respectively.

In (3.51),

$$
\left[\bar{\mathbf{Z}}_{\mathbf{sc}}^{\mathbf{LLLGm}}\right] = \begin{bmatrix} \mathbf{BCBV}_{\mathbf{i}}^{\mathbf{rad,a}} \\ \mathbf{BCBV}_{\mathbf{i}}^{\mathbf{rad,b}} \\ \mathbf{BCBV}_{\mathbf{i}}^{\mathbf{rad,c}} \\ \cdots \\ \mathbf{BCBV}^{\mathbf{loop}} \end{bmatrix} \begin{bmatrix} \mathbf{BIBC}_{\mathbf{i}}^{\mathbf{rad,a}} & \mathbf{BIBC}_{\mathbf{i}}^{\mathbf{rad,b}} & \mathbf{BIBC}_{\mathbf{i}}^{\mathbf{rad,c}} & \vdots & \mathbf{BIBC}^{\mathbf{loop}} \end{bmatrix}
+ \begin{bmatrix} \bar{Z}_f & \bar{Z}_f & \bar{Z}_f & \vdots & \\ \bar{Z}_f & \bar{Z}_f & \bar{Z}_f & \vdots & \\ \bar{Z}_f & \bar{Z}_f & \bar{Z}_f & \vdots & \\ \cdots & \cdots & \cdots & \vdots & \cdots \\ & & & \vdots & 0 \end{bmatrix}
$$

3.3.4 Line to line fault

When an LL fault occurs through a fault impedance $\bar{Z}_f$ on phases a and b of ith bus (shown in Figure 3.1d), the boundary conditions are identical to those described in (3.31) and (3.32).

Equation (3.35) can be modified for a weakly meshed distribution system as

$$
\begin{bmatrix} \bar{\mathbf{I}}_{\mathbf{br,f}}^{\mathbf{rad}} \\ \cdots \\ \bar{\mathbf{I}}_{\mathbf{br,f}}^{\mathbf{loop}} \end{bmatrix} = \begin{bmatrix} \mathbf{BIBC}_{\mathbf{i}}^{\mathbf{rad,a}} & \mathbf{BIBC}_{\mathbf{i}}^{\mathbf{rad,b}} & \vdots & \mathbf{BIBC}^{\mathbf{loop}} \end{bmatrix} \begin{bmatrix} \bar{I}_{i,f}^{a} \\ -\bar{I}_{i,f}^{a} \\ \cdots \\ \bar{\mathbf{I}}_{\mathbf{br,f}}^{\mathbf{loop}} \end{bmatrix} \tag{3.52}
$$

The voltage drops for radial and meshed parts of the network can be written as [11]

$$
\begin{bmatrix} \Delta\bar{V}_{i,f}^{a} \\ \Delta\bar{V}_{i,f}^{b} \\ \cdots \\ \mathbf{0} \end{bmatrix} = \begin{bmatrix} \mathbf{BCBV}_{\mathbf{i}}^{\mathbf{rad,a}} \\ \mathbf{BCBV}_{\mathbf{i}}^{\mathbf{rad,b}} \\ \cdots \\ \mathbf{BCBV}^{\mathbf{loop}} \end{bmatrix} \begin{bmatrix} \mathbf{BIBC}_{\mathbf{i}}^{\mathbf{rad,a}} & \mathbf{BIBC}_{\mathbf{i}}^{\mathbf{rad,b}} & \vdots & \mathbf{BIBC}^{\mathbf{loop}} \end{bmatrix} \begin{bmatrix} \bar{I}_{i,f}^{a} \\ -\bar{I}_{i,f}^{a} \\ \cdots \\ \bar{\mathbf{I}}_{\mathbf{br,f}}^{\mathbf{loop}} \end{bmatrix} \tag{3.53}
$$

Equation (3.53) can be simplified as

$$\begin{bmatrix} \Delta\bar{V}^{a}_{i,f} - \Delta\bar{V}^{b}_{i,f} \\ \cdots \\ \mathbf{0} \end{bmatrix} = \begin{bmatrix} \mathbf{BCBV}_{\mathbf{i}}^{\mathbf{rad,a}} - \mathbf{BCBV}_{\mathbf{i}}^{\mathbf{rad,b}} \\ \cdots \\ \mathbf{BCBV}^{\mathbf{loop}} \end{bmatrix} \begin{bmatrix} \mathbf{BIBC}_{\mathbf{i}}^{\mathbf{rad,a}} - \mathbf{BIBC}_{\mathbf{i}}^{\mathbf{rad,b}} \\ \cdots \\ \mathbf{BIBC}^{\mathbf{loop}} \end{bmatrix} \begin{bmatrix} \bar{I}^{a}_{i,f} \\ \cdots \\ \bar{\mathbf{I}}^{\mathbf{loop}}_{\mathbf{br,f}} \end{bmatrix} \tag{3.54}$$

Substituting (3.33) into (3.54), and after simplification, we get

$$\begin{bmatrix} \bar{V}^{a}_{s} - \bar{V}^{b}_{s} \\ \cdots \\ \mathbf{0} \end{bmatrix} = \left(\begin{bmatrix} \mathbf{BCBV}_{\mathbf{i}}^{\mathbf{rad,a}} - \mathbf{BCBV}_{\mathbf{i}}^{\mathbf{rad,b}} \\ \cdots \\ \mathbf{BCBV}^{\mathbf{loop}} \end{bmatrix} \begin{bmatrix} \mathbf{BIBC}_{\mathbf{i}}^{\mathbf{rad,a}} - \mathbf{BIBC}_{\mathbf{i}}^{\mathbf{rad,b}} \\ \cdots \\ \mathbf{BIBC}^{\mathbf{loop}} \end{bmatrix} + \begin{bmatrix} \bar{Z}_{f} & \vdots & \\ \cdots & \vdots & \cdots \\ & \vdots & 0 \end{bmatrix} \right) \begin{bmatrix} \bar{I}^{a}_{i,f} \\ \cdots \\ \bar{\mathbf{I}}^{\mathbf{loop}}_{\mathbf{br,f}} \end{bmatrix} \tag{3.55}$$

Finally, the fault current and meshed branch currents can be calculated as

$$\begin{bmatrix} \bar{I}^{a}_{i,f} \\ \cdots \\ \bar{\mathbf{I}}^{\mathbf{loop}}_{\mathbf{br,f}} \end{bmatrix} = \left[\bar{\mathbf{Z}}^{\mathbf{LLm}}_{\mathbf{sc}}\right]^{-1} \begin{bmatrix} \bar{V}^{a}_{s} - \bar{V}^{b}_{s} \\ \cdots \\ \mathbf{0} \end{bmatrix} \tag{3.56}$$

where

$$\left[\bar{\mathbf{Z}}^{\mathbf{LLm}}_{\mathbf{sc}}\right] = \begin{bmatrix} \mathbf{BCBV}_{\mathbf{i}}^{\mathbf{rad,a}} - \mathbf{BCBV}_{\mathbf{i}}^{\mathbf{rad,b}} \\ \cdots \\ \mathbf{BCBV}^{\mathbf{loop}} \end{bmatrix} \begin{bmatrix} \mathbf{BIBC}_{\mathbf{i}}^{\mathbf{rad,a}} - \mathbf{BIBC}_{\mathbf{i}}^{\mathbf{rad,b}} \\ \cdots \\ \mathbf{BIBC}^{\mathbf{loop}} \end{bmatrix} + \begin{bmatrix} \bar{Z}_{f} & \vdots & \\ \cdots & \vdots & \cdots \\ & \vdots & 0 \end{bmatrix}$$

For a distribution network with ℓ number of three-phase loops, the size of the $[\bar{\mathbf{Z}}^{\mathbf{LLm}}_{\mathbf{sc}}]$ matrix is $(3\ell + 1) \times (3\ell + 1)$.

After calculating the fault current and the branch currents for the meshed part of the network from (3.56), the branch currents and post-fault bus voltages are obtained with the help of (3.52) and (3.45), respectively.

3.4 Example of short-circuit analysis of unbalanced radial distribution system

To illustrate the short-circuit analysis procedure, a test radial distribution system, as shown in Figure 3.2, has been considered. The distribution system has six buses, three 3-phase feeders, one 2-phase feeder and one 1-phase feeder. The switches S are open in the radial configuration of the distribution system. The feeder data are given in Table 3.2.

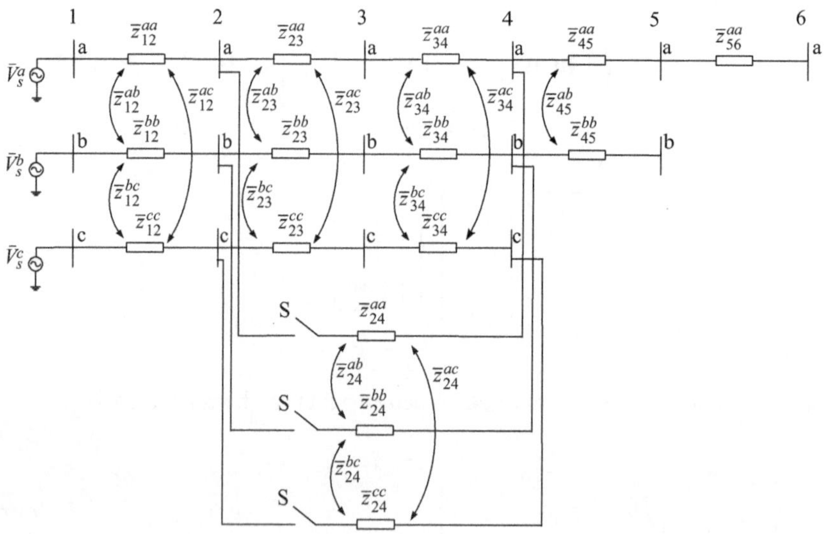

Figure 3.2 6-Bus distribution system for short-circuit analysis

Table 3.2 Feeder data of 6-bus radial distribution system

Feeder no.	From bus	To bus	Length (ft)	Type
1	1	2	20,064	Three phase
2	2	3	14,784	Three phase
3	3	4	22,704	Three phase
4	4	5	17,424	Two phase
5	5	6	16,972	Single phase

The impedance parameters (in Ω/mile) for the feeders are given below.
For three-phase elements (branches 1-2, 2-3 and 3-4):

$$\bar{\mathbf{Z}}_{3-\Phi} = \begin{matrix} a \\ b \\ c \end{matrix} \overset{\begin{matrix} a & \quad\quad\quad\quad b & \quad\quad\quad\quad c \end{matrix}}{\begin{bmatrix} 0.3465+j1.0179 & 0.1560+j0.5017 & 0.1580+j0.4236 \\ 0.1560+j0.5017 & 0.3375+j1.0478 & 0.1535+0.3849 \\ 0.1580+j0.4236 & 0.1535+j0.3849 & 0.3414+j1.0348 \end{bmatrix}}$$

For two-phase element (branch 4-5):

$$\bar{\mathbf{Z}}_{2-\Phi} = \begin{matrix} a \\ b \end{matrix} \overset{\begin{matrix} a & \quad\quad\quad\quad b \end{matrix}}{\begin{bmatrix} 0.3465+j1.0179 & 0.1560+j0.5017 \\ 0.1560+j0.5017 & 0.3375+j1.0478 \end{bmatrix}}$$

For single-phase element (branch 5-6):

$$\bar{\mathbf{Z}}_{1-\Phi} = a \overset{a}{[0.3465+j1.0179]}$$

The LG fault is assumed to occur on *phase a*, while LL and LLG faults are assumed to occur on *phases a and b* of the faulted bus. The fault impedance is taken as $\bar{Z}_f = 0.001$ pu for all the types of faults [10] and the fault is assumed to occur at bus no. 4. The substation bus voltage is 11.4 kV.

3.4.1 Results for 6-bus radial distribution system

Following the algorithm in Chapter 2, the [**BIBC**] matrix of the test radial distribution system of Figure 3.2 is given below:

$$[\mathbf{BIBC}] = \begin{matrix} & 2a & 2b & 2c & 3a & 3b & 3c & 4a & 4b & 4c & 5a & 5b & 6a \\ 1a\text{–}2a & 1 & 0 & 0 & 1 & 0 & 0 & 1 & 0 & 0 & 1 & 0 & 1 \\ 1b\text{–}2b & 0 & 1 & 0 & 0 & 1 & 0 & 0 & 1 & 0 & 0 & 1 & 0 \\ 1c\text{–}2c & 0 & 0 & 1 & 0 & 0 & 1 & 0 & 0 & 1 & 0 & 0 & 0 \\ 2a\text{–}3a & 0 & 0 & 0 & 1 & 0 & 0 & 1 & 0 & 0 & 1 & 0 & 1 \\ 2b\text{–}3b & 0 & 0 & 0 & 0 & 1 & 0 & 0 & 1 & 0 & 0 & 1 & 0 \\ 2c\text{–}3c & 0 & 0 & 0 & 0 & 0 & 1 & 0 & 0 & 1 & 0 & 0 & 0 \\ 3a\text{–}4a & 0 & 0 & 0 & 0 & 0 & 0 & 1 & 0 & 0 & 1 & 0 & 1 \\ 3b\text{–}4b & 0 & 0 & 0 & 0 & 0 & 0 & 0 & 1 & 0 & 0 & 1 & 0 \\ 3c\text{–}4c & 0 & 0 & 0 & 0 & 0 & 0 & 0 & 0 & 1 & 0 & 0 & 0 \\ 4a\text{–}5a & 0 & 0 & 0 & 0 & 0 & 0 & 0 & 0 & 0 & 1 & 0 & 1 \\ 4b\text{–}5b & 0 & 0 & 0 & 0 & 0 & 0 & 0 & 0 & 0 & 0 & 1 & 0 \\ 5a\text{–}6a & 0 & 0 & 0 & 0 & 0 & 0 & 0 & 0 & 0 & 0 & 0 & 1 \end{matrix}$$

In the above **[BIBC]** matrix, the rows correspond to the branches while the columns correspond to the buses. The **[BCBV]** matrix for the same system is given as

$$
[\mathbf{BCBV}] = \begin{array}{c} \\ 2a \\ 2b \\ 2c \\ 3a \\ 3b \\ 3c \\ 4a \\ 4b \\ 4c \\ 5a \\ 5b \\ 6a \end{array}
\begin{array}{cccccccccccc}
1a{-}2a & 1b{-}2b & 1c{-}2c & 2a{-}3a & 2b{-}3b & 2c{-}3c & 3a{-}4a & 3b{-}4b & 3c{-}4c & 4a{-}5a & 4b{-}5b & 5a{-}6a \\
\bar{Z}_{12}^{aa} & \bar{Z}_{12}^{ab} & \bar{Z}_{12}^{ac} & 0 & 0 & 0 & 0 & 0 & 0 & 0 & 0 & 0 \\
\bar{Z}_{12}^{ba} & \bar{Z}_{12}^{bb} & \bar{Z}_{12}^{bc} & 0 & 0 & 0 & 0 & 0 & 0 & 0 & 0 & 0 \\
\bar{Z}_{12}^{ca} & \bar{Z}_{12}^{cb} & \bar{Z}_{12}^{cc} & 0 & 0 & 0 & 0 & 0 & 0 & 0 & 0 & 0 \\
\bar{Z}_{12}^{aa} & \bar{Z}_{12}^{ab} & \bar{Z}_{12}^{ac} & \bar{Z}_{23}^{aa} & \bar{Z}_{23}^{ab} & \bar{Z}_{23}^{ac} & 0 & 0 & 0 & 0 & 0 & 0 \\
\bar{Z}_{12}^{ba} & \bar{Z}_{12}^{bb} & \bar{Z}_{12}^{bc} & \bar{Z}_{23}^{ba} & \bar{Z}_{23}^{bb} & \bar{Z}_{23}^{bc} & 0 & 0 & 0 & 0 & 0 & 0 \\
\bar{Z}_{12}^{ca} & \bar{Z}_{12}^{cb} & \bar{Z}_{12}^{cc} & \bar{Z}_{23}^{ca} & \bar{Z}_{23}^{cb} & \bar{Z}_{23}^{cc} & 0 & 0 & 0 & 0 & 0 & 0 \\
\bar{Z}_{12}^{aa} & \bar{Z}_{12}^{ab} & \bar{Z}_{12}^{ac} & \bar{Z}_{23}^{aa} & \bar{Z}_{23}^{ab} & \bar{Z}_{23}^{ac} & \bar{Z}_{34}^{aa} & \bar{Z}_{34}^{ab} & \bar{Z}_{34}^{ac} & 0 & 0 & 0 \\
\bar{Z}_{12}^{ba} & \bar{Z}_{12}^{bb} & \bar{Z}_{12}^{bc} & \bar{Z}_{23}^{ba} & \bar{Z}_{23}^{bb} & \bar{Z}_{23}^{bc} & \bar{Z}_{34}^{ba} & \bar{Z}_{34}^{bb} & \bar{Z}_{34}^{bc} & 0 & 0 & 0 \\
\bar{Z}_{12}^{ca} & \bar{Z}_{12}^{cb} & \bar{Z}_{12}^{cc} & \bar{Z}_{23}^{ca} & \bar{Z}_{23}^{cb} & \bar{Z}_{23}^{cc} & \bar{Z}_{34}^{ca} & \bar{Z}_{34}^{cb} & \bar{Z}_{34}^{cc} & 0 & 0 & 0 \\
\bar{Z}_{12}^{aa} & \bar{Z}_{12}^{ab} & \bar{Z}_{12}^{ac} & \bar{Z}_{23}^{aa} & \bar{Z}_{23}^{ab} & \bar{Z}_{23}^{ac} & \bar{Z}_{34}^{aa} & \bar{Z}_{34}^{ab} & \bar{Z}_{34}^{ac} & \bar{Z}_{45}^{aa} & \bar{Z}_{45}^{ab} & 0 \\
\bar{Z}_{12}^{ba} & \bar{Z}_{12}^{bb} & \bar{Z}_{12}^{bc} & \bar{Z}_{23}^{ba} & \bar{Z}_{23}^{bb} & \bar{Z}_{23}^{bc} & \bar{Z}_{34}^{ba} & \bar{Z}_{34}^{bb} & \bar{Z}_{34}^{bc} & \bar{Z}_{45}^{ba} & \bar{Z}_{45}^{bb} & 0 \\
\bar{Z}_{12}^{aa} & \bar{Z}_{12}^{ab} & \bar{Z}_{12}^{ac} & \bar{Z}_{23}^{aa} & \bar{Z}_{23}^{ab} & \bar{Z}_{23}^{ac} & \bar{Z}_{34}^{aa} & \bar{Z}_{34}^{ab} & \bar{Z}_{34}^{ac} & \bar{Z}_{45}^{aa} & \bar{Z}_{45}^{ab} & \bar{Z}_{56}^{aa}
\end{array}
$$

In the above **[BCBV]** matrix, the rows correspond to the buses while the columns correspond to the branches.

The required portions of the **[BIBC]** and **[BCBV]** matrices for calculating the short-circuit impedance matrices along with the expressions for the short-circuit impedance matrices of the 6-bus radial distribution system under various types of fault are given next.

LG fault

$$
[\mathbf{BCBV}]_{\mathbf{4}}^{\mathbf{a}} = \begin{array}{c} \\ 4a \end{array}
\begin{array}{cccccccccccc}
1a{-}2a & 1b{-}2b & 1c{-}2c & 2a{-}3a & 2b{-}3b & 2c{-}3c & 3a{-}4a & 3b{-}4b & 3c{-}4c & 4a{-}5a & 4b{-}5b & 5a{-}6a \\
\bar{Z}_{12}^{aa} & \bar{Z}_{12}^{ab} & \bar{Z}_{12}^{ac} & \bar{Z}_{23}^{aa} & \bar{Z}_{23}^{ab} & \bar{Z}_{23}^{ac} & \bar{Z}_{34}^{aa} & \bar{Z}_{34}^{ab} & \bar{Z}_{34}^{ac} & 0 & 0 & 0
\end{array}
$$

$$
\left([\mathbf{BIBC}]_{\mathbf{4}}^{\mathbf{a}}\right)^{T} = \begin{array}{c} \\ 4a \end{array}
\begin{array}{cccccccccccc}
1a{-}2a & 1b{-}2b & 1c{-}2c & 2a{-}3a & 2b{-}3b & 2c{-}3c & 3a{-}4a & 3b{-}4b & 3c{-}4c & 4a{-}5a & 4b{-}5b & 5a{-}6a \\
1 & 0 & 0 & 1 & 0 & 0 & 1 & 0 & 0 & 0 & 0 & 0
\end{array}
$$

From (3.9), the short-circuit impedance matrix for an LG fault can be calculated as

$$
\left[\bar{Z}_{\mathrm{sc}}^{\mathrm{LG}}\right] = \bar{Z}_{12}^{aa} + \bar{Z}_{23}^{aa} + \bar{Z}_{34}^{aa} + \bar{Z}_{f}
$$

LLG fault

$$\begin{bmatrix}[\mathbf{BCBV}]_4^{\mathbf{a}}\\ [\mathbf{BCBV}]_4^{\mathbf{b}}\end{bmatrix} = \begin{array}{c|cccccccccccc} & 1a{-}2a & 1b{-}2b & 1c{-}2c & 2a{-}3a & 2b{-}3b & 2c{-}3c & 3a{-}4a & 3b{-}4b & 3c{-}4c & 4a{-}5a & 4b{-}5b & 5a{-}6a \\ \hline 4a & \bar{Z}_{12}^{aa} & \bar{Z}_{12}^{ab} & \bar{Z}_{12}^{ac} & \bar{Z}_{23}^{aa} & \bar{Z}_{23}^{ab} & \bar{Z}_{23}^{ac} & \bar{Z}_{34}^{aa} & \bar{Z}_{34}^{ab} & \bar{Z}_{34}^{ac} & 0 & 0 & 0 \\ 4b & \bar{Z}_{12}^{ba} & \bar{Z}_{12}^{bb} & \bar{Z}_{12}^{bc} & \bar{Z}_{23}^{ba} & \bar{Z}_{23}^{bb} & \bar{Z}_{23}^{bc} & \bar{Z}_{34}^{ba} & \bar{Z}_{34}^{bb} & \bar{Z}_{34}^{bc} & 0 & 0 & 0 \end{array}$$

$$\begin{bmatrix}[\mathbf{BIBC}]_4^{\mathbf{a}}\\ [\mathbf{BIBC}]_4^{\mathbf{b}}\end{bmatrix}^T = \begin{array}{c|cccccccccccc} & 1a{-}2a & 1b{-}2b & 1c{-}2c & 2a{-}3a & 2b{-}3b & 2c{-}3c & 3a{-}4a & 3b{-}4b & 3c{-}4c & 4a{-}5a & 4b{-}5b & 5a{-}6a \\ \hline 4a & 1 & 0 & 0 & 1 & 0 & 0 & 1 & 0 & 0 & 0 & 0 & 0 \\ 4b & 0 & 1 & 0 & 0 & 1 & 0 & 0 & 1 & 0 & 0 & 0 & 0 \end{array}$$

From (3.20), the short-circuit impedance matrix for an LLG fault can be calculated as

$$[\bar{\mathbf{Z}}_{\mathbf{sc}}^{\mathbf{LLG}}] = \begin{bmatrix} \bar{Z}_{12}^{aa}+\bar{Z}_{23}^{aa}+\bar{Z}_{34}^{aa}+\bar{Z}_f & \bar{Z}_{12}^{ab}+\bar{Z}_{23}^{ab}+\bar{Z}_{34}^{ab}+\bar{Z}_f \\ \bar{Z}_{12}^{ba}+\bar{Z}_{23}^{ba}+\bar{Z}_{34}^{ba}+\bar{Z}_f & \bar{Z}_{12}^{bb}+\bar{Z}_{23}^{bb}+\bar{Z}_{34}^{bb}+\bar{Z}_f \end{bmatrix}$$

LLLG fault

$$\begin{bmatrix}[\mathbf{BCBV}]_4^{\mathbf{a}}\\ [\mathbf{BCBV}]_4^{\mathbf{b}}\\ [\mathbf{BCBV}]_4^{\mathbf{c}}\end{bmatrix} = \begin{array}{c|cccccccccccc} & 1a{-}2a & 1b{-}2b & 1c{-}2c & 2a{-}3a & 2b{-}3b & 2c{-}3c & 3a{-}4a & 3b{-}4b & 3c{-}4c & 4a{-}5a & 4b{-}5b & 5a{-}6a \\ \hline 4a & \bar{Z}_{12}^{aa} & \bar{Z}_{12}^{ab} & \bar{Z}_{12}^{ac} & \bar{Z}_{23}^{aa} & \bar{Z}_{23}^{ab} & \bar{Z}_{23}^{ac} & \bar{Z}_{34}^{aa} & \bar{Z}_{34}^{ab} & \bar{Z}_{34}^{ac} & 0 & 0 & 0 \\ 4b & \bar{Z}_{12}^{ba} & \bar{Z}_{12}^{bb} & \bar{Z}_{12}^{bc} & \bar{Z}_{23}^{ba} & \bar{Z}_{23}^{bb} & \bar{Z}_{23}^{bc} & \bar{Z}_{34}^{ba} & \bar{Z}_{34}^{bb} & \bar{Z}_{34}^{bc} & 0 & 0 & 0 \\ 4c & \bar{Z}_{12}^{ca} & \bar{Z}_{12}^{cb} & \bar{Z}_{12}^{cc} & \bar{Z}_{23}^{ca} & \bar{Z}_{23}^{cb} & \bar{Z}_{23}^{cc} & \bar{Z}_{34}^{ca} & \bar{Z}_{34}^{cb} & \bar{Z}_{34}^{cc} & 0 & 0 & 0 \end{array}$$

$$\begin{bmatrix}[\mathbf{BIBC}]_4^{\mathbf{a}}\\ [\mathbf{BIBC}]_4^{\mathbf{b}}\\ [\mathbf{BIBC}]_4^{\mathbf{c}}\end{bmatrix}^T = \begin{array}{c|cccccccccccc} & 1a{-}2a & 1b{-}2b & 1c{-}2c & 2a{-}3a & 2b{-}3b & 2c{-}3c & 3a{-}4a & 3b{-}4b & 3c{-}4c & 4a{-}5a & 4b{-}5b & 5a{-}6a \\ \hline 4a & 1 & 0 & 0 & 1 & 0 & 0 & 1 & 0 & 0 & 0 & 0 & 0 \\ 4b & 0 & 1 & 0 & 0 & 1 & 0 & 0 & 1 & 0 & 0 & 0 & 0 \\ 4c & 0 & 0 & 1 & 0 & 0 & 1 & 0 & 0 & 1 & 0 & 0 & 0 \end{array}$$

From (3.30), the short-circuit impedance matrix for an LLLG fault can be calculated as

$$[\bar{\mathbf{Z}}_{\mathbf{sc}}^{\mathbf{LLLG}}] = \begin{bmatrix} \bar{Z}_{12}^{aa}+\bar{Z}_{23}^{aa}+\bar{Z}_{34}^{aa}+\bar{Z}_f & \bar{Z}_{12}^{ab}+\bar{Z}_{23}^{ab}+\bar{Z}_{34}^{ab}+\bar{Z}_f & \bar{Z}_{12}^{ac}+\bar{Z}_{23}^{ac}+\bar{Z}_{34}^{ac}+\bar{Z}_f \\ \bar{Z}_{12}^{ba}+\bar{Z}_{23}^{ba}+\bar{Z}_{34}^{ba}+\bar{Z}_f & \bar{Z}_{12}^{bb}+\bar{Z}_{23}^{bb}+\bar{Z}_{34}^{bb}+\bar{Z}_f & \bar{Z}_{12}^{bc}+\bar{Z}_{23}^{bc}+\bar{Z}_{34}^{bc}+\bar{Z}_f \\ \bar{Z}_{12}^{ca}+\bar{Z}_{23}^{ca}+\bar{Z}_{34}^{ca}+\bar{Z}_f & \bar{Z}_{12}^{cb}+\bar{Z}_{23}^{cb}+\bar{Z}_{34}^{cb}+\bar{Z}_f & \bar{Z}_{12}^{cc}+\bar{Z}_{23}^{cc}+\bar{Z}_{34}^{cc}+\bar{Z}_f \end{bmatrix}$$

Table 3.3 Fault currents (kA) for different types of faults at bus no. 4 of 6-bus radial distribution system

	LG fault		LL fault		LLG fault		LLLG fault	
	BIBC	**PSCAD**	**BIBC**	**PSCAD**	**BIBC**	**PSCAD**	**BIBC**	**PSCAD**
Phase *a*	0.55954	0.55952	0.92592	0.92589	0.94461	0.94458	1.0307	1.0306
Phase *b*			0.925921	0.92589	0.94180	0.94184	1.0197	1.0198
Phase *c*							0.8752	0.8751

LL fault

From (3.38), the elements of $[\bar{\mathbf{Z}}_{\mathbf{sc}}^{\mathbf{LL}}]$ can be calculated as

$$\bar{Z}_4^{aa} = \bar{Z}_{12}^{aa} + \bar{Z}_{23}^{aa} + \bar{Z}_{34}^{aa} + \bar{Z}_f$$

$$\bar{Z}_4^{ab} = \bar{Z}_{12}^{ab} + \bar{Z}_{23}^{ab} + \bar{Z}_{34}^{ab} + \bar{Z}_f$$

$$\bar{Z}_4^{ba} = \bar{Z}_{12}^{ba} + \bar{Z}_{ba}^{ab} + \bar{Z}_{34}^{ba} + \bar{Z}_f$$

$$\bar{Z}_4^{bb} = \bar{Z}_{12}^{bb} + \bar{Z}_{23}^{bb} + \bar{Z}_{34}^{bb} + \bar{Z}_f$$

From (3.40), the short-circuit impedance matrix $\left[\bar{\mathbf{Z}}_{\mathbf{sc}}^{\mathbf{LL}}\right]$ for a double line fault at bus no. 4 can be calculated as

$$\bar{Z}_{\text{sc}}^{\text{LL}} = \bar{Z}_{12}^{aa} - 2\bar{Z}_{12}^{ab} + \bar{Z}_{12}^{bb} + \bar{Z}_{23}^{aa} - 2\bar{Z}_{23}^{ab} + \bar{Z}_{23}^{bb} + \bar{Z}_{34}^{aa} - 2\bar{Z}_{34}^{ab} + \bar{Z}_{34}^{bb} + \bar{Z}_f$$

The fault currents, for the four types of faults (considered one at a time), at the fault point are tabulated in Table 3.3, and the *phase a* voltage of the faulted bus is given in Table 3.4. The results obtained through BIBC-based method have also been compared with PSCAD simulation results. It can be seen from these two tables that the results of BIBC method are in excellent agreement with those obtained through PSCAD simulation. The voltage profiles of the phases *a*, *b* and *c* of the distribution system, for these four types of faults, obtained through BIBC method and PSCAD simulations are shown in Figure 3.3. As can be observed from these figures, the voltage profiles obtained from these two methods are also in excellent agreement.

3.4.2 Results for 6-bus meshed distribution system

When switches S, as shown in Figure 3.2, are closed, a new feeder from bus no. 2 to bus no. 4 is added to the distribution system and a mesh is created. The system now operates in a weakly meshed configuration. The length of the added feeder is 20,064 ft.

Table 3.4 Phase a *voltage in kV for different types of faults at bus no. 4 of 6-bus radial distribution system*

Bus no.	LG fault			LLG fault			LLLG fault			LL fault		
	PSCAD	**BIBC**	**%Error**	**PSCAD**	**BIBC**	**%Error**	**PSCAD**	**BIBC**	**%Error**	**PSCAD**	**BIBC**	**%Error**
1	6.58165	6.58179	0.00217	6.58165	6.58179	0.00217	6.58165	6.58179	0.00217	6.58165	6.58179	0.00217
2	4.29562	4.29571	0.00208	4.31427	4.31435	0.00188	4.30655	4.30663	0.00195	4.99111	4.99124	0.00262
3	2.61139	2.61144	0.00197	2.64393	2.64401	0.00309	2.63008	2.63087	0.03009	4.01805	4.01817	0.00301
4	0.07274	0.07272	0.02625	0.12279	0.12276	0.02574	0.13399	0.13395	0.02817	3.25672	3.25680	0.00245
5	0.07274	0.07272	0.02762	0.12279	0.12276	0.02452	0.13399	0.13395	0.02519	3.25672	3.25680	0.00245
6	0.07274	0.07272	0.02762	0.12280	0.12276	0.02737	0.13399	0.13395	0.02519	3.25672	3.25680	0.00245

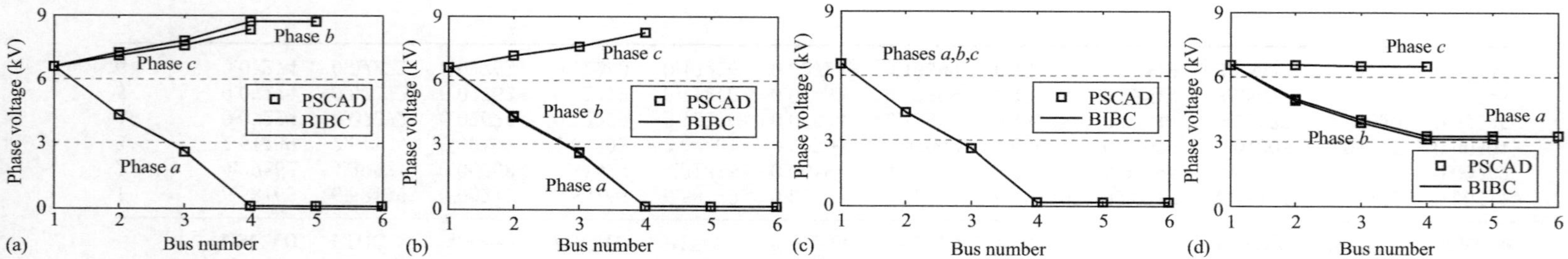

Figure 3.3 Voltage profile of 6-bus radial distribution system under fault conditions: (a) LG fault; (b) LLG fault; (c) LLLG fault; and (d) LL fault

The **[BIBC]** matrix of the 6-bus meshed distribution system of Figure 3.2 is as given below:

$$
[\mathbf{BIBC}] = \begin{array}{c|cccccccccccc|ccc}
 & 2a & 2b & 2c & 3a & 3b & 3c & 4a & 4b & 4c & 5a & 5b & 6a & 2a\!-\!4a & 2b\!-\!4b & 2c\!-\!4c \\
\hline
1a\!-\!2a & 1 & 0 & 0 & 1 & 0 & 0 & 1 & 0 & 0 & 1 & 0 & 1 & 0 & 0 & 0 \\
1b\!-\!2b & 0 & 1 & 0 & 0 & 1 & 0 & 0 & 1 & 0 & 0 & 1 & 0 & 0 & 0 & 0 \\
1c\!-\!2c & 0 & 0 & 1 & 0 & 0 & 1 & 0 & 0 & 1 & 0 & 0 & 0 & 0 & 0 & 0 \\
2a\!-\!3a & 0 & 0 & 0 & 1 & 0 & 0 & 1 & 0 & 0 & 1 & 0 & 1 & -1 & 0 & 0 \\
2b\!-\!3b & 0 & 0 & 0 & 0 & 1 & 0 & 0 & 1 & 0 & 0 & 1 & 0 & 0 & -1 & 0 \\
2c\!-\!3c & 0 & 0 & 0 & 0 & 0 & 1 & 0 & 0 & 1 & 0 & 0 & 0 & 0 & 0 & -1 \\
3a\!-\!4a & 0 & 0 & 0 & 0 & 0 & 0 & 1 & 0 & 0 & 1 & 0 & 1 & -1 & 0 & 0 \\
3b\!-\!4b & 0 & 0 & 0 & 0 & 0 & 0 & 0 & 1 & 0 & 0 & 1 & 0 & 0 & -1 & 0 \\
3c\!-\!4c & 0 & 0 & 0 & 0 & 0 & 0 & 0 & 0 & 1 & 0 & 0 & 0 & 0 & 0 & -1 \\
4a\!-\!5a & 0 & 0 & 0 & 0 & 0 & 0 & 0 & 0 & 0 & 1 & 0 & 1 & 0 & 0 & 0 \\
4b\!-\!5b & 0 & 0 & 0 & 0 & 0 & 0 & 0 & 0 & 0 & 0 & 1 & 0 & 0 & 0 & 0 \\
5a\!-\!6a & 0 & 0 & 0 & 0 & 0 & 0 & 0 & 0 & 0 & 0 & 0 & 1 & 0 & 0 & 0 \\
\hline
2a\!-\!4a & 0 & 0 & 0 & 0 & 0 & 0 & 0 & 0 & 0 & 0 & 0 & 0 & 1 & 0 & 0 \\
2b\!-\!4b & 0 & 0 & 0 & 0 & 0 & 0 & 0 & 0 & 0 & 0 & 0 & 0 & 0 & 1 & 0 \\
2c\!-\!4c & 0 & 0 & 0 & 0 & 0 & 0 & 0 & 0 & 0 & 0 & 0 & 0 & 0 & 0 & 1
\end{array}
$$

From the structure of **[BIBC]** matrix of the meshed system, it can be seen that additional rows and columns, corresponding to the created loop, are added to the **[BIBC]** matrix found earlier for the radial distribution system. Hence, the **[BIBC]** matrix of meshed system can be represented in a partitioned form as

$$
[\mathbf{BIBC}] = \begin{array}{c} \\ \text{radial} \\ \text{loop} \end{array} \begin{array}{c|c} \text{radial} & \text{loop} \\ [\mathbf{BIBC}]_{\mathbf{radial}} & [\mathbf{BIBC}]_{\mathbf{loop}} \\ \hline [\mathbf{0}] & [\mathbf{I}] \end{array}
$$

Combining the rows for the two partitions, **[BIBC]** matrix may be written as

$$
[\mathbf{BIBC}] = \left[\, [\mathbf{BIBC}]_{\mathbf{radial}} \ \vdots\ [\mathbf{BIBC}]_{\mathbf{loop}} \right]
$$

The **[BCBV]** matrix for the same system is given as

BCBV =

	1a–2a	1b–2b	1c–2c	2a–3a	2b–3b	2c–3c	3a–4a	3b–4b	3c–4c	4a–5a	4b–5b	5a–6a	2a–4a	2b–4b	2c–4c
2a	$\bar{Z}_{12}^{aa}$	$\bar{Z}_{12}^{ab}$	$\bar{Z}_{12}^{ac}$	0	0	0	0	0	0	0	0	0	0	0	0
2b	$\bar{Z}_{12}^{ba}$	$\bar{Z}_{12}^{bb}$	$\bar{Z}_{12}^{bc}$	0	0	0	0	0	0	0	0	0	0	0	0
2c	$\bar{Z}_{12}^{ca}$	$\bar{Z}_{12}^{cb}$	$\bar{Z}_{12}^{cc}$	0	0	0	0	0	0	0	0	0	0	0	0
3a	$\bar{Z}_{12}^{aa}$	$\bar{Z}_{12}^{ab}$	$\bar{Z}_{12}^{ac}$	$\bar{Z}_{23}^{aa}$	$\bar{Z}_{23}^{ab}$	$\bar{Z}_{23}^{ac}$	0	0	0	0	0	0	0	0	0
3b	$\bar{Z}_{12}^{ba}$	$\bar{Z}_{12}^{bb}$	$\bar{Z}_{12}^{bc}$	$\bar{Z}_{23}^{ba}$	$\bar{Z}_{23}^{bb}$	$\bar{Z}_{23}^{bc}$	0	0	0	0	0	0	0	0	0
3c	$\bar{Z}_{12}^{ca}$	$\bar{Z}_{12}^{cb}$	$\bar{Z}_{12}^{cc}$	$\bar{Z}_{23}^{ca}$	$\bar{Z}_{23}^{cb}$	$\bar{Z}_{23}^{cc}$	0	0	0	0	0	0	0	0	0
4a	$\bar{Z}_{12}^{aa}$	$\bar{Z}_{12}^{ab}$	$\bar{Z}_{12}^{ac}$	$\bar{Z}_{23}^{aa}$	$\bar{Z}_{23}^{ab}$	$\bar{Z}_{23}^{ac}$	$\bar{Z}_{34}^{aa}$	$\bar{Z}_{34}^{ab}$	$\bar{Z}_{34}^{ac}$	0	0	0	0	0	0
4b	$\bar{Z}_{12}^{ba}$	$\bar{Z}_{12}^{bb}$	$\bar{Z}_{12}^{bc}$	$\bar{Z}_{23}^{ba}$	$\bar{Z}_{23}^{bb}$	$\bar{Z}_{23}^{bc}$	$\bar{Z}_{34}^{ba}$	$\bar{Z}_{34}^{bb}$	$\bar{Z}_{34}^{bc}$	0	0	0	0	0	0
4c	$\bar{Z}_{12}^{ca}$	$\bar{Z}_{12}^{cb}$	$\bar{Z}_{12}^{cc}$	$\bar{Z}_{23}^{ca}$	$\bar{Z}_{23}^{cb}$	$\bar{Z}_{23}^{cc}$	$\bar{Z}_{34}^{ca}$	$\bar{Z}_{34}^{cb}$	$\bar{Z}_{34}^{cc}$	0	0	0	0	0	0
5a	$\bar{Z}_{12}^{aa}$	$\bar{Z}_{12}^{ab}$	$\bar{Z}_{12}^{ac}$	$\bar{Z}_{23}^{aa}$	$\bar{Z}_{23}^{ab}$	$\bar{Z}_{23}^{ac}$	$\bar{Z}_{34}^{aa}$	$\bar{Z}_{34}^{ab}$	$\bar{Z}_{34}^{ac}$	$\bar{Z}_{45}^{aa}$	$\bar{Z}_{45}^{ab}$	0	0	0	0
5b	$\bar{Z}_{12}^{ba}$	$\bar{Z}_{12}^{bb}$	$\bar{Z}_{12}^{bc}$	$\bar{Z}_{23}^{ba}$	$\bar{Z}_{23}^{bb}$	$\bar{Z}_{23}^{bc}$	$\bar{Z}_{34}^{ba}$	$\bar{Z}_{34}^{bb}$	$\bar{Z}_{34}^{bc}$	$\bar{Z}_{45}^{ba}$	$\bar{Z}_{45}^{bb}$	0	0	0	0
6a	$\bar{Z}_{12}^{aa}$	$\bar{Z}_{12}^{ab}$	$\bar{Z}_{12}^{ac}$	$\bar{Z}_{23}^{aa}$	$\bar{Z}_{23}^{ab}$	$\bar{Z}_{23}^{ac}$	$\bar{Z}_{34}^{aa}$	$\bar{Z}_{34}^{ab}$	$\bar{Z}_{34}^{ac}$	$\bar{Z}_{45}^{aa}$	$\bar{Z}_{45}^{ab}$	$\bar{Z}_{56}^{aa}$	0	0	0
2a–4a	0	0	0	$-\bar{Z}_{23}^{aa}$	$-\bar{Z}_{23}^{ab}$	$-\bar{Z}_{23}^{ac}$	$-\bar{Z}_{34}^{aa}$	$-\bar{Z}_{34}^{ab}$	$-\bar{Z}_{34}^{ac}$	0	0	0	$\bar{Z}_{24}^{aa}$	$\bar{Z}_{24}^{ab}$	$\bar{Z}_{24}^{ac}$
2b–4b	0	0	0	$-\bar{Z}_{23}^{ba}$	$-\bar{Z}_{23}^{bb}$	$-\bar{Z}_{23}^{bc}$	$-\bar{Z}_{34}^{ba}$	$-\bar{Z}_{34}^{bb}$	$-\bar{Z}_{34}^{bc}$	0	0	0	$\bar{Z}_{24}^{ba}$	$\bar{Z}_{24}^{bb}$	$\bar{Z}_{24}^{bc}$
2c–4c	0	0	0	$-\bar{Z}_{23}^{ca}$	$-\bar{Z}_{23}^{cb}$	$-\bar{Z}_{23}^{cc}$	$-\bar{Z}_{34}^{ca}$	$-\bar{Z}_{34}^{cb}$	$-\bar{Z}_{34}^{cc}$	0	0	0	$\bar{Z}_{24}^{ca}$	$\bar{Z}_{24}^{cb}$	$\bar{Z}_{24}^{cc}$

From the [**BCBV**] matrix of the meshed system, it can also be observed that additional rows and columns, corresponding to the loop created, are added to the [**BCBV**] matrix of the radial distribution system.

The [**BCBV**] matrix can then be represented in a concise form as

$$
\left[\mathbf{BCBV}\right] = \begin{array}{c} \\ \text{radial} \\ \text{loop} \end{array} \begin{array}{c} \begin{array}{cc} \text{radial} & \qquad\qquad \text{loop} \end{array} \\ \left[\begin{array}{c|c} [\mathbf{BCBV}]_{\mathbf{radial}} & [\mathbf{0}] \\ \hline [\mathbf{BCBV}]_{\mathbf{loop}} & [\bar{\mathbf{Z}}]_{\text{loop branch}} \end{array}\right] \end{array}
$$

Now combining the columns of the two partitions, [**BCBV**] may be written as

$$
\left[\mathbf{BCBV}\right] = \left[\begin{array}{c} [\mathbf{BCBV}]_{\mathbf{radial}} \\ \hline [\mathbf{BCBV}]_{\mathbf{loop}} \end{array}\right]
$$

From the above [**BCBV**] matrix of the meshed system, it can also be observed that additional rows and columns, corresponding to the loop created, are added to the [**BCBV**] matrix of the radial distribution system.

For faults at bus no. 4 of the 6-bus meshed distribution system, the [**BIBC**] matrix for the three phases $\left[\mathbf{BIBC_4^a} \quad \mathbf{BIBC_4^b} \quad \mathbf{BIBC_4^c}\right]^T$ is given as

$$
\begin{array}{c} \\ 4a \\ 4b \\ 4c \end{array}
\begin{array}{c}
\begin{array}{ccccccccccccccc} 1a\text{–}2a & 1b\text{–}2b & 1c\text{–}2c & 2a\text{–}3a & 2b\text{–}3b & 2c\text{–}3c & 3a\text{–}4a & 3b\text{–}4b & 3c\text{–}4c & 4a\text{–}5a & 4b\text{–}5b & 5a\text{–}6a & 2a\text{–}4a & 2b\text{–}4b & 2c\text{–}4c \end{array} \\
\left[\begin{array}{ccccccccccccccc}
1 & 0 & 0 & 1 & 0 & 0 & 1 & 0 & 0 & 0 & 0 & 0 & 0 & 0 & 0 \\
0 & 1 & 0 & 0 & 1 & 0 & 0 & 1 & 0 & 0 & 0 & 0 & 0 & 0 & 0 \\
0 & 0 & 1 & 0 & 0 & 1 & 0 & 0 & 1 & 0 & 0 & 0 & 0 & 0 & 0
\end{array}\right]
\end{array}
$$

For the LG fault, only the row corresponding to phase *a* of bus no. 4 is required. For the LLG and LL faults, the rows corresponding to phases *a* and *b* of bus no. 4 are needed; while for the LLLG fault, the [**BIBC**] matrix corresponding to all the three phases is used. The $\left[\mathbf{BCBV_{loop}}\right]$ part of the [**BCBV**] matrix is

$$
\begin{array}{c} \\ 2a\text{–}4a \\ 2b\text{–}4b \\ 2c\text{–}4c \end{array}
\begin{array}{c}
\begin{array}{ccccccccccccccc} 1a\text{–}2a & 1b\text{–}2b & 1c\text{–}2c & 2a\text{–}3a & 2b\text{–}3b & 2c\text{–}3c & 3a\text{–}4a & 3b\text{–}4b & 3c\text{–}4c & 4a\text{–}5a & 4b\text{–}5b & 5a\text{–}6a & 2a\text{–}4a & 2b\text{–}4b & 2c\text{–}4c \end{array} \\
\left[\begin{array}{ccccccccccccccc}
0 & 0 & 0 & -\bar{Z}_{23}^{aa} & -\bar{Z}_{23}^{ab} & -\bar{Z}_{23}^{ac} & -\bar{Z}_{34}^{aa} & -\bar{Z}_{34}^{ab} & -\bar{Z}_{34}^{ac} & 0 & 0 & 0 & \bar{Z}_{24}^{aa} & \bar{Z}_{24}^{ab} & \bar{Z}_{24}^{ac} \\
0 & 0 & 0 & -\bar{Z}_{23}^{ba} & -\bar{Z}_{23}^{bb} & -\bar{Z}_{23}^{bc} & -\bar{Z}_{34}^{ba} & -\bar{Z}_{34}^{bb} & -\bar{Z}_{34}^{bc} & 0 & 0 & 0 & \bar{Z}_{24}^{ba} & \bar{Z}_{24}^{bb} & \bar{Z}_{24}^{bc} \\
0 & 0 & 0 & -\bar{Z}_{23}^{ca} & -\bar{Z}_{23}^{cb} & -\bar{Z}_{23}^{cc} & -\bar{Z}_{34}^{ca} & -\bar{Z}_{34}^{cb} & -\bar{Z}_{34}^{cc} & 0 & 0 & 0 & \bar{Z}_{24}^{ca} & \bar{Z}_{24}^{cb} & \bar{Z}_{24}^{cc}
\end{array}\right]
\end{array}
$$

Following the procedure outlined in Section 3.3.1, the final expression for the short-circuit impedance matrix for an LG fault in 6-bus meshed distribution system is as follows:

$$
\left[\bar{Z}_{\text{sc}}^{\text{LGm}}\right] = \left[\begin{array}{cccc}
\bar{Z}_{12}^{aa} + \bar{Z}_{23}^{aa} + \bar{Z}_{34}^{aa} + \bar{Z}_f & -\bar{Z}_{23}^{aa} - \bar{Z}_{34}^{aa} & -\bar{Z}_{23}^{ab} - \bar{Z}_{34}^{ab} & -\bar{Z}_{23}^{ac} - \bar{Z}_{34}^{ac} \\
-\bar{Z}_{23}^{aa} - \bar{Z}_{34}^{aa} & \bar{Z}_{23}^{aa} + \bar{Z}_{24}^{aa} + \bar{Z}_{34}^{aa} & \bar{Z}_{23}^{ab} + \bar{Z}_{24}^{ab} + \bar{Z}_{34}^{ab} & \bar{Z}_{23}^{ac} + \bar{Z}_{24}^{ac} + \bar{Z}_{34}^{ac} \\
-\bar{Z}_{23}^{ba} - \bar{Z}_{34}^{ba} & \bar{Z}_{23}^{ba} + \bar{Z}_{24}^{ba} + \bar{Z}_{34}^{ba} & \bar{Z}_{23}^{bb} + \bar{Z}_{24}^{bb} + \bar{Z}_{34}^{bb} & \bar{Z}_{23}^{bc} + \bar{Z}_{24}^{bc} + \bar{Z}_{34}^{bc} \\
-\bar{Z}_{23}^{ca} - \bar{Z}_{34}^{ca} & \bar{Z}_{23}^{ca} + \bar{Z}_{24}^{ca} + \bar{Z}_{34}^{ca} & \bar{Z}_{23}^{cb} + \bar{Z}_{24}^{cb} + \bar{Z}_{34}^{cb} & \bar{Z}_{23}^{cc} + \bar{Z}_{24}^{cc} + \bar{Z}_{34}^{cc}
\end{array}\right]
$$

Table 3.5 *Short-circuit impedance matrices for different types of faults at bus no. 4 of 6-bus meshed distribution system*

Fault	Short-circuit impedance matrix
LLG	$\left[\bar{\mathbf{Z}}_{sc}^{\mathbf{LLGm}}\right] = \begin{bmatrix} \bar{Z}_{12}^{aa}+\bar{Z}_{23}^{aa}+\bar{Z}_{34}^{aa}+\bar{Z}_f & \bar{Z}_{12}^{ab}+\bar{Z}_{23}^{ab}+\bar{Z}_{34}^{ab} & -\bar{Z}_{23}^{aa}-\bar{Z}_{34}^{aa} & -\bar{Z}_{23}^{ab}-\bar{Z}_{34}^{ab} & -\bar{Z}_{23}^{ac}-\bar{Z}_{34}^{ac} \\ \bar{Z}_{12}^{ba}+\bar{Z}_{23}^{ba}+\bar{Z}_{34}^{ba} & \bar{Z}_{12}^{bb}+\bar{Z}_{23}^{bb}+\bar{Z}_{34}^{bb}+\bar{Z}_f & -\bar{Z}_{23}^{ba}-\bar{Z}_{34}^{ba} & -\bar{Z}_{23}^{bb}-\bar{Z}_{34}^{bb} & -\bar{Z}_{23}^{bc}-\bar{Z}_{34}^{bc} \\ -\bar{Z}_{23}^{aa}-\bar{Z}_{34}^{aa} & -\bar{Z}_{23}^{ab}-\bar{Z}_{34}^{ab} & \bar{Z}_{23}^{aa}+\bar{Z}_{24}^{aa}+\bar{Z}_{34}^{aa} & \bar{Z}_{23}^{ab}+\bar{Z}_{24}^{ab}+\bar{Z}_{34}^{ab} & \bar{Z}_{23}^{ac}+\bar{Z}_{24}^{ac}+\bar{Z}_{34}^{ac} \\ -\bar{Z}_{23}^{ba}-\bar{Z}_{34}^{ba} & -\bar{Z}_{23}^{bb}-\bar{Z}_{34}^{bb} & \bar{Z}_{23}^{ba}+\bar{Z}_{24}^{ba}+\bar{Z}_{34}^{ba} & \bar{Z}_{23}^{bb}+\bar{Z}_{24}^{bb}+\bar{Z}_{34}^{bb} & \bar{Z}_{23}^{bc}+\bar{Z}_{24}^{bc}+\bar{Z}_{34}^{bc} \\ -\bar{Z}_{23}^{ca}-\bar{Z}_{34}^{ca} & -\bar{Z}_{23}^{cb}-\bar{Z}_{34}^{cb} & \bar{Z}_{23}^{ca}+\bar{Z}_{24}^{ca}+\bar{Z}_{34}^{ca} & \bar{Z}_{23}^{cb}+\bar{Z}_{24}^{cb}+\bar{Z}_{34}^{cb} & \bar{Z}_{23}^{cc}+\bar{Z}_{24}^{cc}+\bar{Z}_{34}^{cc} \end{bmatrix}$
LLLG	$\left[\bar{\mathbf{Z}}_{sc}^{\mathbf{LLLGm}}\right] = \begin{bmatrix} \bar{Z}_{12}^{aa}+\bar{Z}_{23}^{aa}+\bar{Z}_{34}^{aa}+\bar{Z}_f & \bar{Z}_{12}^{ab}+\bar{Z}_{23}^{ab}+\bar{Z}_{34}^{ab} & \bar{Z}_{12}^{ac}+\bar{Z}_{23}^{ac}+\bar{Z}_{34}^{ac} & -\bar{Z}_{23}^{aa}-\bar{Z}_{34}^{aa} & -\bar{Z}_{23}^{ab}-\bar{Z}_{34}^{ab} & -\bar{Z}_{23}^{ac}-\bar{Z}_{34}^{ac} \\ \bar{Z}_{12}^{ba}+\bar{Z}_{23}^{ba}+\bar{Z}_{34}^{ba} & \bar{Z}_{12}^{bb}+\bar{Z}_{23}^{bb}+\bar{Z}_{34}^{bb}+\bar{Z}_f & \bar{Z}_{12}^{bc}+\bar{Z}_{23}^{bc}+\bar{Z}_{34}^{bc} & -\bar{Z}_{23}^{ba}-\bar{Z}_{34}^{ba} & -\bar{Z}_{23}^{bb}-\bar{Z}_{34}^{bb} & -\bar{Z}_{23}^{bc}-\bar{Z}_{34}^{bc} \\ \bar{Z}_{12}^{ca}+\bar{Z}_{23}^{ca}+\bar{Z}_{34}^{ca} & \bar{Z}_{12}^{cb}+\bar{Z}_{23}^{cb}+\bar{Z}_{34}^{cb} & \bar{Z}_{12}^{cc}+\bar{Z}_{23}^{cc}+\bar{Z}_{34}^{cc}+\bar{Z}_f & -\bar{Z}_{23}^{ca}-\bar{Z}_{34}^{ca} & -\bar{Z}_{23}^{cb}-\bar{Z}_{34}^{cb} & -\bar{Z}_{23}^{cc}-\bar{Z}_{34}^{cc} \\ -\bar{Z}_{23}^{aa}-\bar{Z}_{34}^{aa} & -\bar{Z}_{23}^{ab}-\bar{Z}_{34}^{ab} & -\bar{Z}_{23}^{ac}-\bar{Z}_{34}^{ac} & \bar{Z}_{23}^{aa}+\bar{Z}_{24}^{aa}+\bar{Z}_{34}^{aa} & \bar{Z}_{23}^{ab}+\bar{Z}_{24}^{ab}+\bar{Z}_{34}^{ab} & \bar{Z}_{23}^{ac}+\bar{Z}_{24}^{ac}+\bar{Z}_{34}^{ac} \\ -\bar{Z}_{23}^{ba}-\bar{Z}_{34}^{ba} & -\bar{Z}_{23}^{bb}-\bar{Z}_{34}^{bb} & -\bar{Z}_{23}^{bc}-\bar{Z}_{34}^{bc} & \bar{Z}_{23}^{ba}+\bar{Z}_{24}^{ba}+\bar{Z}_{34}^{ba} & \bar{Z}_{23}^{bb}+\bar{Z}_{24}^{bb}+\bar{Z}_{34}^{bb} & \bar{Z}_{23}^{bc}+\bar{Z}_{24}^{bc}+\bar{Z}_{34}^{bc} \\ -\bar{Z}_{23}^{ca}-\bar{Z}_{34}^{ca} & -\bar{Z}_{23}^{cb}-\bar{Z}_{34}^{cb} & -\bar{Z}_{23}^{cc}-\bar{Z}_{34}^{cc} & \bar{Z}_{23}^{ca}+\bar{Z}_{24}^{ca}+\bar{Z}_{34}^{ca} & \bar{Z}_{23}^{cb}+\bar{Z}_{24}^{cb}+\bar{Z}_{34}^{cb} & \bar{Z}_{23}^{cc}+\bar{Z}_{24}^{cc}+\bar{Z}_{34}^{cc} \end{bmatrix}$
LL	$\left[\bar{\mathbf{Z}}_{sc}^{\mathbf{LLm}}\right] = \begin{bmatrix} \bar{Z}_{12}^{aa}-2\bar{Z}_{12}^{ab}+\bar{Z}_{12}^{bb}+\bar{Z}_{23}^{aa}-2\bar{Z}_{23}^{ab}+\bar{Z}_{23}^{bb}+\bar{Z}_{34}^{aa}-2\bar{Z}_{34}^{ab}+\bar{Z}_{34}^{bb}+\bar{Z}_f & \bar{Z}_{23}^{ab}-\bar{Z}_{23}^{aa}-\bar{Z}_{34}^{aa}+\bar{Z}_{34}^{ab} & \bar{Z}_{23}^{bb}-\bar{Z}_{23}^{ab}-\bar{Z}_{34}^{ab}+\bar{Z}_{34}^{bb} & \bar{Z}_{23}^{bc}-\bar{Z}_{23}^{ac}-\bar{Z}_{34}^{ac}+\bar{Z}_{34}^{bc} \\ \bar{Z}_{23}^{ab}-\bar{Z}_{23}^{aa}-\bar{Z}_{34}^{aa}+\bar{Z}_{34}^{ab} & \bar{Z}_{23}^{aa}+\bar{Z}_{24}^{aa}+\bar{Z}_{34}^{aa} & \bar{Z}_{23}^{ab}+\bar{Z}_{24}^{ab}+\bar{Z}_{34}^{ab} & \bar{Z}_{23}^{ab}+\bar{Z}_{24}^{ac}+\bar{Z}_{34}^{ac} \\ \bar{Z}_{23}^{bb}-\bar{Z}_{23}^{ba}-\bar{Z}_{34}^{ba}+\bar{Z}_{34}^{bb} & \bar{Z}_{23}^{ba}+\bar{Z}_{24}^{ba}+\bar{Z}_{34}^{ba} & \bar{Z}_{23}^{bb}+\bar{Z}_{24}^{bb}+\bar{Z}_{34}^{bb} & \bar{Z}_{23}^{bc}+\bar{Z}_{24}^{bc}+\bar{Z}_{34}^{bc} \\ \bar{Z}_{23}^{cb}-\bar{Z}_{23}^{ca}-\bar{Z}_{34}^{ca}+\bar{Z}_{34}^{cb} & \bar{Z}_{23}^{ca}+\bar{Z}_{24}^{ca}+\bar{Z}_{34}^{ca} & \bar{Z}_{23}^{cb}+\bar{Z}_{24}^{cb}+\bar{Z}_{34}^{cb} & \bar{Z}_{23}^{cc}+\bar{Z}_{24}^{cc}+\bar{Z}_{34}^{cc} \end{bmatrix}$

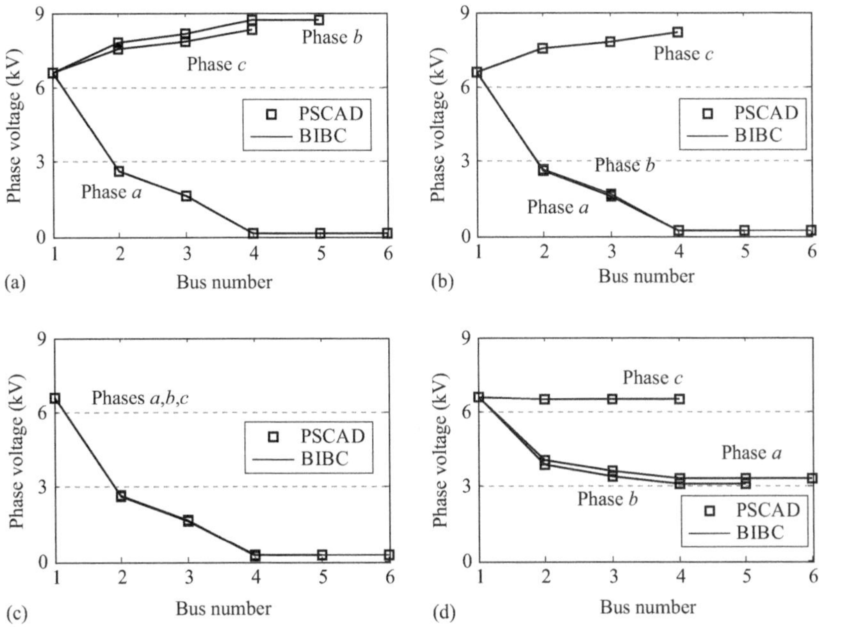

Figure 3.4 Voltage profile of 6-bus meshed distribution system under fault conditions: (a) LG fault; (b) LLG fault; (c) LLLG fault; and (d) LL fault

Table 3.6 Fault currents (kA) for different types of faults at bus no. 4 of 6-bus meshed system

	LG fault		**LL fault**		**LLG fault**		**LLLG fault**	
	BIBC	**PSCAD**	**BIBC**	**PSCAD**	**BIBC**	**PSCAD**	**BIBC**	**PSCAD**
Phase *a*	0.96926	0.96926	1.6040	1.6039	1.6289	1.6288	1.7793	1.7792
Phase *b*			1.6040	1.6039	1.6303	1.6302	1.7648	1.7647
Phase *c*							1.5134	1.5133

Using the same procedure, the short-circuit impedance matrices for LLG, LLLG and LL faults for the 6-bus meshed distribution system are given in Table 3.5. The voltage profiles for the four types of faults at bus no. 4 of the 6-bus meshed distribution system are shown in Figure 3.4, while the fault currents at the fault point are given in Table 3.6. Further, in Table 3.7, the phase *a* voltage at bus no. 4 for different types of faults (obtained by the short-circuit analysis method and PSCAD simulation studies) are shown.

From Tables 3.6 and 3.7 and Figure 3.4, it can be concluded that the BIBC-based short-circuit analysis method performs equally well for the meshed distribution systems also.

Table 3.7 Phase a voltages (kV) for different types of faults at bus no. 4 of 6-bus meshed system

Bus no.	LG fault			LLG fault			LLLG fault			LL fault		
	PSCAD	BIBC	%Error	PSCAD	BIBC	%Error	PSCAD	BIBC	%Error	PSCAD	BIBC	%Error
1	6.5817	6.5818	0.00151	6.5817	6.5818	0.0015	6.5817	6.5818	0.00151	6.5817	6.5818	0.00151
2	2.623	2.623	0	2.6801	2.6802	0.0037	2.6585	2.6585	0	4.043	4.0431	0.00247
3	1.6074	1.6074	0	1.6802	1.6802	0	1.6546	1.6546	0	3.6124	3.6124	0
4	0.126	0.126	0	0.2118	0.2117	0.0472	0.2313	0.2312	0.04323	3.2979	3.298	0.00303
5	0.126	0.126	0	0.2118	0.2117	0.0472	0.2313	0.2312	0.04323	3.2979	3.298	0.00303
6	0.126	0.126	0	0.2118	0.2117	0.0472	0.2313	0.2312	0.04323	3.2979	3.298	0.00303

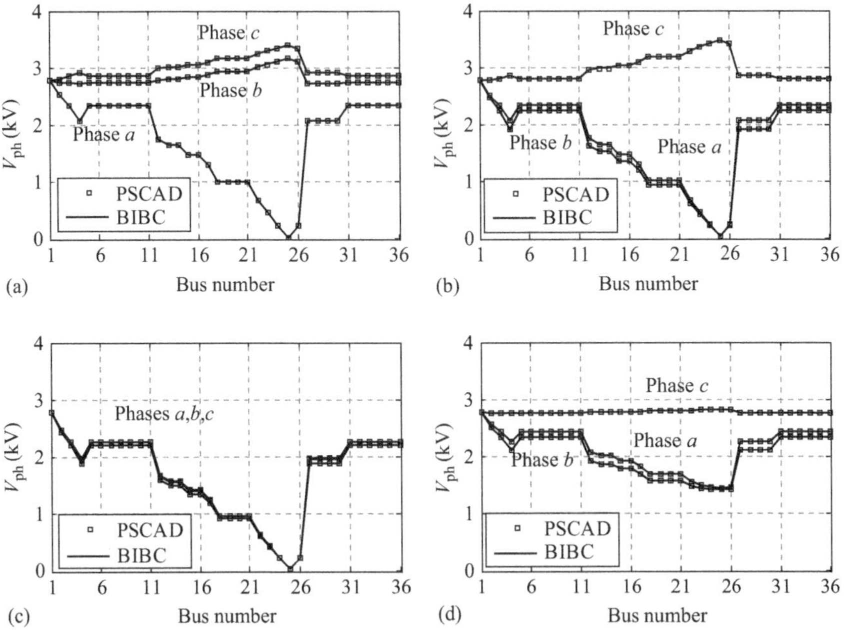

Figure 3.5 Voltage profile of 36-bus radial distribution system under fault conditions: (a) LG fault; (b) LLG fault; (c) LLLG fault; and (d) LL fault

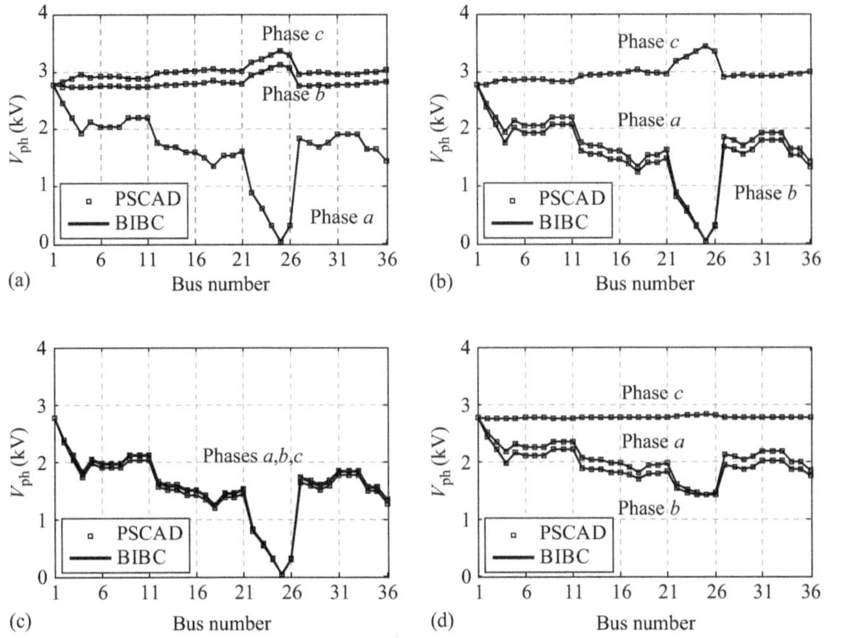

Figure 3.6 Voltage profile of 36-bus meshed distribution system under fault conditions: (a) LG fault; (b) LLG fault; (c) LLLG fault; and (d) LL fault

Table 3.8 Fault currents (kA) for different types of faults at bus no. 25 of 36-bus radial distribution system

	LG fault		LL fault		LLG fault		LLLG fault	
	BIBC	**PSCAD**	**BIBC**	**PSCAD**	**BIBC**	**PSCAD**	**BIBC**	**PSCAD**
Phase *a*	1.9346	1.9343	2.4957	2.4953	2.6708	2.6704	2.6654	2.6650
Phase *b*			2.4957	2.4953	2.5002	2.4999	2.9478	2.9473
Phase *c*							2.7734	2.7730

Table 3.9 Fault currents (kA) for different types of faults at bus no. 25 of 36-bus meshed distribution system

	LG fault		LL fault		LLG fault		LLLG fault	
	BIBC	**PSCAD**	**BIBC**	**PSCAD**	**BIBC**	**PSCAD**	**BIBC**	**PSCAD**
Phase *a*	2.5886	2.5881	3.2595	3.2589	3.4961	3.4954	3.4937	3.4930
Phase *b*			3.2595	3.2589	3.2744	3.2737	3.8436	3.8429
Phase *c*							3.6077	3.6070

3.4.3 Results for 36-bus radial distribution system

The method was also tested on a larger system to test its efficacy. The voltage profiles of the 36-bus distribution system, for all the four types of faults at bus no. 25, obtained through BIBC-based method and PSCAD simulation studies are shown in Figure 3.5 for radial configuration and in Figure 3.6 for weakly meshed configuration. The fault currents for radial and meshed distribution systems are tabulated in Tables 3.8 and 3.9, respectively. As can be observed from these tables and figures, the results of both the methods are in very good agreement.

3.5 Conclusion

In this chapter, a method for the short-circuit analysis of radial and meshed distribution systems has been explained in details. The results obtained from the discussed method have also been compared with PSCAD simulation results to establish their accuracy. The values of the calculated short-circuit currents can be used for the selection of equipment ratings and protection coordination.

Acknowledgements

The author wishes to thank Mr. Akhilesh Mathur, Senior Research Fellow, Department of Electrical Engineering, IIT Roorkee for his help in computer and PSCAD simulation results.

References

[1] J. J. Grainger and W. D. Stevenson Jr., *Power System Analysis*, McGraw-Hill, New York, NY, 1994.

[2] M. R. Patel, *Introduction to Electrical Power and Power Electronics*, CRC Press – Taylor & Francis Group, Boca Raton, FL, 2013.

[3] J. Arrillaga and N. R. Watson, *Computer Modeling of Electrical Power Systems*, 2nd ed., Wiley, New York, NY, 2001.

[4] M. A. Akher and K. M. Nor, 'Fault analysis of multi-phase distribution systems using symmetrical components', *IEEE Transactions on Power Delivery*, Vol. 25, No. 4, pp. 2931–2939, September 2010.

[5] K. Gampa, Srinaga A. C. Vemprala, and Sukumar M. Brahma, 'Errors in fault analysis of power distribution systems using sequence component approach', *2010 IEEE PES Transmission and Distribution Conference and Exposition*, April 2010, New Orleans, LA, pp. 1–6.

[6] T. H. Chen, M. S. Chen, W. J. Lee, P. Kotas, and P. Van Olinda, 'Distribution system short-circuit analysis – a rigid approach', *IEEE Transactions on Power Systems*, Vol. 7, No. 1, pp. 444–450, February 1992.

[7] S. M. Halpin, L.L. Grigsby, C. A. Gross, and R. M. Nelms, 'An improved fault analysis algorithm for unbalanced multi-phase power distribution systems', *IEEE Transactions on Power Delivery*, Vol. 9, No. 3, pp. 1332–1338, July 1994.

[8] X. Zhang, F. Soudi, D. Shirmohammadi and C. S. Cheng, 'A distribution short-circuit analysis using hybrid compensation method', *IEEE Transactions on Power Systems*, Vol. 10, No. 4, pp. 2053–2059, November 1995.

[9] J. H. Teng, 'Fast short-circuit analysis method for unbalanced distribution systems', *IEEE Power Engineering Society General Meeting*, Vol. 1, pp. 240–245, July 2003.

[10] J. H. Teng, 'Systematic short-circuit analysis method for unbalanced distribution systems', *IEE Proceedings of Generation, Transmission and Distribution*, Vol. 152, No. 4, pp. 549–555, July 2005.

[11] J. H. Teng, 'Unsymmetrical short-circuit fault analysis for weakly meshed distribution systems', *IEEE Transactions on Power Systems*, Vol. 25, No. 1, pp. 96–105, March 2005.

Chapter 4

Choice of solver for distribution system state estimation

Ravindra Singh[1], *Bikash C. Pal*[2] *and Rabih A. Jabr*[3]

Abstract

In this chapter, a statistical framework is introduced to assess the suitability of various state estimation methodologies for the purpose of distribution system state estimation. The existing algorithms adopted in the transmission system state estimation are re-evaluated for the distribution system. The performance of three-state estimation algorithms has been examined and discussed in standard 12-bus and 95-bus United Kingdom-Generic Distribution System network models.

Keywords

Distribution system state estimation, Distribution management system, Error statistics, χ^2-Distribution

List of symbols

m, n	Number of measurements and state variables
$\boldsymbol{x}_t, \hat{\boldsymbol{x}}$	True state and estimated state vectors, respectively $(n \times 1)$
$P_x, \hat{P}_x$	Numerically computed and estimated error covariance matrices, respectively $(n \times n)$
$E[\cdot]$	Expectation operator
ϵ	Normalised state error squared variable
$\boldsymbol{z}$	Measurement vector $(m \times 1)$
$\boldsymbol{h}(\boldsymbol{x})$	Expectation of measurement vector $(m \times 1)$
σ_{zi}	Standard deviation of the ith measurement

[1]Scientist Power Technologies, ABB Inc. (Corporate Research), 940 Main Campus Drive, STE 200, Raleigh, NC, USA
[2]Department of Electrical and Electronic Engineering, Imperial College London, London, UK
[3]Department of Electrical & Computer Engineering, American University of Beirut, Riad El-Solh, Beirut, Lebanon

R_z Measurement error covariance matrix ($m \times m$)
$\boldsymbol{e}_z$ Measurement error vector ($m \times 1$)
r_i Normalised residual of ith measurement

4.1 Introduction

The deregulation of power systems and the introduction of distributed generation (DG) to distribution networks have challenged the operational philosophy of the distribution systems. The passive nature of the network can only accommodate restricted amount of DG capacity. This means that significant network reinforcement will be necessary to accommodate DG and load growth in the future. An alternative would be to change the approach to network operation such as introducing control to distribution network operation. A range of technology innovations is needed to change the way distribution systems operate. The innovations must pin down on new architecture for distribution network control centre with performance critical software functions like state estimation (SE), optimal power flow, and network-specific sensor placement and integration.

In transmission systems, SE is a fairly routine task and a host of established methodologies exist [1]. These cannot simply be adopted to distribution systems because the planning, design and operation philosophy of distribution networks are different from those in the transmission networks. The distribution network topology and characteristics are different and most importantly the amount of available network measurements is very limited. The SE methodologies adopted in transmission systems start showing their limitations when exposed to the specifics of distribution networks [2].

Furthermore, the potential benefits of using SE technologies in distribution network control have not been explored mainly due to the absence of adequate network measurements and also the lack of rigorous methodology and tools that could be applied on restricted measurements. The development of new distribution system state estimation (DSSE) is a challenging task as the tools to evaluate the quality of SE must consider a number of issues relating to measurement types, locations and numbers.

Methodologies on which such tools could be built are not available at present. However, some interesting research has been done in DSSE [3–10]. Lu et al. [3] propose a three-phase DSSE algorithm. The algorithm uses a current-based formulation of the weighted least squares (WLS) method in which the power measurements, current measurements and voltage measurements are converted to their equivalent currents, and the Jacobian terms are constant and equal to the admittance matrix elements. The observability analysis of the proposed distribution system is also discussed. Lin and Teng [4] have proposed a new fast decoupled state estimator with equality constraints. The proposed method is based on the equivalent current measurement in rectangular coordinates. Baran and Kelley [5] have introduced a computationally efficient algorithm based on branch currents as state variables. The method is demonstrated to work well in radial and weakly meshed systems. This concept is further refined by Wang and Schulz [6]. Wang and Schulz [6] have presented a revised branch current-based DSSE algorithm. In this algorithm, the load estimated at every node from an automated meter reading system is used as a pseudo measurement.

Li [7] has presented a distribution system state estimator based on WLS approach and three-phase modelling techniques. Li has also demonstrated the impact of the measurement placement and measurement accuracy on the estimated results. A rule-based approach for measurement placement is presented by Baran et al. [8]. Ghosh et al. [9] have presented an alternative approach to DSSE using a probabilistic extension of the radial load flow algorithm treating the real measurements as solution constraints. The algorithm that accounts for non-normally distributed loads incorporates the concept of load diversity and can interact with a load allocation routine. The field results are discussed in Reference 10.

The DSSE literature is either based on the probabilistic load flow or direct adaptation of transmission system SE algorithms (particularly WLS). The issue of measurement inadequacy is addressed through pseudo measurements which are stochastic in nature. However, the performance of the SE algorithms under the stochastic behaviour of pseudo measurements is not addressed in the DSSE literature.

The work presented in this chapter and also reported in References 11, 12 investigates the existing transmission system SE techniques and algorithms and assesses their suitability to the DSSE problem. The selected algorithms are tested on the 12-bus and 95-bus United Kingdom-Generic Distribution System (UK-GDS) network models against some statistical measures like *Bias*, *Consistency* and overall *Quality* of the estimates. Unlike many other distribution systems, the UK distribution network is fairly balanced and that has prompted us to go with a single-phase approach although the method is generic. Furthermore, the statistical measures utilised in this chapter mainly depend on the probability distribution of the measurements and not on the line model of the network. Following this introduction, a theoretical framework for the statistical measures is established in Section 4.2. The consistency and the quality of the estimates utilise the asymptotic state error covariance matrix. The various SE techniques along with the details of their state error covariance matrices are discussed in Section 4.3. The efficacy of the algorithms is examined on standard test systems and discussed in Section 4.4.

4.2 Statistical measures

In distribution systems, measurements are predominantly of pseudo type, which are statistical in nature, so the performance of a state estimator should be based on some statistical measures. Various statistical measures such as bias, consistency and quality have been adopted for assessing the effectiveness of SE in other technology areas such as target tracking [13]. We explore these for the DSSE applications. Briefly, we describe the statistical measures as follows:

4.2.1 Bias

A state estimator is said to be unbiased if the expected value of error in the state estimate is zero. Mathematically an unbiased estimator can be defined as:

$$E\left[\left(\boldsymbol{x}_t - \hat{\boldsymbol{x}}\right)\right] = \mathbf{0} \tag{4.1}$$

4.2.2 Consistency

If the error in an estimate statistically corresponds to the corresponding covariance matrix then the estimate (and hence the technique generating this estimate) is said to be consistent. One measure of consistency is the normalised state error squared variable:

$$\epsilon = (\boldsymbol{x}_t - \hat{\boldsymbol{x}})^T \hat{P}_x^{-1} (\boldsymbol{x}_t - \hat{\boldsymbol{x}}) \tag{4.2}$$

where $\hat{P}_x$ denotes the estimated state error covariance matrix.

For the estimator to be consistent, ϵ should be within its confidence bounds, which can be obtained from the error statistics.

4.2.2.1 Choice of confidence regions

In the univariate case when the estimation error is represented by a normal distribution with zero mean and known variance, one can use the tables of normal distribution to compute the confidence intervals. However, in the multivariate case when the estimation error is represented by a normal distribution with zero mean vector and known covariance matrix, such confidence intervals are difficult to compute because tables are available only for the bivariate case. Alternatively, one could setup limits for each component on the basis of distribution, but this procedure has the disadvantages that the choice of limit is somewhat arbitrary and in some cases leads to tests that may be poor against some alternatives. Moreover, such limits are difficult to compute. The procedure given below, which is based on χ^2-statistics, can be easily computed and applied in the multivariate case. Furthermore, it can be theoretically justified based on the following lemma. The proof of the lemma can be found in Reference 14.

Lemma 4.1. *If an n-component vector* $\boldsymbol{v}$ *is distributed according to normal distribution* $\mathcal{N}(\boldsymbol{0}, T)$ *(non-singular), then* $\boldsymbol{v}^T T^{-1} \boldsymbol{v}$ *is distributed according to chi-square* (χ^2)*-distribution with n degrees of freedom.*

4.2.2.2 χ^2-Statistics

It can be shown that if the errors in measurements are normally distributed, the SE error corresponding to these measurements will be normally distributed with zero mean vector and covariance matrix given by $E[(\boldsymbol{x}_t - \hat{\boldsymbol{x}})(\boldsymbol{x}_t - \hat{\boldsymbol{x}})^T]$. Utilising this fact and Lemma 4.1, the normalised squared error ϵ (4.2) should follow a χ^2-distribution with n degrees of freedom for a consistent estimator, where n is the number of states. In other words, for the estimator to be consistent, ϵ should lie within its confidence bounds that can be obtained from the standard χ^2-table for a chosen confidence level α. Lower and upper bounds for this confidence level can be given by $\chi_n^2((1-\alpha)/2)$ and $\chi_n^2((1+\alpha)/2)$, respectively. In statistics, a 95% confidence is considered to be adequate.

4.2.2.3 χ^2-Test over Monte Carlo simulations

In practice, statistical tests are performed using a number of Monte Carlo simulations. Consider that the system has n number of states and M is the number of Monte Carlo simulations, then the normalised squared error follows a χ^2-distribution with Mn degrees of freedom. Mathematically:

$$E[\epsilon] = \frac{\chi^2_{Mn}(\alpha)}{M} \tag{4.3}$$

For large number of Monte Carlo runs $\chi^2_{Mn}(\alpha) \approx Mn$, which results in:

$$E[\epsilon] = n \tag{4.4}$$

Hence, the mean of ϵ should approach to the number of states with the increase in the number of simulations.

4.2.3 Quality

Quality of an estimate is inversely related to its variance. For the multivariate case, the square root of the determinant of the error covariance matrix measures the volume of 1-σ (one standard deviation) ellipsoid and is used here to quantify the total variance of an estimate. Hence, the quality of the estimate can be defined as:

$$Q_{\text{det}} = \log\left(\frac{1}{\sqrt{\det(P_x)}}\right) \tag{4.5}$$

Sometimes, in large networks, it becomes difficult to compute the determinant of the error covariance matrix numerically due to precision limits of the solver. In this situation, an alternate way to define the quality is to use the *trace* of the error covariance matrix. However, this ignores the off-diagonal information. The quality as function of the *trace* of the error covariance matrix can be written as:

$$Q_{trace} = \log\left(\frac{1}{\text{tr}(P_{\text{x}})}\right) \tag{4.6}$$

4.3 SE techniques

Various algorithms have been suggested for transmission system state estimation [1]. All these algorithms work well in transmission systems because there is high redundancy in the measurements. However, in distribution systems, due to sparsity of measurements, there is less or no redundancy in the measurements. Hence, when these algorithms are exposed to distribution systems they start showing their limitations. For example, in transmission systems, weighted least absolute value (WLAV) eliminates bad data out of redundant measurements, but in distribution systems, it fails to work because it treats every pseudo measurement as bad data and there is no redundancy to eliminate these pseudo measurements.

This section briefly explains the most common SE techniques to examine their suitability for the DSSE problem under stochastic behaviour of the pseudo measurements and limited or no redundancy. All these techniques use the following measurement model.

4.3.1 Measurement model

$$z = \boldsymbol{h}(\boldsymbol{x}) + \boldsymbol{e}_z \tag{4.7}$$

where $\boldsymbol{e}_z \sim \mathcal{N}(\boldsymbol{0}, R_z)$ is zero mean Gaussian noise with error covariance matrix $R_z(=\text{diag}\{\sigma_{z1}^2, \sigma_{z2}^2, \ldots, \sigma_{zm}^2\})$. We define the normalised residual of ith measurement r_i as:

$$r_i = \frac{z_i - h_i(\boldsymbol{x})}{\sigma_{zi}} \tag{4.8}$$

where $r_i \sim \mathcal{N}(0, 1)$. The class of estimators discussed in this section is based on maximum likelihood theory. They rely on a priori knowledge of the distribution of the measurement error (Gaussian in this case, with zero mean and known covariance). A generalised estimation problem seeks to minimise the following objective:

$$J = \sum_{i=1}^{m} \rho(r_i) \tag{4.9}$$

The different estimators can be characterized based on the choice of the ρ function.

4.3.2 WLS estimation

WLS is a quadratic form of the maximum likelihood estimation problem. The WLS problem can be stated as the minimisation of the following objective function:

$$\frac{1}{2}\,[z - \boldsymbol{h}(\boldsymbol{x})]^T R_z^{-1}\,[z - \boldsymbol{h}(\boldsymbol{x})] \tag{4.10}$$

The above objective takes the form given in (4.9) for:

$$\rho(r_i) = \frac{1}{2} r_i^2 \tag{4.11}$$

An estimate of state was obtained iteratively using the Newton method according to:

$$\hat{\boldsymbol{x}}_{k+1} = \hat{\boldsymbol{x}}_k + \left(H^T(\hat{\boldsymbol{x}}_k)\, R_z^{-1} H(\hat{\boldsymbol{x}}_k)\right)^{-1} H^T(\hat{\boldsymbol{x}}_k)\, R_z^{-1}\left[z - \boldsymbol{h}(\hat{\boldsymbol{x}}_k)\right] \tag{4.12}$$

where

$$H(\hat{\boldsymbol{x}}_k) = \left[\frac{\partial \boldsymbol{h}(\boldsymbol{x})}{\partial \boldsymbol{x}}\right]_{x=h(\hat{x}_k)} \tag{4.13}$$

4.3.3 WLAV estimator

WLAV estimator is based on the minimisation of the L_1 norm of weighted measurement residual, and can be expressed as:

$$\left\| R_z^{-\frac{1}{2}} \left[z - \boldsymbol{h}(\boldsymbol{x}) \right] \right\|_1 \tag{4.14}$$

which is equivalent to (4.9) when,

$$\rho(r_i) = |r_i| \tag{4.15}$$

Existing techniques use linear programming or interior point methods to solve this problem. In this chapter, we have used a primal dual interior point (PDIP) method [15].

4.3.4 Schweppe Huber generalized M estimator

This estimator combines both WLS and WLAV estimators. The ρ function for Schweppe Huber generalized M (SHGM) estimator is given by:

$$\rho(r_i) = \begin{cases} \frac{1}{2} r_i^2 & \text{if } |r_i| \le a\omega_i \\ a\omega_i |r_i| - \frac{1}{2} a^2 \omega_i^2 & \text{otherwise} \end{cases} \tag{4.16}$$

The performance of this estimator highly depends upon the weight factor ω_i and tuning parameter a. In this chapter, the solution to this problem was obtained using iteratively re-weighted least squares (IRLS) method [1]. The parameter $a = 1.5$ was used in simulations.

4.3.4.1 State error covariance matrix

An estimate of the asymptotic covariance matrix at convergence can be expressed as [16, 17]:

$$\hat{P}_x = \alpha \left(H^T(\hat{\boldsymbol{x}}) \, R_z^{-1} H(\hat{\boldsymbol{x}}) \right)^{-1} \tag{4.17}$$

where $\hat{\boldsymbol{x}} = \lim_{k \to \infty} \hat{\boldsymbol{x}}_k$. The value of α depends on the choice of the estimator. An expression for α is given by [16]:

$$\alpha = \frac{E\left[\psi^2(r)\right]}{\left(E\left[\psi'(r)\right]\right)^2} \tag{4.18}$$

where $\psi(r) = \frac{\partial \rho(r)}{\partial r}$ and $\psi'(r) = \frac{\partial \psi(r)}{\partial r}$.

The numerical computation of α for various estimators is given in Appendix 4.6. Table 4.1 summarises the various estimators used in this chapter. Table 4.1 indicates a typical value of α for SHGM considering $a\omega_i = 1.5$. Since the IRLS method is used for SHGM, the value of α changes during the estimation process depending upon the weight ω_i.

Table 4.1 State estimators: summary

	Solution for $\hat{x}$	Asymptotic error covariance $\hat{P}_x$
WLS	Newton	$(H^T(\hat{x})\,R_z^{-1}H(\hat{x}))^{-1}$
WLAV	PDIP	$\frac{\pi}{2}(H^T(\hat{x})\,R_z^{-1}H(\hat{x}))^{-1}$
SHGM	IRLS	$^*1.037(H^T(\hat{x})\,R_z^{-1}H(\hat{x}))^{-1}$

$^*a = 1.5, w_i = 1.$

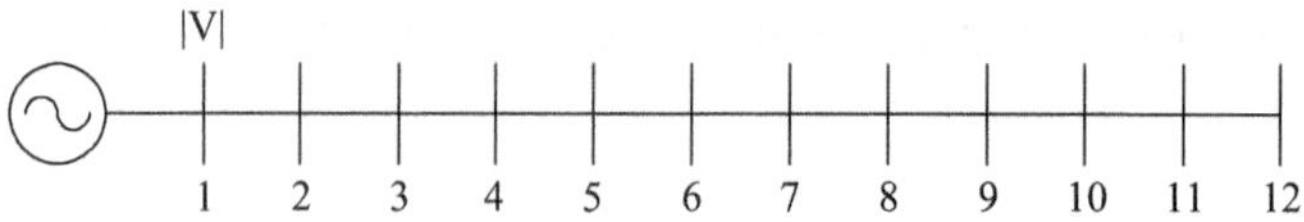

Figure 4.1 12-Bus test system

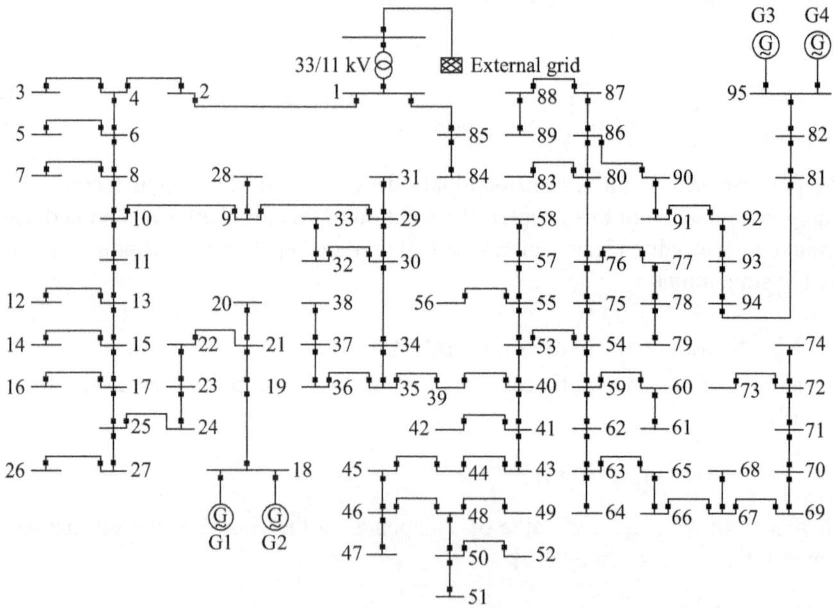

Figure 4.2 UK-GDS: 95-bus test system

4.4 Case study

The algorithms discussed in the previous section were applied on a 12-bus radial distribution network model and on a part of the UK-GDS model (95-bus UK-GDS). Figures 4.1 and 4.2 show the schematic diagrams of the test systems. Network and load data for these networks can be found in References 18 and 19, respectively.

4.4.1 State variables

The bus voltage magnitudes and angles were considered as state variables except at the reference bus (bus #1) for which the bus angle was assumed to be zero. Hence, the number of states to be evaluated was 23 and 189 for the 12-bus test system and the UK-GDS, respectively.

4.4.2 Measurements

It was assumed that the errors associated with the measurements are independent identically distributed. Three types of measurements were taken into consideration. The telemetered measurements were utilised as real measurements. Zero injections with a very low variance (10^{-8}) were modelled as virtual measurements. Loads were modelled as pseudo measurements. Various scenarios considering the errors in real measurements as 1% and 3%, while 20% and 50% in pseudo measurements were examined. The range of error in pseudo measurements was chosen on the basis of errors in load estimates of various classes of customers, such as industrial, domestic and commercial. The loads of the industrial customers can be estimated more accurately than the domestic and commercial, thus they have less error. On the other hand, loads of domestic customers are difficult to estimate; hence, they have large error. The error in commercial load estimates lies between these two cases. It was also taken into consideration that with this choice of range, the maximum demand limits at various buses are not violated and the condition of linear approximation is valid. The mean value for these measurements was obtained using distribution system load flow. Table 4.2 summarises the measurements and their redundancy level for the two test network models.

4.4.3 Measurement variance

A $\pm 3\sigma$ deviation around the mean covers more than 99.7% area of the Gaussian curve. Hence, for a given percentage of maximum error about mean μ_{zi}, the standard deviation of error was computed as follows:

$$\sigma_{zi} = \frac{\mu_{zi} \times \% \text{ error}}{3 \times 100} \tag{4.19}$$

The square of standard deviation gives the variance of the measurement.

4.4.4 Simulation results

The performance of the estimators was evaluated for the following cases:

Case 1: Error in real measurement 1% and pseudo measurement 20%.
Case 2: Error in real measurement 1% and pseudo measurement 50%.
Case 3: Error in real measurement 3% and pseudo measurement 20%.
Case 4: Error in real measurement 3% and pseudo measurement 50%.

Table 4.2 Measurements used in study

Test system	**Real measurements (m_r)**	**Virtual and pseudo measurements (m_p)**	**Redundancy $\frac{(m_r + m_p)}{n}$**
12-Bus	3 (V_1, P_{1-2}, Q_{1-2})	22 (Loads only no zero injections)	$\frac{25}{23} = 1.09$
UK-GDS **(a) Limited redundancy**	5 ($V_1, P_{1-2}, Q_{1-2}, P_{1-85}, Q_{1-85}$)	188 (Loads and zero injections)	$\frac{193}{189} = 1.02$
UK-GDS **(b) Increased redundancy**	21 ($V_1, V_{18}, V_{19}, V_{20}, V_{21}, V_{95}, P_{1-2}, Q_{1-2}, P_{1-85}, Q_{1-85}, P_{18-19}, Q_{18-19}, P_{82-95}, Q_{82-95}$ $P_{15-17}, Q_{15-17}, P_{34-35}, Q_{34-35}, \delta_{19}, \delta_{20}, \delta_{21}$)	188 (Loads and zero injections)	$\frac{209}{189} = 1.11$

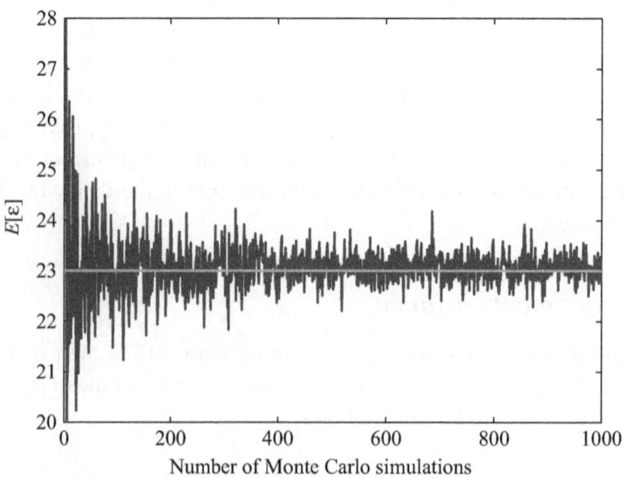

Figure 4.3 Variation of $E[\epsilon]$ with different Monte Carlo steps

4.4.4.1 12-Bus system

In the 12-bus test system, the voltage magnitude measurement at bus #1 and power flow measurement in lines #1–2 were considered as real measurement. Figure 4.3 shows the variation of the expected value of the normalised state error squared with various Monte Carlo steps for the 12-bus distribution system. It is clear from the figure that as the number of Monte Carlo steps increases, the expected value of normalised state error square variable approaches to the number of states, which agrees with (4.4). Also after 400 Monte Carlo steps, the error in $E[\epsilon]$ is within 1% of the number of

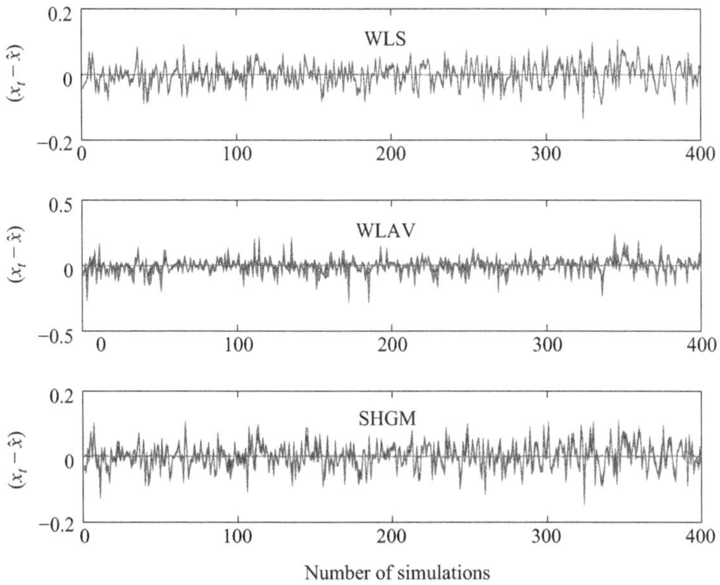

Figure 4.4 12-Bus system estimation error plot for all state variables: error in true measurements = 3%, error in pseudo measurements = 50%

states. Hence, we chose 400 Monte Carlo steps for the simulations. A larger number of Monte Carlo steps give slightly better results but it increases the computation time.

Figure 4.4 shows the error plots with the number of simulations for the three estimators. The plots shown are for the worst-case scenario (Case 4), i.e. the error associated with real measurements is 3% and that with pseudo measurements is 50%. The estimation errors for all the states are displayed in Figure 4.4; however, they are indistinguishable because of the overlaps. It is evident from the figure that the error varies about zero mean. This indicates that all the three estimators are unbiased. It was also found that for all other cases, the three estimators were unbiased. Figures 4.5–4.8 show the consistency plots for the estimators for Cases 1–4. A 95% confidence level was used to define the confidence bounds. It was found that WLS shows consistent results in all test cases. On the other hand, WLAV is inconsistent in all the cases. It is interesting to note that SHGM is inconsistent for small errors in pseudo measurements and consistent for large errors in pseudo measurements. The reason is that the measurement set considered for study is predominantly comprised of the pseudo measurements, and large error in pseudo measurements increases the measurement variance (4.19). Also the computation of variance in (4.19) is based on the maximum error. This results in low normalised residual ($|r_i|$) for pseudo measurements. Due to this fact the normalised residual becomes less than the cut-off value $a\omega_i$ (4.16), and the estimator behaves like WLS. However, this is not always true. Whenever the normalised residual exceeds the cut-off value, the estimator becomes inconsistent. It will be shown, that for the 95-bus UK-GDS system, SHGM becomes inconsistent for these cases of large errors too.

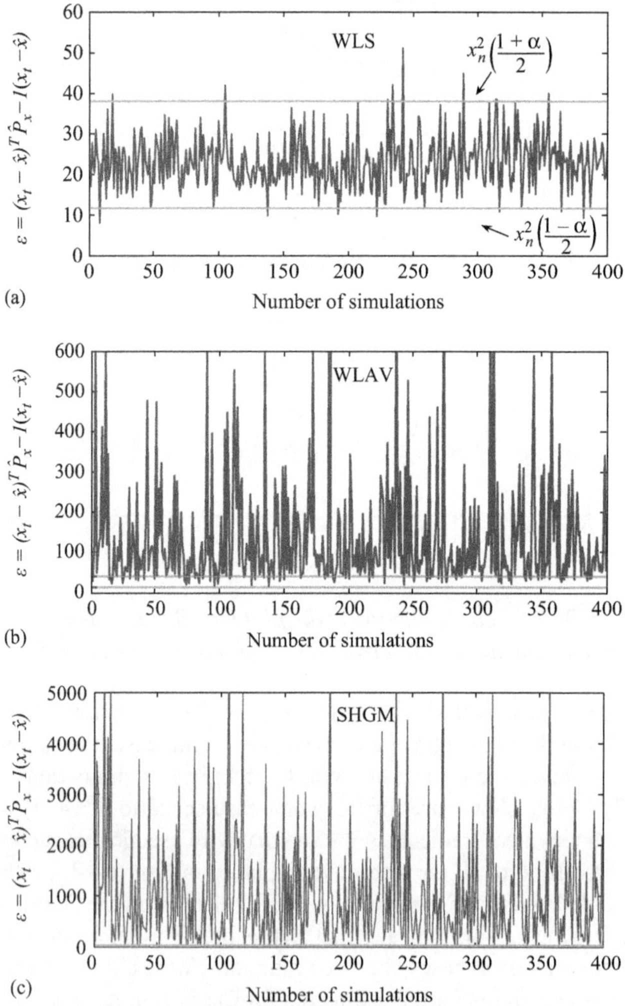

Figure 4.5 12-Bus system consistency plot: error in true measurements = 1%, error in pseudo measurements = 20%

Table 4.3 shows the performance summary of the 12-bus test system. Two types of qualities are shown. As expected the quality of the estimates decreases with the increase in the error in measurements. This decrease is significant with the increase in the error in the real measurements as compared to the pseudo measurements.

4.4.4.2 95-Bus UK-GDS

The performance of the estimators was also evaluated on the 95-bus test system model for all the test cases analysed in the 12-bus test system. It was observed that in the

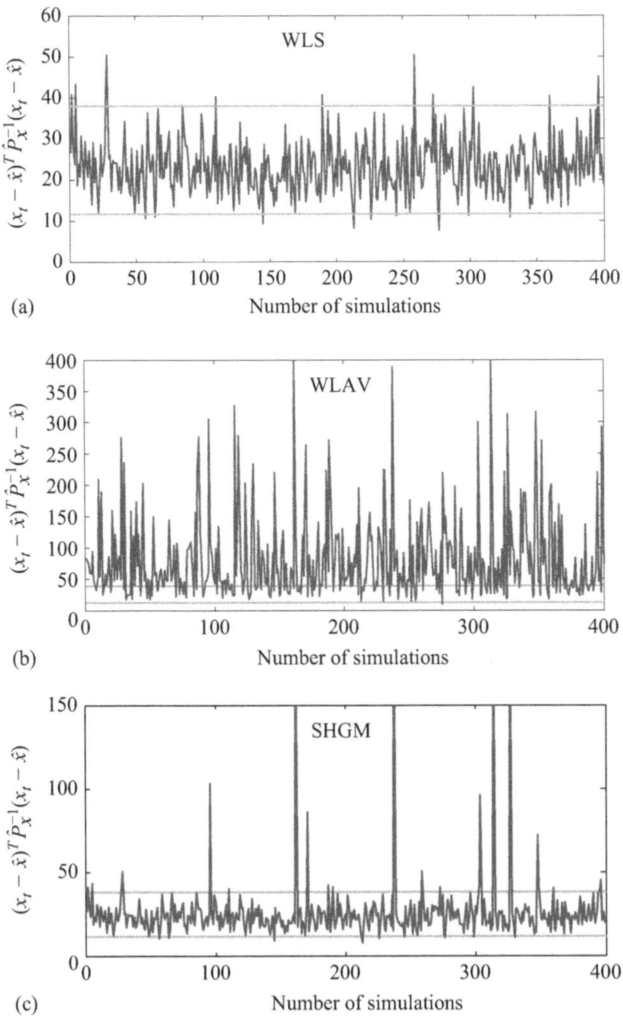

Figure 4.6 12-Bus system consistency plot: error in true measurements = 1%, error in pseudo measurements = 50%

95-bus test system also, 400 Monte Carlo steps are sufficient to bring down the error in $E[\epsilon]$ within 1% of the number of states. The following two cases were considered.

(a) Limited redundancy

In this case, the real measurements were considered to be available at the main substation. Hence, the voltage magnitude measurement at bus #1 and power flow measurements in lines #1–2 and #1–85 were taken as real measurements. It was observed that in the 95-bus test system all the estimators were unbiased. However, only WLS was found to be consistent in all the test cases. Hence, the consistency

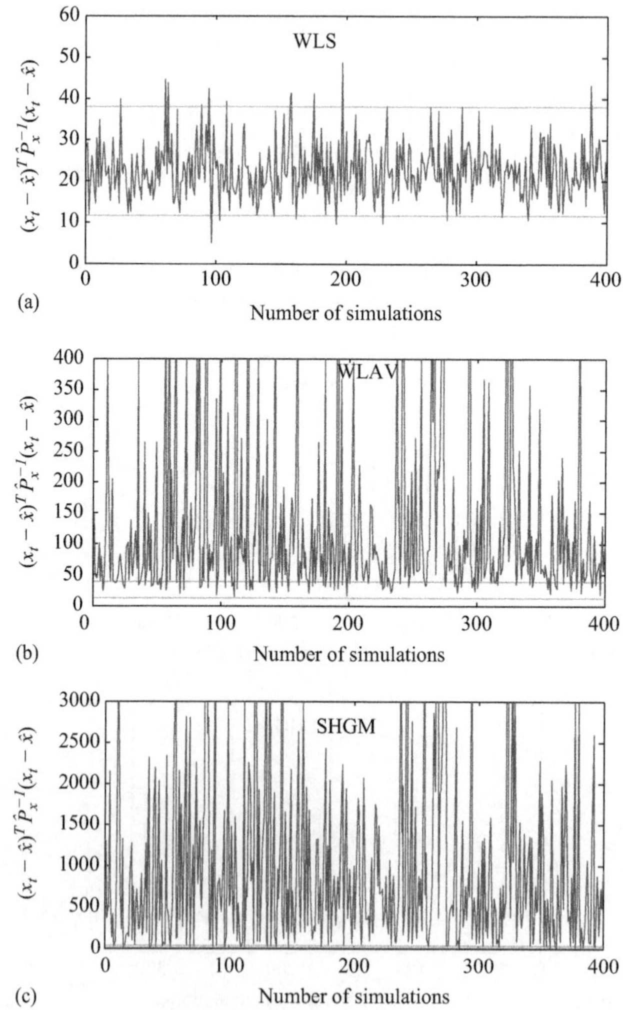

Figure 4.7 12-Bus system consistency plot: error in true measurements = 3%, error in pseudo measurements = 20%

plots of WLS in all four test cases are displayed in Figure 4.9. The consistency plot for SHGM is also shown in Figure 4.10 for the test Case 2. It is clear from Figure 4.10 that the SHGM which was consistent in Case 2 in the 12-bus system no longer remains consistent in larger systems.

(b) Increased redundancy

In this case, the redundancy was increased by placing the measurements at DG locations first and then measurements were placed at optimal locations. The optimality criterion and details of the measurement placement appear in Reference 20.

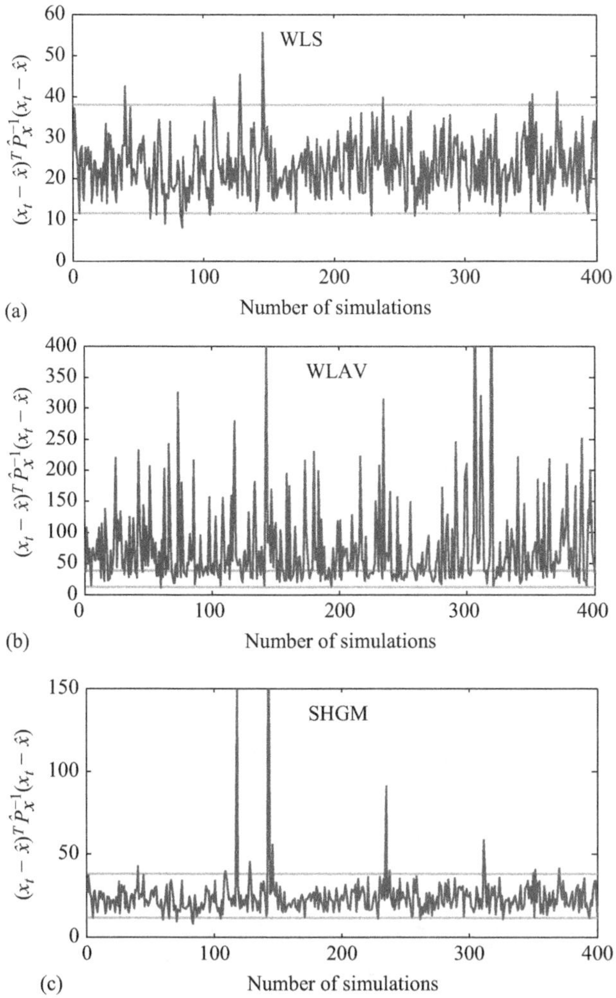

Figure 4.8 12-Bus system consistency plot: error in true measurements = 3%, error in pseudo measurements = 50%

Furthermore, the phasor measurements were also deployed at optimally selected buses. The real measurement set in this study consists of following measurements:

1. Voltage measurements at buses #1, #18, #19, #20, #21 and #95.
2. Line flow measurements in lines #1–2, #1–85, #82–95, #18–19, #15–17 and #34–35.
3. Phasor measurements at buses #19, #20 and #21.

The consistency plots for WLS and SHGM with increased redundancy are shown in Figures 4.11 and 4.12, respectively. The WLS shows the consistent performance

Table 4.3 12-Bus system performance summary

Estimator	**Real 1%, pseudo 20%**			**Real 1%, pseudo 50%**			**Real 3%, pseudo 20%**			**Real 3%, pseudo 50%**		
	Bias	**Consistency/ $E[\epsilon]$**	**Quality *tr/det***	**Bias**	**Consistency/ $E[\epsilon]$**	**Quality *tr/det***	**Bias**	**Consistency/ $E[\epsilon]$**	**Quality *tr/det***	**Bias**	**Consistency/ $E[\epsilon]$**	**Quality *tr/det***
WLS	Unbiased	Consistent/ 23.13	4.08/199.24	Unbiased	Consistent/ 23.25	3.7/179.01	Unbiased	Consistent/ 22.85	1.89/198.4	Unbiased	Consistent/ 22.97	1.81/178.02
WLAV	Unbiased	Inconsistent/ 136.7	3.17/191.46	Unbiased	Inconsistent/ 80.67	2.26/173.98	Unbiased	Inconsistent/ 130.91	1.68/190.53	Unbiased	Inconsistent/ 72.81	1.45/173.07
SHGM	Unbiased	Inconsistent/ 1033.1	3.86/193.1	Unbiased	**Consistent/ 26.37**	3.7/178.28	Unbiased	Inconsistent/ 953.8	1.86/192.37	Unbiased	**Consistent/ 24.1**	1.80/177.63

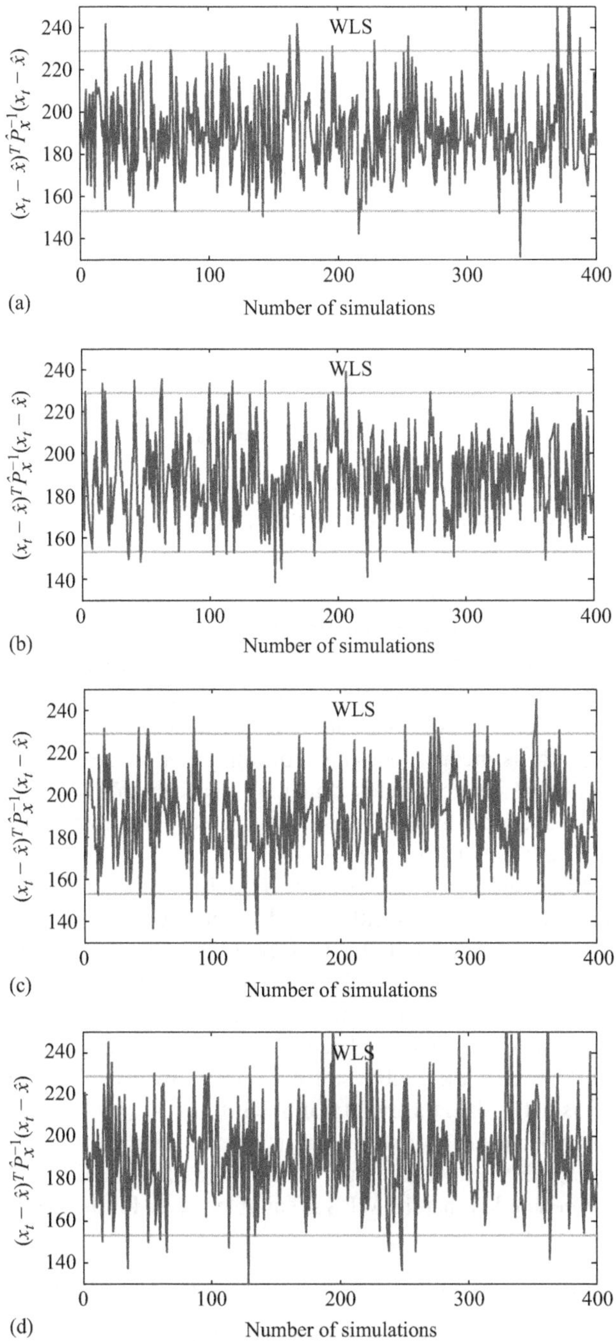

Figure 4.9 95-Bus system consistency plot with limited redundancy: WLS shows consistency in all test cases. (a) Case 1; (b) Case 2; (c) Case 3; and (d) Case 4

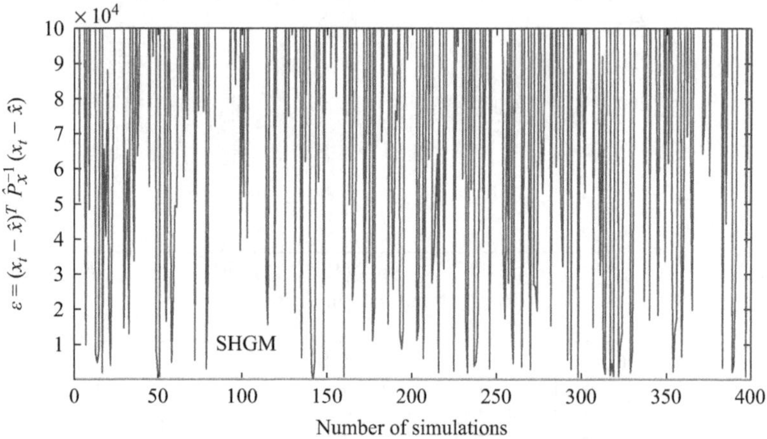

Figure 4.10 95-Bus system consistency plot with limited redundancy: error in true measurements = 1%, error in pseudo measurements = 50%

whereas the SHGM shows the inconsistency in all the simulated cases. A very high degree of inconsistency was observed in WLAV, which is difficult to show graphically.

The performance summaries for both of these cases are shown in Tables 4.4(a) and 4.4(b). In both the cases, the quality defined in (4.5) gives numerical instability in computations; hence, it does not appear in the tables. Furthermore, the quality for WLAV estimator is inconsistent and shows negative values. This is because of very high variance of state estimates which are unacceptable for SE. In WLS and SHGM, as expected the qualities decrease with increase in errors in real and pseudo measurements. The value of $E[\epsilon]$ in case of SHGM does not converge to the number of states (i.e. 189), which numerically confirms its inconsistency.

It is also important to note that with limited redundancy the *trace qualities* defined in (4.6) are close for both WLS and SHGM in cases 2 and 4. This gives the impression that SHGM should be consistent for these cases. Since *trace* captures the diagonal information of the error covariance matrix, it can be attributed that inconsistency in SHGM is mainly due to off-diagonal elements. In case of increased redundancy, there is significant difference in the qualities of WLS and SHGM in all the test cases. The quality of WLS is better than the quality of SHGM.

In all the simulated cases only WLS satisfies the three statistical criteria (Bias, Consistency and Quality) under the assumption of normal distribution of measurement errors. It can be concluded that the WLS is suitable solver for the DSSE problem.

4.4.5 Comments on error distribution and choice of solver

The statistical criteria discussed in this chapter depend on the characteristics of the distribution of measurement errors. The results presented are based on the assumption that the measurement errors are normally distributed. Under this assumption, the WLS

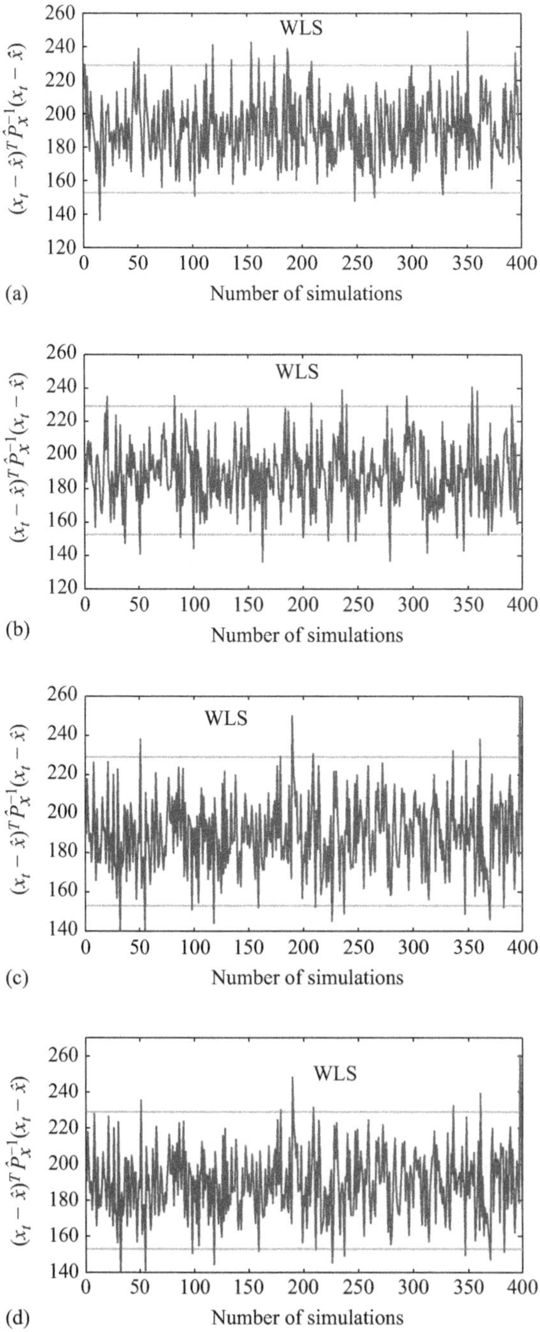

Figure 4.11 95-Bus system consistency plot with increased redundancy: WLS shows consistency in all the test cases. (a) Case 1; (b) Case 2; (c) Case 3; and (d) Case 4

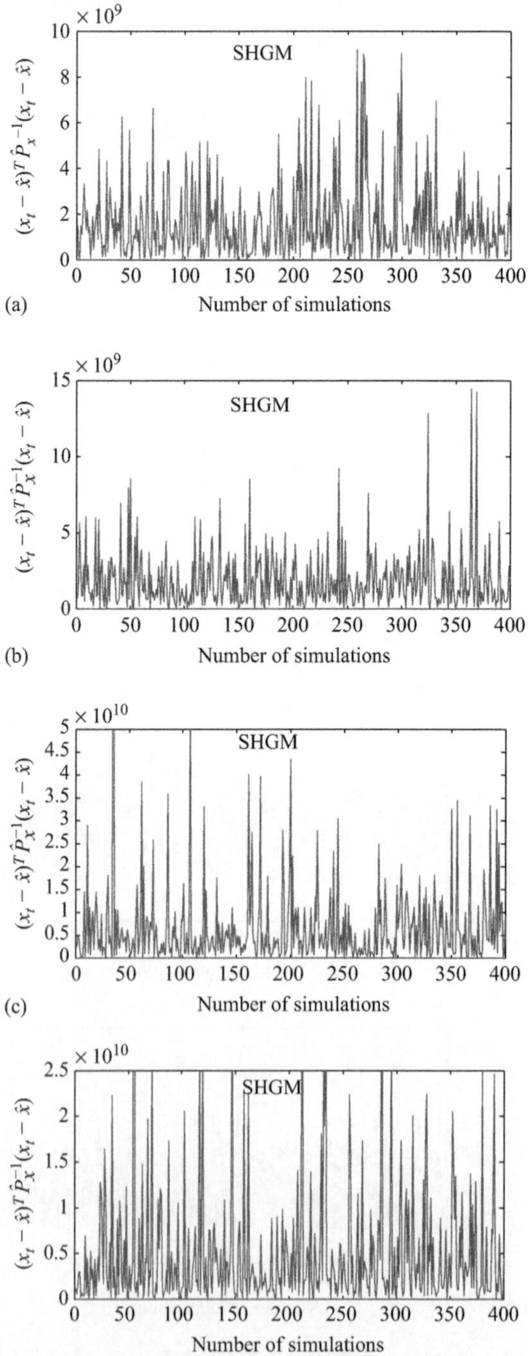

Figure 4.12 95-Bus system consistency plot with increased redundancy: SHGM shows inconsistency in all the test cases. (a) Case 1; (b) Case 2; (c) Case 3; and (d) Case 4

Table 4.4 95-Bus UK-GDS performance summary

Estimator	Real 1%, pseudo 20%			Real 1%, pseudo 50%			Real 3%, pseudo 20%			Real 3%, pseudo 50%		
	Bias	**Consistency/** $E[\epsilon]$	**Quality** *tr/det*	**Bias**	**Consistency/** $E[\epsilon]$	**Quality** *tr/det*	**Bias**	**Consistency/** $E[\epsilon]$	**Quality** *tr/det*	**Bias**	**Consistency/** $E[\epsilon]$	**Quality** *tr/det*
(a) Limited redundancy												
WLS	Unbiased	Consistent/ 190.02	6.63/–	Unbiased	Consistent/ 188.16	6.24/–	Unbiased	Consistent/ 189.84	4.61/–	Unbiased	Consistent/ 190.23	4.41/–
WLAV	Unbiased	Inconsistent/ ∞	−52/–	Unbiased	Inconsistent/ ∞	−44.42/–	Unbiased	Inconsistent/ ∞	−45.18/–	Unbiased	Inconsistent/ ∞	−41.12/–
SHGM	Unbiased	Inconsistent/ 3.06×10^4	6.46/–	Unbiased	**Inconsistent/** 2.53×10^5	6.16/–	Unbiased	Inconsistent/ 2.89×10^4	4.75/–	Unbiased	**Inconsistent/** 2.27×10^5	4.4/–
(b) Increased redundancy												
WLS	Unbiased	Consistent/ 190	8.86/–	Unbiased	Consistent/ 188.3	8.75/–	Unbiased	Consistent/ 188.65	6.85/–	Unbiased	Consistent/ 189.23	6.75
WLAV	Unbiased	Inconsistent/ ∞	−55.65/–	Unbiased	Inconsistent/ ∞	−63.74/–	Unbiased	Inconsistent/ ∞	−43.73/–	Unbiased	Inconsistent/ ∞	−49.88/–
SHGM	Unbiased	Inconsistent/ 1.65×10^9	6.70/–	Unbiased	Inconsistent/ 1.82×10^9	6.35/–	Unbiased	Inconsistent/ 6.58×10^9	4.32/–	Unbiased	Inconsistent/ 1.0×10^{10}	4.28/–

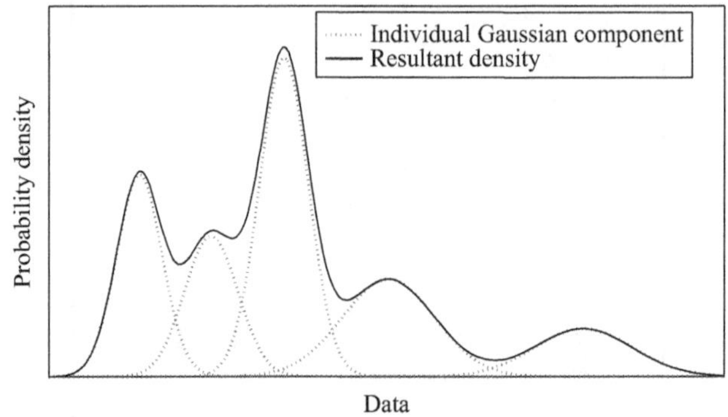

Figure 4.13 Gaussian mixture approximation of the density

satisfies the statistical criteria and hence was found to be the suitable solver for the SE. However, this may not be true if the measurement errors are not normally distributed. For instance if the errors follow the Laplace distribution [21], the WLAV estimator gives better performance than WLS and SHGM. The reason for this is that the WLAV is consistent with the Laplace distribution and maximisation of log-likelihood of the Laplace density function results in the WLAV formulation. Hence, depending on the distribution of the errors, the corresponding statistical criterion discussed in Section 4.2.2 can be modified in order to identify the consistent solver for that distribution.

In reality, different probabilistic load distributions exist in the distribution networks and no standard distribution can fit all of them. Furthermore, the large size of the distribution network having various probability distributions at different buses makes accommodating them in a single state estimator impractical. A more practical approach is to model the actual probability distributions as a mixture of several Gaussian distributions (Figure 4.13) and apply the WLS state estimator which is consistent with the normal distribution. This requires the modelling of the distribution of errors through Gaussian mixture model (GMM) [22–24]. As shown in Figure 4.13, the GMM represents an arbitrary distribution as a weighted combination of several Gaussian components. Mathematically, a GMM having M_c mixture components with mean and variance of kth component as μ_k and σ_k^2 can be written as:

$$f(x) = \sum_{k=1}^{M_c} w_k \mathcal{N}\left(\mu_k, \sigma_k^2\right)(x) \quad \text{and} \quad \sum_{k=1}^{M_c} w_k = 1 \tag{4.20}$$

The Expectation Maximisation algorithm [22–24] is used to obtain the parameters $\left(w_k, \mu_k, \sigma_k^2\right)$ of the GMM.

In transmission systems, all the estimators work well because of very high redundancy and thus the statistical measures to evaluate the performance are not required. For example, a highly erroneous measurement is treated as a bad data by the WLAV estimator and a redundant measurement is always available to replace this. But in distribution systems, the measurements are mainly the pseudo measurements with very limited redundancy. Since pseudo measurements are derived from the historical load profiles and customer demand pattern, they are highly erroneous. This is why the statistical framework is required to identify the suitable solver for the DSSE.

4.5 Conclusion

The performance evaluation of SE techniques shows that the existing solution methodology of WLAV and SHGM cannot be applied to the distribution systems. In order to obtain the consistent and good quality estimate, significant modifications are required in these algorithms. WLS gives consistent and better quality performance when applied to distribution systems. Hence, WLS is found to be a suitable solver for the DSSE problem.

The WLS works well if the noise characteristics are known. In the absence of this knowledge, either the WLS needs to be modified or a new class of algorithms needs to be introduced. Furthermore, with growing interest in the distribution automation, new DSSE techniques are expected to be introduced in the future. However, any modification in existing techniques or introduction of new algorithms should qualify some statistical criteria due to limited number of measurements. This chapter highlights some important statistical criteria against which a SE algorithm should be tested to assess its suitability to DSSE.

Acknowledgements

The authors would like to thank Peter D. Lang of UK Power Networks (UKPN) for his valuable suggestions and discussions.

4.6 Appendix

4.6.1 Computation of α for various estimators

The fact that normalised measurement residual r is normally distributed with zero mean and unit variance can be used to compute the α for the state estimators discussed in Section 4.3.

4.6.1.1 Weighted least squares

$$\psi(r) = r, \qquad \psi'(r) = 1 \tag{4.21}$$

$$E\left[\psi^2(r)\right] = \frac{1}{\sqrt{2\pi}} \int_{-\infty}^{\infty} r^2 e^{-\frac{1}{2}r^2}\, dr = \mathrm{Var}(r) = 1 \tag{4.22}$$

$$E\left[\psi'(r)\right] = \frac{1}{\sqrt{2\pi}} \int_{-\infty}^{\infty} e^{-\frac{1}{2}r^2}\, dr = 1 \tag{4.23}$$

$$\alpha = \frac{E\left[\psi^2(r)\right]}{\left(E\left[\psi'(r)\right]\right)^2} = 1 \tag{4.24}$$

4.6.1.2 Weighted least absolute value

$$\psi(r) = \text{sgn}(r), \qquad \psi'(r) = 2\delta(r) \tag{4.25}$$

$$E[\psi^2(r)] = \frac{1}{\sqrt{2\pi}} \int_{-\infty}^{\infty} (\text{sgn}(r))^2 e^{-\frac{1}{2}r^2}\, dr = 1 \tag{4.26}$$

We use the fact that $\int_{-\infty}^{\infty} \delta(t - t_0)\, f(t)\, dt = f(t_0)$ in the following expressions:

$$E[\psi'(r)] = \frac{1}{\sqrt{2\pi}} \int_{-\infty}^{\infty} 2\delta(r) e^{-\frac{1}{2}r^2}\, dr = \sqrt{\frac{2}{\pi}} \tag{4.27}$$

$$\alpha = \frac{E[\psi^2(r)]}{(E[\psi'(r)])^2} = \frac{\pi}{2} \tag{4.28}$$

4.6.1.3 Schweppe Huber generalized M

$$\psi(r) = \begin{cases} r & \text{if } |r| \le a\omega \\ a\omega\, \text{sgn}(r) & \text{otherwise} \end{cases} \tag{4.29}$$

$$\psi'(r) = \begin{cases} 1 & \text{if } |r| \le a\omega \\ 2a\omega\, \delta(r) = 0 & \text{otherwise} \end{cases} \tag{4.30}$$

$$\begin{aligned} E[\psi^2(r)] = {} & \frac{1}{\sqrt{2\pi}} \int_{-\infty}^{-a\omega} (a\omega\, \text{sgn}(r))^2 e^{-\frac{1}{2}r^2}\, dr \\ & + \frac{1}{\sqrt{2\pi}} \int_{-a\omega}^{a\omega} r^2 e^{-\frac{1}{2}r^2}\, dr \\ & + \frac{1}{\sqrt{2\pi}} \int_{a\omega}^{\infty} (a\omega\, \text{sgn}(r))^2 e^{-\frac{1}{2}r^2}\, dr \end{aligned} \tag{4.31}$$

By symmetry of the distribution the above equation can be expressed as:

$$E[\psi^2(r)] = \frac{2}{\sqrt{2\pi}} \int_{-\infty}^{-a\omega} (a\omega\, \text{sgn}(r))^2 e^{-\frac{1}{2}r^2}\, dr + \frac{2}{\sqrt{2\pi}} \int_{0}^{a\omega} r^2 e^{-\frac{1}{2}r^2}\, dr \tag{4.32}$$

$$= \left(2a^2\omega^2 \Phi(-a\omega) + \frac{2}{\sqrt{2\pi}} \int_{0}^{a\omega} r^2 e^{-\frac{1}{2}r^2}\, dr \right) \tag{4.33}$$

where Φ is the cumulative probability function. The integral term in the above equation is given by:

$$\frac{2}{\sqrt{2\pi}} \int_{0}^{a\omega} r^2 e^{-\frac{1}{2}r^2}\, dr = -\sqrt{\frac{2}{\pi}} a\omega e^{-\frac{a^2\omega^2}{2}} + (2\Phi(a\omega) - 1) \tag{4.34}$$

Using the relation $\Phi(-a\omega) = 1 - \Phi(a\omega)$ and substituting (4.34) in (4.33), we get:

$$E[\psi^2(r)] = 1 - \sqrt{\frac{2}{\pi}} a\omega e^{-\frac{a^2\omega^2}{2}} + 2\left(a^2\omega^2 - 1\right)(1 - \Phi(a\omega)) \tag{4.35}$$

$$E[\psi'(r)] = \frac{1}{\sqrt{2\pi}} \int_{-a\omega}^{a\omega} e^{-\frac{1}{2}r^2}\, dr = 2\Phi(a\omega) - 1 \tag{4.36}$$

In this case, α depends on parameters 'a' and 'ω', i.e. if $a = 1.5$ and $\omega = 1$, the value of α is:

$$\alpha = \frac{E[\psi^2(r)]}{(E[\psi'(r)])^2} = \frac{0.7785}{(0.8664)^2} = 1.0371 \tag{4.37}$$

References

[1] Abur, A., and Exposito, A. G.: 'Power system state estimation: theory and implementation' (New York, NY: Marcel Dekker, Inc., 2004).

[2] Shafiu, A., Thornley, V., Jenkins, and Strbac, G.: 'Control of active networks', *CIRED*, 2005.

[3] Lu, C. N., Teng, J. H., and Liu, W.-H. E.: 'Distribution system state estimation', *IEEE Trans. Power Syst.*, 1995, **10** (1), pp. 229–240.

[4] Lin, W.-M., and Teng, J.-H.: 'Distribution fast decoupled state estimation by measurement pairing', *IEE Proc. Gener. Transm. Distrib.*, 1996, **143** (1), pp. 43–48.

[5] Baran, M. E., and Kelley, A. W.: 'A branch current based state estimation method for distribution systems', *IEEE Trans. Power Syst.*, 1995, **10** (1), pp. 483–491.

[6] Wang, H., and Schulz, N. N.: 'A revised branch current based distribution system state estimation algorithm and meter placement impact', *IEEE Trans. Power Syst.*, 2004, **19** (1), pp. 207–213.

[7] Li, K.: 'State estimation for power distribution system and measurement impacts', *IEEE Trans. Power Syst.*, 1996, **11** (2), pp. 911–916.

[8] Baran, M. E., Zhu, J. X., and Kelley, A. W.: 'Meter placement for real time monitoring of distribution feeders', *IEEE Trans. Power Syst.*, 1996, **11** (1), pp. 332–337.

[9] Ghosh, A. K., Lubkeman, D. L., Downey, M. J., and Jones, R. H.: 'Distribution circuit state estimation using a probabilistic approach', *IEEE Trans. Power Syst.*, 1997, **12** (1), pp. 45–51.

[10] Lubkeman, D. L., Zhang, J., Ghosh, A. K., and Jones, R. H.: 'Field results for a distribution circuit state estimator implementation', *IEEE Trans. Power Deliv.*, 2000, **15** (1), pp. 399–406.

[11] Singh, R., Pal, B. C., and Jabr, R. A.: 'Choice of estimator for distribution system state estimation', *IET Gener. Transm. Distrib.*, 2009, **3** (7), pp. 666–678.

[12] Singh, R.: 'State estimation in power distribution network operation' (Ph.D. Thesis, Imperial College, London, 2009).
[13] Blackman, S. S.: 'Multiple target tracking with radar applications' (Norwood, MA: Artech House, Inc., 1986).
[14] Anderson, T. W.: 'An introduction to multivariate statistical analysis' (New York, NY: John Wiley & Sons, Inc. seventh printing, 1966).
[15] Jabr, R. A.: 'Primal dual interior point approach to compute the L_1 solution of the state estimation problem', *IEE Proc. Gener. Transm. Distrib.*, 2005, **152** (3), pp. 313–320.
[16] Huber, P. J.: 'Robust statistics' (New York, NY: John Wiley & Sons, Inc., 1981).
[17] Celik, M. K., and Liu, W.-H. E.: 'An incremental measurement placement algorithm for state estimation', *IEEE Trans. Power Syst.*, 1995, **10** (3), pp. 1698–1703.
[18] Das, D., Nagi, H. S., and Kothari, D. P.: 'Novel method for solving radial distribution networks', *IEE Proc. Gener. Transm. Distrib.*, 1994, **141** (4), pp. 291–298.
[19] 'United Kingdom Generic Distribution Network (UK-GDS)' [online]. Available from: http://monaco.eee.strath.ac.uk/ukgds/ [Accessed June 2008].
[20] Singh, R., Pal, B. C., and Vinter R. B.: 'Measurement placement in distribution system state estimation', *IEEE Trans. Power Syst.*, 2009, **24** (2), pp. 668–675.
[21] Kotz, S., Kozubowski, T. J., and Podgórski, K.: 'The Laplace distribution and generalizations' (Boston, MA: Birkhäuser Boston, 2001).
[22] Dempster, A. P., Laird, N. M., and Rubin, D. B.: 'Maximum-likelihood from incomplete data via the EM algorithm', *J. Royal Statist. Soc. Ser. B*, 1977, **39** (1), pp. 1–38.
[23] Redner, R. A., and Walker, H. F.: 'Mixture densities, maximum likelihood and the EM algorithm', *SIAM Rev.*, 1984, **26** (2), pp. 195–239.
[24] Bilmes, J. A.: 'A gentle tutorial on the EM algorithm and its application to parameter estimation for Gaussian mixture and hidden Markov models', ICSI-TR-97-021, Technical Report, International Computer Science Institute, 1998.

Chapter 5

Feeder reconfiguration for loss reduction

Debapriya Das[1]

5.1 Introduction

Distribution networks are generally built as meshed networks, while they are operated radially. Their configurations may be varied with manual or automatic switching operations so that all of the loads are supplied and reduce power loss, increase system security, and enhance power quality. Reconfiguration also relieves the overloading of the network components. The change in network configuration is performed by opening sectionalizing (normally closed) and closing tie (normally open) switches of the network. These switching are performed in such a way that the radiality of the network is maintained and all of the loads are energized. Obviously, the greater the number of switches is, the greater the possibilities are for reconfiguration and the better the effects are. In recent years, considerable attention has been conducted for loss minimization in the area of network reconfiguration of distribution systems [1–19] using heuristics as well as artificial intelligence techniques. In this chapter, the author has considered multiple objectives and reduction of real power loss is one of the objectives. Hence, the work formulates the network reconfiguration problem as a multiple objective problem subject to operational and electric constraints.

5.2 Definitions of different indices and objective function

The problem formulation considers four objectives related to:

- minimization of the system power loss
- minimization of the deviations of the nodes voltage
- minimization of branch current constraint violation
- minimization of feeders currents imbalance

Multiobjective optimization problems have many objectives [20–23] and a trade-off between the objectives exist, and we never have a situation in which all the

[1] Department of Electrical Engineering, Indian Institute of Technology, Kharagpur, India

objectives can be in the best possible way be satisfied simultaneously. The four objectives, described as follows, are integrated into an objective function J through appropriate weighting factors.

Real power loss reduction index (X_i) may defined as:

$$X_i = \frac{Ploss(i)}{Ploss^0} \quad \forall i = 1, 2, \ldots, N_k \tag{5.1}$$

where $Ploss(i)$ = total real power loss when ith branch in the loop is opened, $Ploss^0$ = total real power loss in the network before network reconfiguration, and N_k = total number of branches in the loop including tie branch, when kth tie switch is closed.

Equation (5.1) indicates that if X_i is low, then power loss reduction is high. For better results, the value of X_i should be less than unity. During iterative process, if X_i is greater than unity, value of objective function is set to very high value.

Maximum voltage deviation index (Y_i) may be defined as:

$$Y_i = \max\left|V_s - V_{i,j}\right| \quad \begin{array}{l} \forall i = 1, 2, \ldots, N_k \\ \forall j = 1, 2, \ldots, NB \end{array} \tag{5.2}$$

where, V_s = voltage at the substation (in p.u.), $V_{i,j}$ = voltage of node j corresponding to the opening of the ith branch in the loop (in p.u.), and NB = total number of nodes in the system.

In this case, if the value of Y_i is less, the system has better voltage profile. For example, if the substation voltage is 1.0 p.u. and system minimum voltage constraint is set to $V^s_{\min} = 0.9$ p.u., then $Y_i = Y_{\max} = 0.10$. Therefore, it is desirable that for better reconfiguration result, Y_i must be less than 0.10. If the value of Y_i is greater than 0.10, the objective function value is set to very high value.

Maximum branch current loading index (Z_i) is defined as:

$$Z_i = \max\left(\frac{|I(i,m)|}{I_c(m)}\right) \quad \begin{array}{l} \forall i = 1, 2, \ldots, N_k \\ \forall m = 1, 2, \ldots, NB - 1 \end{array} \tag{5.3}$$

where, $|I(i,m)|$ = magnitude of current of branch m when the ith branch in the loop is opened and $I_c(m)$ = maximum current carrying capacity of branch m.

It is desirable that the branch currents of the system are less than or equal to their respective rated capacity. However, a 15% overloading [21, 22] is allowed for each branch. During iterative process, if any branch is overloaded more than 15%, i.e., if Z_i is greater than 1.15, the objective function value is set to very high value.

Feeder load balancing index (U_i) is defined as:

$$U_i = \max\left(\frac{IFF_i^{\max} - IF_{i,j}}{IFF_i^{\max}}\right) \quad \begin{array}{l} \forall i = 1, 2, \ldots, N_k \\ \forall j = 1, 2, \ldots, NF \end{array} \tag{5.4}$$

where, NF = total number of feeders, $IF_{i,j}$ = current of feeder j corresponding to opening of the ith branch in the loop, and IFF_i^{max} = maximum of all the feeder currents corresponding to opening of the ith branch in the loop = max($IF_{i,j}$), for $j = 1, 2, \ldots, NF$.

Equation (5.4) indicates that a better feeder load balancing can be achieved if the value of U_i is low. In this case a limit [21] is imposed on U_i, i.e., $U_i = U_{max} = 0.25$. $U_{max} = 0.25$ indicates that the maximum deviation of feeder currents will be 25% with respect to the maximum value of feeder current. During iterative process, if U_i is greater than 0.25, the objective function value is set to very high value.

The four different indices described above are combined through appropriate weighting factors to form the objective function as follows:

$$\text{Min } J_i = w_1 X_i + w_2 Y_i + w_3 Z_i + w_4 U_i \quad \forall i = 1, 2, \ldots, N_k \tag{5.5}$$

The reconfiguration depends on the proper selection of weighting factors. One has to select proper values of weighting factors such that each objective may be given preference as desired by the operator. Three cases for different weighting factors are considered.

5.3 Explanation of the reconfiguration technique

Here, all the tie switches are considered and a heuristic rule is incorporated for selecting the tie switch one at a time. This heuristic rule is explained below:

In the first iteration, compute the voltage difference across all the open tie switches by running a load flow and detect the open tie switch across which the voltage difference is maximum and consider this tie switch first for closing, as it is expected that this switching will cause maximum loss reduction, minimum nodes voltage deviation, minimum branch current constraint violation, and better feeder load balancing. The reason for using this heuristic is that a good portion of the load will be transferred from lower voltage side to higher voltage side. In the next iteration, the same procedure is repeated for the remaining tie switches and so on.

For the purpose of explanation of the proposed algorithm, consider the sample radial distribution system as shown in Figure 5.1. It is assumed that every branch has a sectionalizing switch. This system has three feeders, four tie branches, and hence four tie switches.

Initially, run the load flow program. Now compute the voltage difference across all the open tie switches and detect the open tie switch across which the voltage difference is maximum. Say the voltage difference across the open tie switch, **tie-4** (Figure 5.1) is maximum, then this tie switch (**tie-4**) will be considered first.

Now if **tie-4** is closed, a loop will be formed (Figure 5.2) and the total number of branches including tie branch (24–13) in this loop will be 10. These branches are 13–12, 12–11, 11–10, 10–26, 27–18, 18–19, 19–22, 22–23, 23–24, and 24–13.

Opening of each branch in this loop is an option. For each option considered, value of objective function is evaluated. Say in this loop, first open sectionalizing

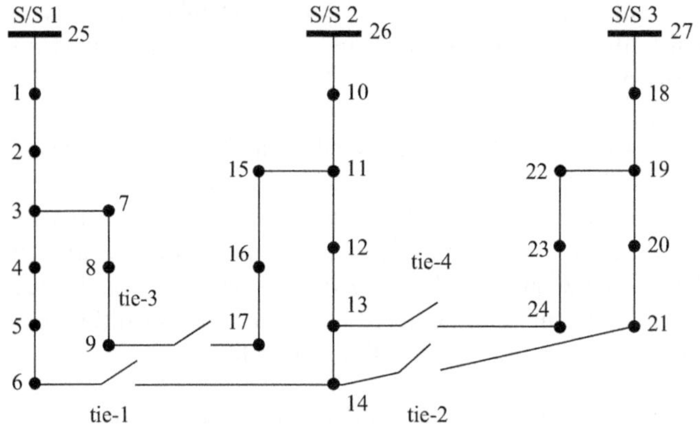

Figure 5.1 Sample distribution network with four tie branches

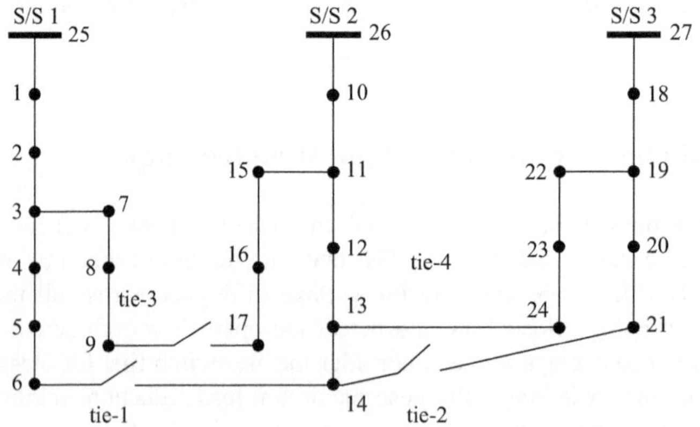

Figure 5.2 Distribution system with tie switch tie-4 closed

switch of branch 13–12 (radial structure is retained) and run the load flow program. Now compute X_1, Y_1, Z_1, and U_1 using (5.1)–(5.4) and then the objective function value J_1 is evaluated using (5.5), i.e.:

$$J_1 = w_1X_1 + w_2Y_1 + w_3Z_1 + w_4U_1 \tag{5.6}$$

Similarly, now close the sectionalizing switch of branch 13–12 and open the sectionalizing switch of branch 12–11 (radial structure is retained) and run the load flow program. Now the objective function value J_2 for this option is computed as:

$$J_2 = w_1X_2 + w_2Y_2 + w_3Z_2 + w_4U_2 \tag{5.7}$$

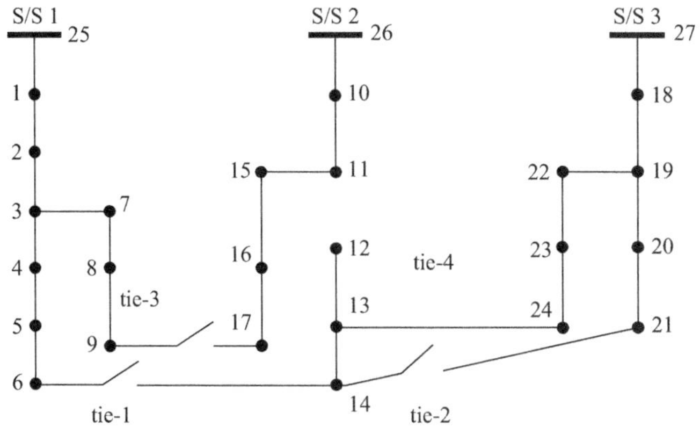

Figure 5.3 Radial configuration after the first switching operation

Similarly, J_3, J_4, …, J_{10} have to be computed. The optimal solution OS_1 for this tie switch (**tie-4**) operation is the minimum of all such values of J_i. Thus, the optimal solution for this tie switch (**tie-4**) operation can be obtained as:

$$OS_1 = \min\{J_1, J_2, \ldots, J_{10}\} \tag{5.8}$$

Suppose $OS_1 = J_2$, which means that optimal solution for this tie switch operation (**tie-4**) can be obtained by opening the sectionalizing switch of branch 12–11 and closing the open tie switch, **tie-4** of the branch 24–13 and the radial structure of the network is retained. Figure 5.3 shows the radial configuration of the network after the first switching operation.

Again, run the load flow program, voltage difference across the remaining open tie switches (**tie-1**, **tie-2**, and **tie-3**) are computed, and say the voltage difference across tie switch, **tie-1** is maximum. Now, this tie switch (**tie-1**) is closed, and a loop will be formed as shown in Figure 5.4.

The total branches in this loop including tie branch (6–14) is 14. These branches are 25–1, 1–2, 2–3, 3–4, 4–5, 5–6, 6–14, 14–13, 13–24, 24–23, 23–22, 22–19, 19–18, and 18–27.

Again by opening sectionalizing switches one by one, objective function values $J_1, J_2, J_3, \ldots, J_{14}$ are evaluated. Now the optimal solution for the tie switch operation (**tie-1**) can be obtained as:

$$OS_2 = \min\{J_1, J_2, \ldots, J_{14}\} \tag{5.9}$$

Say $OS_2 = J_8$, then, the optimal solution for tie switch (**tie-1**) can be obtained by opening sectionalizing switch of branch 14–13 and closing this tie switch (**tie-1**) of the branch 6–14. Figure 5.5 shows the radial configuration of the network after the second switching operation. The same procedure is repeated till all tie switches are considered.

A complete algorithm for the network reconfiguration technique is given below.

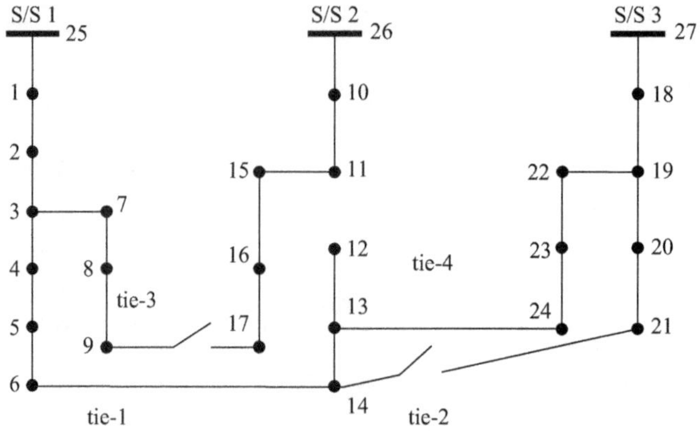

*Figure 5.4 Distribution system with tie switch **tie-1** closed*

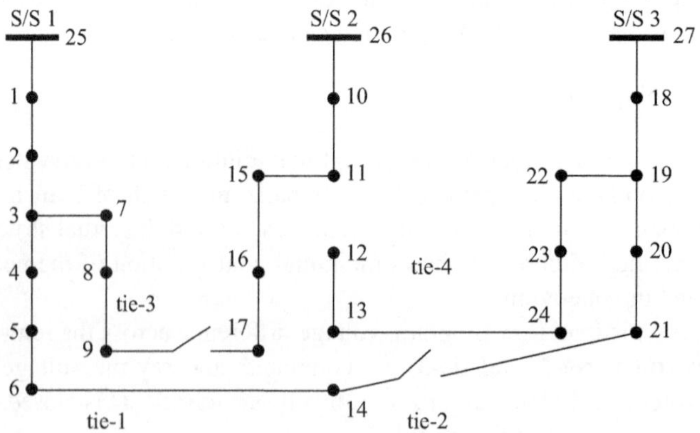

Figure 5.5 Radial configuration after the second switching operation

5.4 Algorithm

STEP-1: Read system data.
STEP-2: Run the load flow program for radial distribution network [24].
STEP-3: Compute the voltage difference across the open tie switches, i.e., $\Delta V_{tie}(i)$, for $i = 1, 2, \ldots, N_{tie}$, where N_{tie} is the total number of tie switches in the system.
STEP-4: Identify the open tie switch across which the voltage difference is maximum and its code k, i.e., $\Delta V_{tie,max} = \Delta V_{tie}(k)$.
STEP-5: Select the tie switch "k" and identify the total number of loop branches (N_k) including the tie branch when the tie switch "k" is closed.

STEP-6: Open one branch at a time in the loop and run the load flow program [24] and evaluate the value for each objective, i.e., for $i = 1$ to N_k, compute X_i, Y_i, Z_i, and U_i using (5.1)–(5.4), respectively, and then compute J_i using (5.5).
STEP-7: Obtain the optimal solution for the operation of tie switch "k," i.e., $OS_k = \min\{J_i\}$, for $i = 1, 2, \ldots, N_k$, and open the branch corresponding to $\min\{J_i\}$. Note that if the value of OS_k and OS_{k-1} are the same (i.e., for kth tie switch operation, there is no improvement of result), the same kth tie switch is made open again (i.e., the operation of the kth tie switch is not carried out).
STEP-8: $N_{\text{tie}} = N_{\text{tie}} - 1$.
STEP-9: Check whether $N_{\text{tie}} = 0$. If Yes, go to STEP-11 Otherwise, go to STEP-10.
STEP-10: Rearrange the coding of the rest of the tie switches, i.e., renumber the remaining tie switches and go to STEP-2.
STEP-11: Print output results and stop.

5.5 Example 5.1

The reconfiguration algorithm is tested with an 11 kV radial distribution system having 2 substations, 4 feeders, 70 nodes, and 79 branches including tie branches as shown in Figure 5.6. This network has 11 tie switches, and these tie switches are open under normal operating condition. Data for this system are given in Appendix 1. Results of load flow before reconfiguration are given in Table 5.1.

Table 5.2 shows the results for various cases after reconfiguration. For Case 1, all the objectives as given in (5.1)–(5.4) are considered. From Table 5.2, it is seen that power loss is 210.79 kW. This means reduction of power loss is 16.74 kW, i.e., 7.35%. Minimum system voltage is $V_{\min} = V_{51} = 0.91841$ p.u., i.e., minimum system voltage has improved from 0.90518 p.u. to 0.91841 p.u. It can be seen that the minimum voltage before reconfiguration occurs at node 69, and that after reconfiguration occurs at node 51. Feeder currents are more balanced as compared to that obtained before reconfiguration (comparing the results for **Case 1** of Table 5.2 with Table 5.1). In this case, only four tie switch operations were required to get the optimum results.

Figure 5.7 shows the plot of optimum objective function value (OS_i) versus order of tie switch operations. In the very first iteration, proposed algorithm considers the tie switch, tie-4, next, it considers tie-1. After that, it considers tie-2 but in this case, closing branch and opening branch is the same (i.e., tie branch 9–40, tie-2). So, there is no improvement of optimum objective function value for this tie switch operation. Next, it considers tie-9, and there was some improvement in the value of OS_i. After that, it considers tie-3, tie-10, tie-6, tie-5, tie-11, and tie-8, but there was no improvement of the value of OS_i. But for tie-7, there was very small change in the value OS_i. In Table 5.2, the last two columns show the branches out and branches in. Change of branches is also in bold in Table 5.2. Final configuration for **Case 1** is given in Figure 5.8.

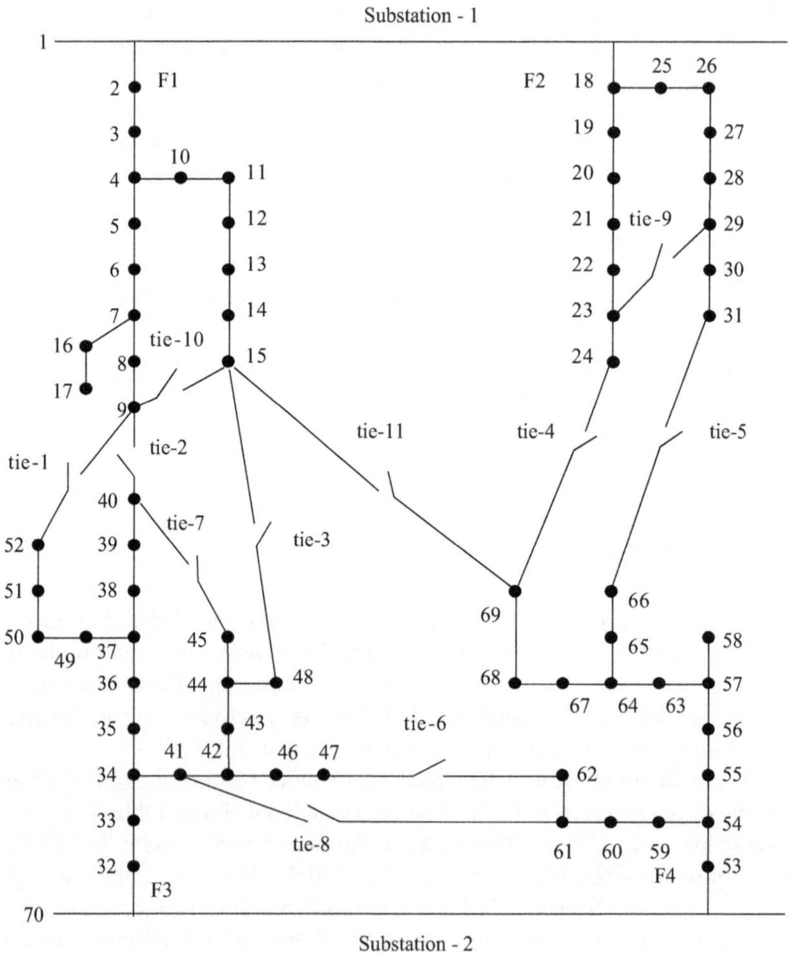

Figure 5.6 Distribution system with 11 tie switches before reconfiguration

Table 5.1 Load flow results before reconfiguration

Power loss (kW)	**Minimum system voltage $V_{\min}$ (p.u.)**	**Feeder current (A)**
227.53	$V_{\min} = V_{69} = 0.90518$	$IF_{1-2} = IF_1 = 99.92$
		$IF_{1-18} = IF_2 = 109.02$
		$IF_{70-32} = IF_3 = 162.30$
		$IF_{70-53} = IF_4 = 148.86$

Table 5.2 Power loss, minimum voltage, and configuration changes for different weighting factors

Different cases	Weighting factors	Ploss (kW)	V_{min} (p.u.)	Feeder current (A)	Changes in configuration	
					Branches out	Branches in
Case 1	$w_1 = 2$	210.79	$V_{min} = V_{51} =$	$IF_1 = 129.89$	**67–68**	**24–69**
	$w_2 = 10$		0.91841	$IF_2 = 131.52$	**50–51**	**9–52**
	$w_3 = 1$			$IF_3 = 132.59$	9–40	9–40
	$w_4 = 1$			$IF_4 = 124.29$	**28–29**	**23–29**
					15–48	15–48
					9–15	9–15
					47–62	47–62
					31–66	31–66
					15–69	15–69
					41–61	41–61
					44–45	**40–45**
Case 2	$w_1 = 1$	203.84	$V_{min} = V_{31} =$	$IF_1 = 112.08$	**67–68**	**24–69**
	$w_2 = 0$		0.92730	$IF_2 = 131.52$	**51–52**	**9–52**
	$w_3 = 0$			$IF_3 = 138.69$	**28–29**	**23–29**
	$w_4 = 0$			$IF_4 = 135.11$	**46–47**	**47–62**
					14–15	**15–48**
					15–69	15–69
					9–15	9–15
					9–40	9–40
					31–66	31–66
					40–45	40–45
					41–61	41–61
Case 3	$w_1 = 2$	204.79	$V_{min} = V_{31} =$	$IF_1 = 112.08$	**67–68**	**24–69**
	$w_2 = 10$		0.92831	$IF_2 = 131.03$	**51–52**	**9–52**
	$w_3 = 0$			$IF_3 = 138.98$	23–29	23–29
	$w_4 = 0$			$IF_4 = 135.11$	15–69	15–69
					46–47	**47–62**
					14–15	**15–48**
					9–15	9–15
					9–40	9–40
					40–45	40–45
					31–66	31–66
					41–61	41–61

For **Case 2**, only power loss reduction index is considered in the objective function, i.e., $w_1 = 1$ and $w_2 = w_3 = w_4 = 0$. From Table 5.2, it can be seen that power loss is 203.84 kW. This means power loss reduction is 23.7 kW, i.e., 10.41% loss reduction. The minimum system voltage has improved from 0.90518 p.u. to 0.92730 p.u. In this case, minimum voltage occurs at node 31. However, the feeder currents are not that balanced as compared to **Case 1**. Moreover, in this case five tie switch

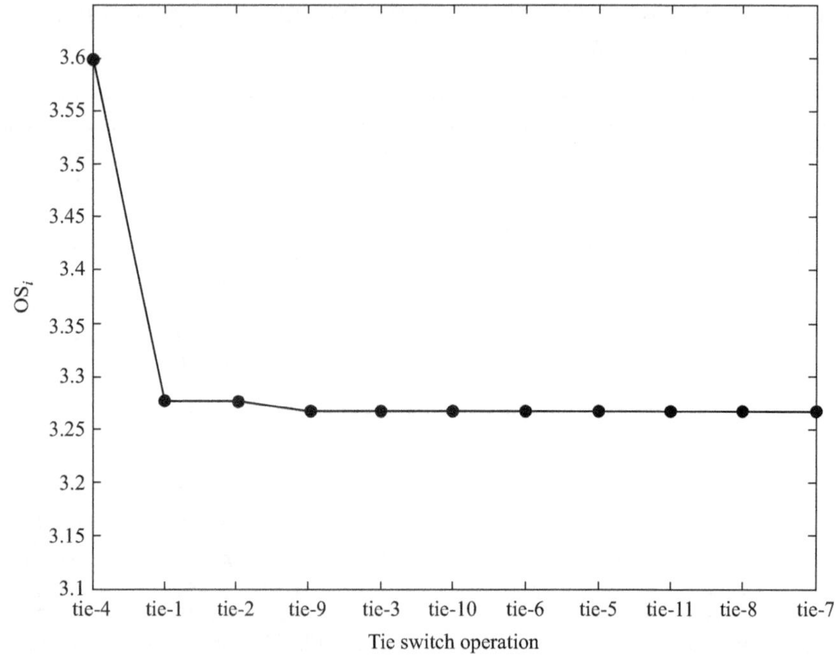

*Figure 5.7 Plot of optimum objective function value V_s tie switch operation (**Case 1**)*

operations are required, i.e., tie-4, tie-1, tie-9, tie-6, and tie-3. Figure 5.9 shows the plot of optimum objective function value versus order of tie switch operations. The final configuration is given in Figure 5.10.

For **Case 3**, two objectives are considered, i.e., power loss reduction and minimization of maximum voltage deviation; and other two objectives are not considered in the objective function. From Table 5.2, it is seen that the power loss is 204.79 kW. This means power loss reduction is 22.74 kW, i.e., 9.99 % loss reduction. Minimum system voltage has improved from 0.90518 p.u. to 0.92831 p.u. In this case, minimum voltage occurs at node 31. However, feeder currents are not that balanced as compared to **Case 1**. Also note that, in this case, feeder currents IF_1 and IF_4 are the same as those in **Case 2**, but IF_2 and IF_3 are slightly different because in this case, four tie switch operations were required for optimal solution, i.e., tie-4, tie-1, tie-6, and tie-3. Figure 5.11 shows the plot of optimum objective function value versus order of tie switch operations. The final configuration is given in Figure 5.12.

From the above discussions, it is clear that each objective has some influence on the tie switch operations and combination of different objectives gives different results.

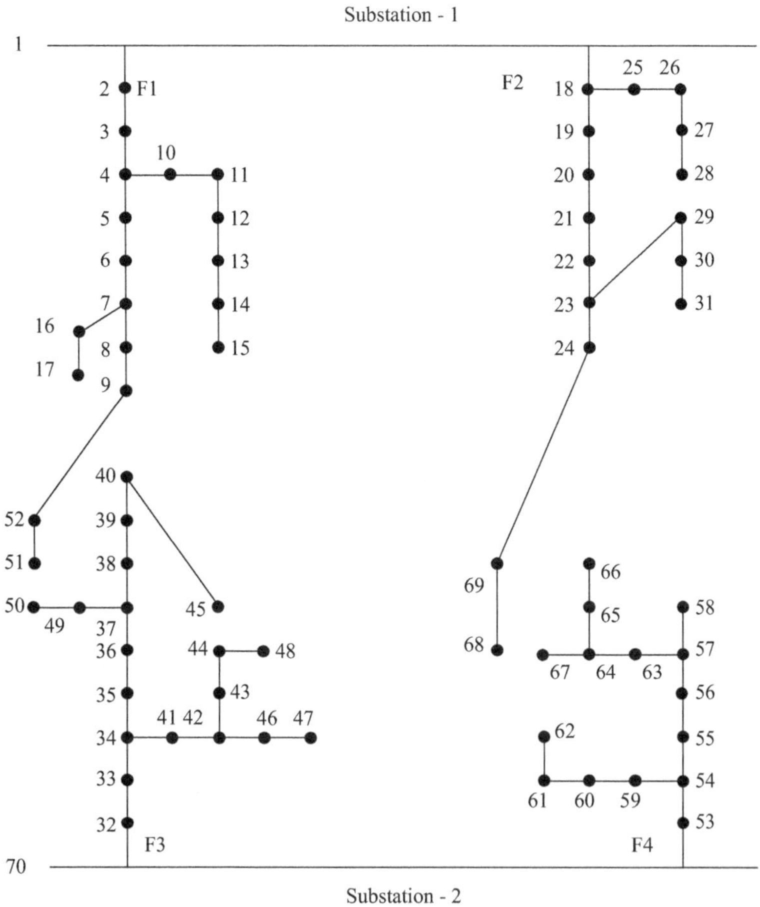

Figure 5.8 Distribution system after reconfiguration ***(Case 1)***

5.6 Reconfiguration using fuzzy multiobjective approach

Here, network reconfiguration problem is considered as a multiple objectives problem subject to operational and electric constraints. The problem formulation considers two different objectives related to:

- minimization of the system's power loss
- minimization of the deviation of nodes voltage.

At the same time, a radial network structure must remain after network reconfiguration in which all loads must be energized. These two objectives are modeled with fuzzy sets to evaluate their imprecise nature. Heuristic rules are also incorporated in the algorithm for minimizing the number of tie switch operations. In Section 5.2,

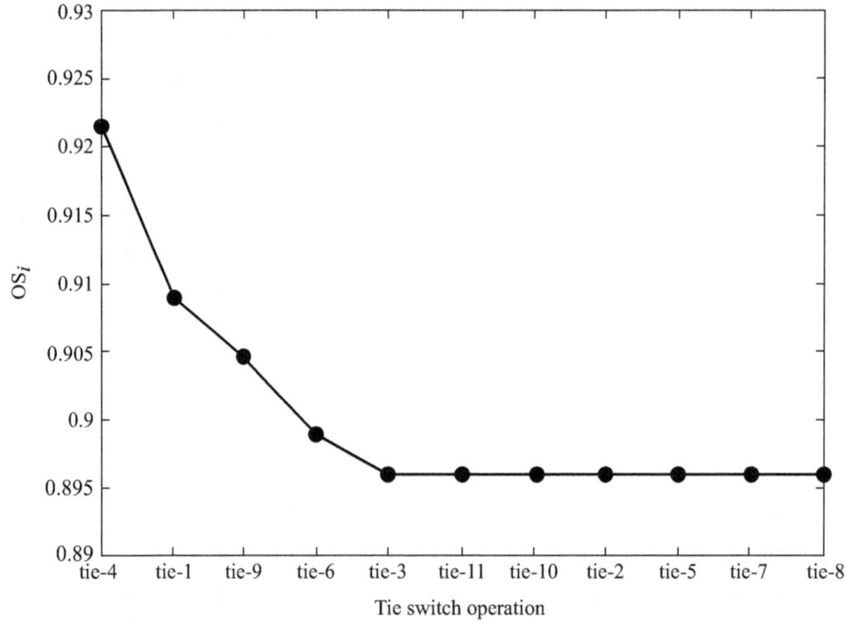

Figure 5.9 Plot of optimum objective function value V_s tie switch operation ***(Case 2)***

different objectives are combined through weighting factors. It is difficult to select best value of the weighting factors. But in the case of fuzzy multiobjective approach, no such weighting factors are involved.

5.7 Membership functions of different objectives

In the fuzzy domain, each objective is associated with a membership function. The membership function indicates the degree of satisfaction of the objective. In the crisp domain, either the objective is satisfied or it is violated, implying membership values of unity and zero, respectively. On the contrary, fuzzy sets entertain varying degrees of membership function values from zero to unity. Thus, fuzzy set theory is an extension of standard set theory [25].

5.7.1 Membership function for real power loss reduction (μL_i)

The basic purpose for this membership function is to reduce the real power loss of the system. Let us define

$$x_i = \frac{PLOSS(i)}{PLOSS^0}, \quad \text{for } i = 1, 2, \ldots, N_k \tag{5.10}$$

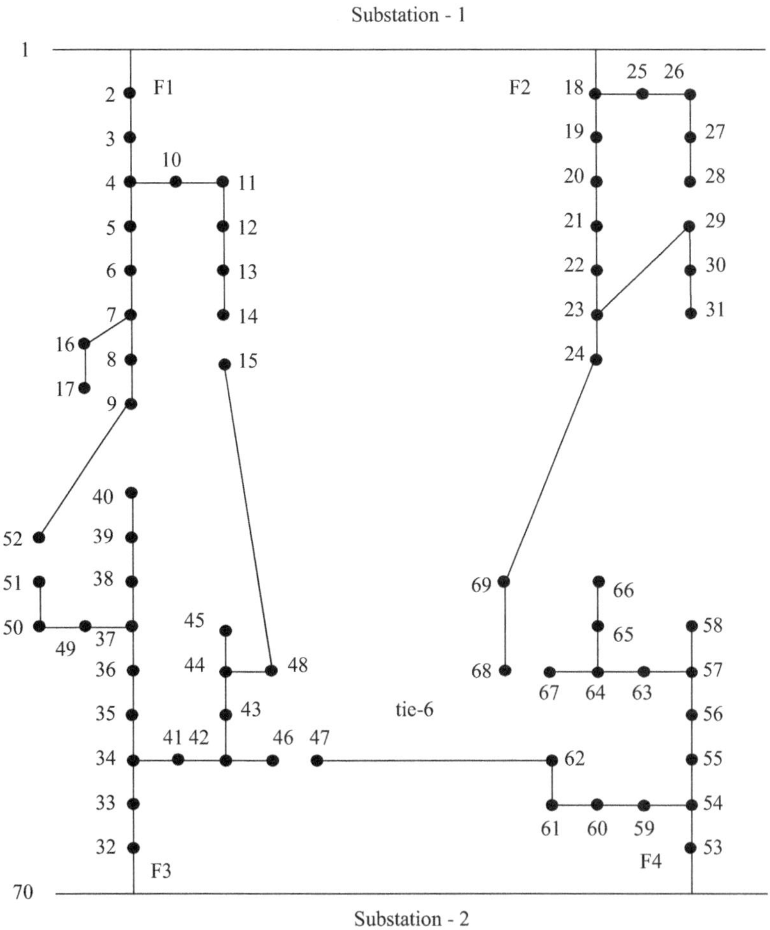

Figure 5.10 Distribution system after reconfiguration ***(Case 2)***

Equation (5.10) indicates that if x_i is high, power loss reduction is low, and hence, a lower membership value is assigned, and if x_i is low, the power loss reduction is high and a higher membership value is assigned. The membership function for real power loss reduction is given in Figure 5.13.

From Figure 5.13, μL_i can be written as

$$\mu L_i = \frac{(x_{\max} - x_i)}{(x_{\max} - x_{\min})} \quad \text{for } x_{\min} < x_i < x_{\max}$$

$$\mu L_i = 1 \quad \text{for } x_i \leq x_{\min}$$

$$\mu L_i = 0 \quad \text{for } x_i \geq x_{\max} \tag{5.11}$$

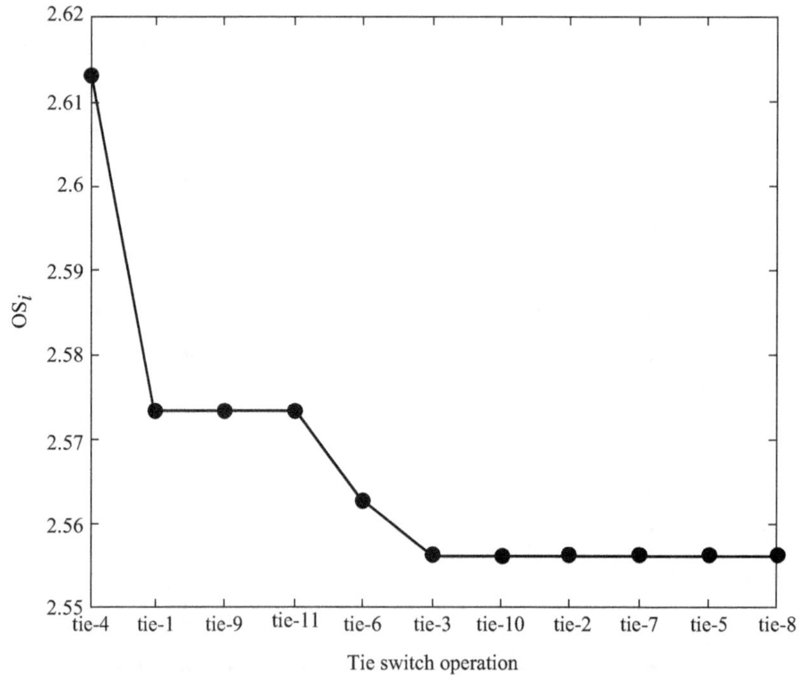

*Figure 5.11 Plot of optimum objective function value V_s tie switch operation **(Case 3)***

Here, it has been assumed that $x_{min} = 0.5$ and $x_{max} = 1.0$, this means if the loss is 50% or less of the $PLOSS^0$, the unity membership value is assigned and if the loss is 100% or more of $PLOSS^0$, the zero membership value is assigned.

5.7.2 Membership function for maximum node voltage deviation (μV_i)

The basic purpose for this membership function is that the deviation of nodes voltage should be less. Let us define

$$y_i = \max\left|V_{i,j} - V_s\right|, \quad \text{for } i = 1, 2, \ldots, N_k; \quad j = 1, 2, \ldots, NB \tag{5.12}$$

If maximum value of nodes voltage deviation is less, then a higher membership value is assigned, and if deviation is more, then a lower membership value is assigned. Figure 5.14 shows the membership function for maximum node voltage deviation.

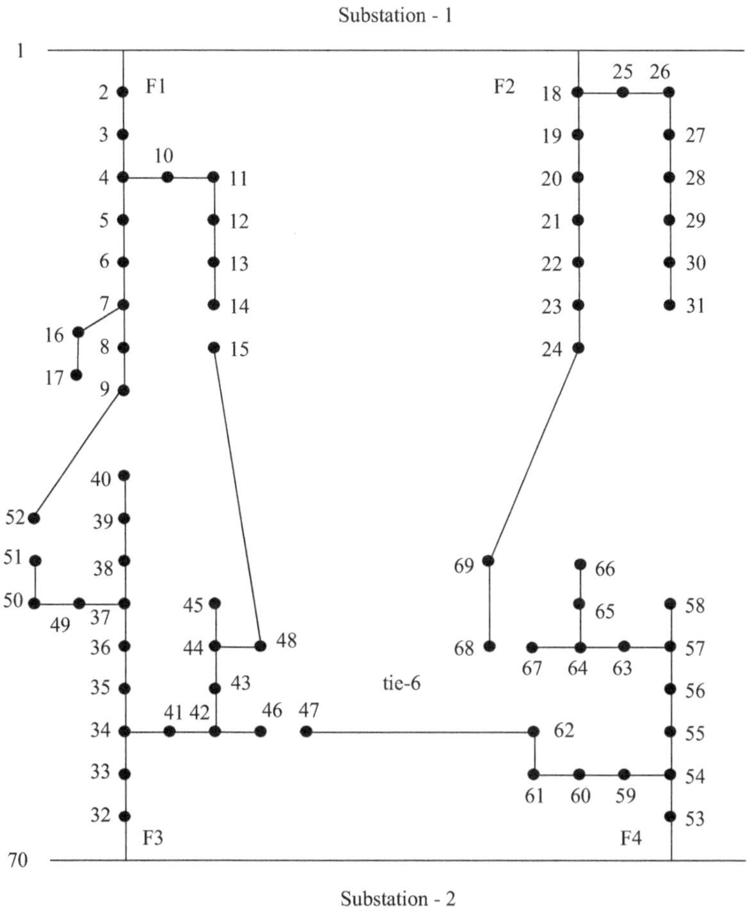

*Figure 5.12 Distribution system after reconfiguration (**Case 3**)*

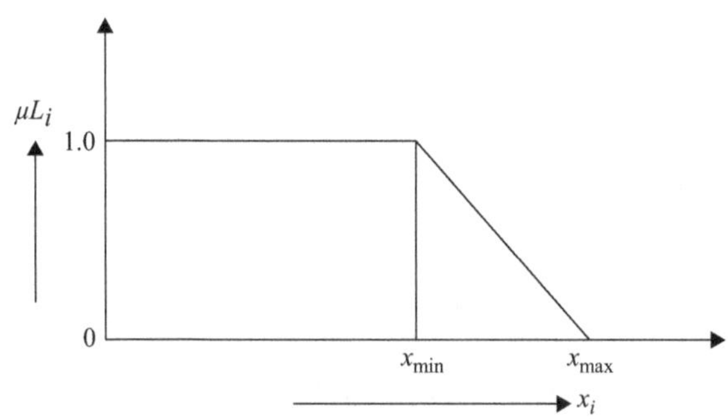

Figure 5.13 Membership function for power loss reduction

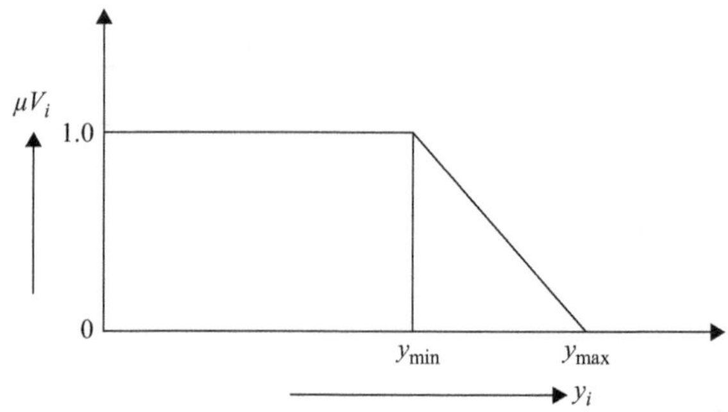

Figure 5.14 Membership function for maximum node voltage deviation

From Figure 5.14, we can write

$$\begin{aligned}\mu V_i &= \frac{(y_{max} - y_i)}{(y_{max} - y_{min})} \text{ for } y_{min} < y_i < y_{max} \\ \mu V_i &= 1 \text{ for } y_i \le y_{min} \\ \mu V_i &= 0 \text{ for } y_i \ge y_{max} \end{aligned} \tag{5.13}$$

Here, it has been assumed that $y_{min} = 0.05$ and $y_{max} = 0.10$. $y_{min} = 0.05$ means if the substation voltage is 1.0 p.u., then the minimum system voltage will be 0.95 p.u., and if the minimum system voltage is greater than or equal to 0.95 p.u., the unity membership value is assigned. Similarly, if $y_{max} = 0.10$, the minimum system voltage will be 0.90 p.u., and if the minimum system voltage is less than or equal to 0.90 p.u., the zero membership value is assigned.

5.7.3 Optimization in fuzzy environment

When there are multiple objectives to be satisfied simultaneously, a compromise has to be made to obtain the best solution. One solution methodology for multiobjective optimization in fuzzy framework is based on max–min principle [21, 23, 25], which is described as follows:

STEP-1: For each option considered, the membership values of all the different objectives are evaluated.

For example, when *k*th tie switch of a distribution system is closed, a loop is formed with N_k number of branches in the loop. Now opening of each branch in this loop is an option. After opening the *i*th branch in this loop (radial structure is retained), load flow run was carried out to compute μL_i and μV_i for $i = 1, 2, \ldots, N_k$.

STEP-2: In this step, the fuzzy decision corresponding to opening of *i*th branch in the loop is computed. This decision may be defined as the choice that satisfies both the objectives and if we interpret this as a logical "and," we can model it with the intersection of the fuzzy sets. Here, classical fuzzy set intersection is used, and the fuzzy decision corresponding to opening of ith branch in the loop is then given by:

$$D_{k,i} = \min\{\mu L_i, \mu V_i\}, \quad \text{for } i = 1, 2, \ldots, N_k \tag{5.14}$$

STEP-3: In this step, the overall fuzzy decision corresponding to the closing of kth tie switch is computed. This overall fuzzy decision may be defined as the choice that maximizes over all the fuzzy decisions corresponding to opening of ith branch in the loop, for $i = 1, 2, \ldots, N_k$. If we interpret this as a logical "or," we can model it with the union of fuzzy sets. Here, classical fuzzy set union is used, and the fuzzy decision for optimal solution is then given by:

$$OS_k = \max\{D_{k,i}\} \text{ for } i = 1, 2, \ldots, N_k \tag{5.15}$$

5.7.4 Heuristic rules to minimize the number of tie switch operations and algorithm

Heuristic rules [21] are considered which minimize the number of tie switch operations. These heuristic rules are explained below.

In the first iteration, compute the voltage difference across all the open tie switches and detect the open tie switch across which the voltage difference is maximum. If this maximum voltage difference is greater than some specified value (ε), then this tie switch is considered first. It is expected that because of the largest voltage difference, this switching will cause maximum loss reduction, improvement of minimum system voltage, and will give better load balancing. In the next iteration, same procedure is repeated for remaining tie switches and so on.

If, in any iteration, this maximum voltage difference is less than the specified value (ε), then this tie switch operation is discarded and automatically other tie switch operations are discarded because voltage difference across all other open tie switches are less than ε.

Actually, number of tie switch operations depends on the value of ε. If ε is large, the number of tie switch operations may be small, but solution may not be satisfactory and if ε is too small, the solution is good but it may consider large number of tie switch operations and computational effort will be more for very small gain. Therefore, by some trial and error, one can select the value of ε to reduce the number of tie switch operations. It was found that $\varepsilon = 0.01$ gives acceptable results.

A complete algorithm for the above network reconfiguration process is given below:

STEP-1: Read system data.
STEP-2: Run the load flow program for radial distribution networks [24].
STEP-3: Compute the voltage difference across the open tie switches, i.e., $\Delta V_{\text{tie}}(\text{i})$, for $i = 1, 2, \ldots, N_{\text{tie}}$.
STEP-4: Identify the open tie switch across which the voltage difference is maximum and its code k, i.e., $\Delta V_{\text{tie,max}} = \Delta V_{\text{tie}}(k)$

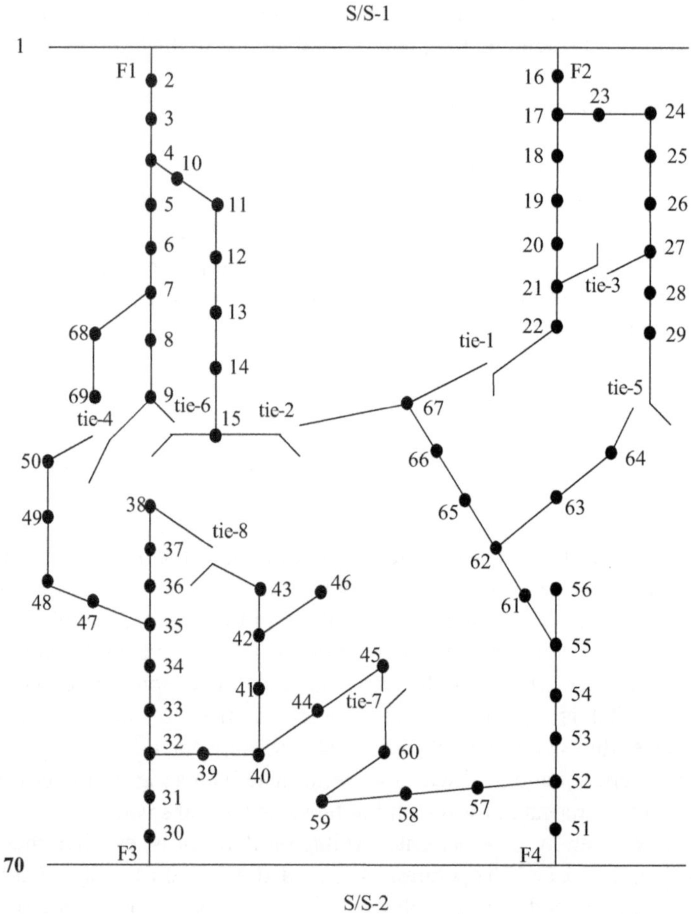

Figure 5.15 Seventy node radial distribution network before reconfiguration

STEP-5: If $\Delta V_{\text{tie,max}} > \varepsilon$, go to **STEP-6;** otherwise, go to **STEP-10**.
STEP-6: Select the tie switch "k" and identify the total number of loop branches (N_k) including the tie-branch when the tie switch "k" is closed.
STEP-7: Open one branch at a time in the loop and run the load flow [24] and evaluate the membership value for each objective and also evaluate overall degree of satisfaction, i.e., for $i = 1 \ldots N_k$, compute μL_i and is μV_i using (5.11) and (5.13), respectively, and evaluate: $D_{k,i} = \min(\mu L_i, \mu V_i)$.
STEP-8: Obtain the optimal solution corresponding to the operation of tie switch "k," i.e., $\text{OS}_k = \max\{D_{k,i}\}$, for $i = 1, 2, \ldots, N_k$.
STEP-9: $N_{\text{tie}} = N_{\text{tie}} - 1$ and rearrange the coding of the rest of the tie switches and go to **STEP-2.**
STEP-10: Print output results and stop.

Table 5.3 Power loss and minimum voltage before and after reconfiguration

Status	Power loss (kW)	Minimum voltage (pu)
Before	341.4	$V_{67} = 0.8839$
After	304.9	$V_{29} = 0.9124$

Table 5.4 Optimal solution for tie switch operation

Tie switch operation (tie-k)	Opening branches	Closing branches	Optimal solution (OS_k)
tie-1	65–66	22–67	0.1617
tie-4	49–50	9–50	0.1917
tie-3	26–27	21–27	0.2020
tie-7	44–45	45–60	0.2139

5.8 Example 5.2

The tested system is an 11 kV radial distribution system having 2 substations, 4 feeders, 70 nodes, and 76 branches (including tie branches) as shown in Figure 5.15. Tie switches of this system are open in normal conditions. Load and tie-branch data for this system are given in Appendix 2. Line resistance and reactance are the same as Example 5.1 as given in Appendix 1.

Power loss of the system and minimum voltage are given in Table 5.3 before and after network reconfiguration. From Table 5.3, it is seen that the loss has decreased and minimum voltage profile has improved.

Table 5.4 shows the optimal solution after each tie switch operation. From Table 5.4, it is noticed that the solution has improved after each tie switch operation, and during the iterative process, the algorithm has considered only four out of eight tie switches and the remaining tie switches have been discarded. Figure 5.16 shows the final radial configuration after network reconfiguration.

In the second example, less number of tie-branches is used to reduce the computational burden. For the second example, line parameters are same but load power are more than the first example to make the network more lossy and without reconfiguration, minimum voltage was 0.8839 p.u. and loss was much higher. But after reconfiguration using fuzzy multiobjective approach minimum voltage is 0.9124 pu. Note that for the first example four objectives are considered and for the second example two objectives are considered. Therefore, no attempt is made to compare the results. But the author feels that these two examples will help the reader to understand the main purpose of reconfiguration problem considering different objectives.

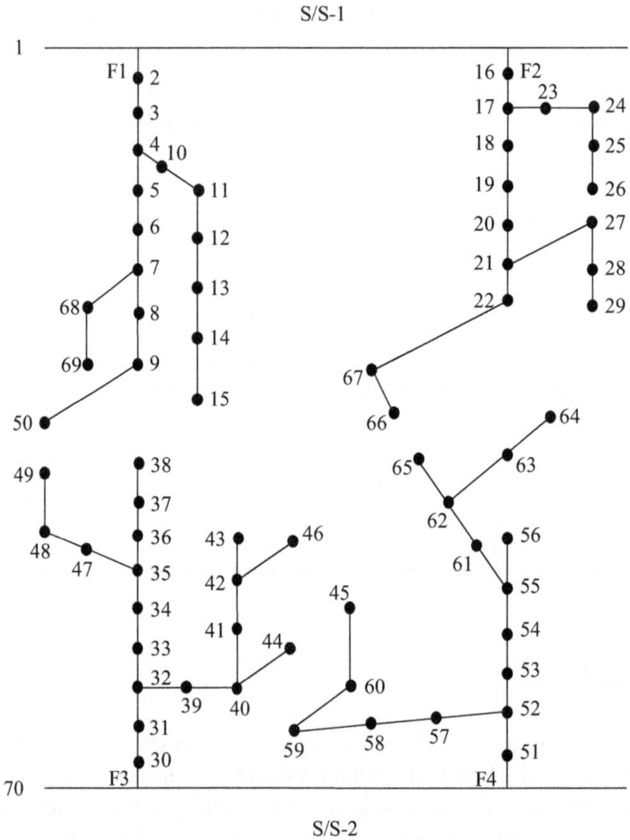

Figure 5.16 Network topology after reconfiguration

References

[1] A. Merlin and H. Back, 'Search for a minimal loss operating spanning tree configuration in an urban power distribution system', Proc. 5th Power System Computation Conf., Cambridge, UK, pp. 1–18, 1975.

[2] D. Shirmohammadi and H.W. Hong, 'Reconfiguration of electric distribution networks for resistive line loss reduction', IEEE Trans. Power Deliv., vol. 4, pp. 1492–1498, 1989.

[3] V. Borozan, D. Rajicic, and R. Ackovski, 'Improved method for loss minimization in distribution networks', IEEE Trans. Power Syst., vol. 10, pp. 1420–1425, 1995.

[4] S. Civanlar, J.J. Grainger, H. Yin, and S.S.H. Lee, 'Distribution feeder reconfiguration for loss reduction', IEEE Trans. Power Deliv., vol. 3, pp. 1217–1223, 1988.

[5] M.E. Baran and F.F. Wu, 'Network reconfiguration in distribution systems for loss reduction and load balancing', IEEE Trans. Power Deliv., vol. 4, pp. 1401–1407, 1989.

[6] C.C. Liu, S.J. Lee, and K. Vu, 'Loss minimization of distribution feeders: optimality and algorithms', IEEE Trans. Power Deliv., vol. 4, pp. 1281–1289, 1989.

[7] H.D. Chiang and R.M. Jean-Jameau, 'Optimal network reconfiguration in distribution systems, part 1: A new formulation and a solution methodology', IEEE Trans. Power Deliv., vol. 5, pp. 1902–1909, 1990.

[8] H.D. Chiang and R.M. Jean-Jameau, 'Optimal network reconfigurations in distribution systems, part 2: solution algorithms and numerical results', IEEE Trans. Power Deliv., vol. 5, pp. 1568–1574, 1990.

[9] Y.J. Jeon, J.C. Kim, J.O. Kim, J.R. Shin, and K.Y. Lee, 'An efficient simulated annealing algorithm for network reconfiguration in large-scale distribution systems', IEEE Trans. Power Deliv., vol. 17, pp. 1070–1078, 2002.

[10] T.P. Wagner, A.Y. Chikhani, and R. Hackam, 'Feeder reconfiguration for loss reduction', IEEE Trans. Power Deliv., vol. 6, pp. 1922–1933, 1991.

[11] S.K. Goswami and S.K. Basu, 'A new algorithm for reconfiguration of distribution feeders for loss minimization', IEEE Trans. Power Deliv., vol. 7, pp. 1484–1491, 1992.

[12] T.E. Lee, M.Y. Cho, and C.S. Chen, 'Distribution system reconfiguration to reduce resistive losses', Int. J. Elec. Power Syst. Res., vol. 30, pp. 25–33, 1994.

[13] D. Jiang and R. Baldick, 'Optimal electric distribution system with switch reconfiguration and switch control', IEEE Trans. Power Syst., vol. 11, pp. 890–897, 1996.

[14] M.A. Kashem, V. Ganapathy, and G.B. Jasmon, 'A geometrical approach for network reconfiguration based loss minimization in distribution systems', Int. J. Elec. Power Syst. Res., vol. 23, pp. 295–304, 2001.

[15] G.B. Jasmon and L.H.C.C. Lee, 'A modified technique for minimization of distribution system losses', Int. J. Elec. Power Syst. Res., vol. 20, pp. 81–88, 1991.

[16] S. Ghosh and D. Das, 'An efficient algorithm for loss minimization via network reconfiguration', Electr. Power Compo. Syst., vol. 31, pp. 791–804, 2003.

[17] E.M. Carreno, R. Romero, and A.P. Feltrin, 'An efficient codification to solve distribution network reconfiguration for loss reduction problem', IEEE Trans. Power Syst., vol. 23(4), pp. 1542–1551, 2008.

[18] G.K.V. Raju and P.R. Bijwe, 'An efficient algorithm for minimum loss reconfiguration of distribution system based on sensitivity and heuristics', IEEE Trans. Power Syst., vol.23, pp.1280–1287, 2008.

[19] A.A.M. Zin, A.K. Ferdavani, A.B. Khairuddin, and M.M. Naeini, 'Reconfiguration of radial electrical distribution network through minimum-current circular-updating-mechanism method', IEEE Trans. Power Syst., vol. 27, pp. 968–974, 2012.

[20] Y.T. Hsiao and C.Y. Chien, 'Multiobjective optimal feeder reconfiguration', IEEE Proc. Genr. Trans. Distrib., vol. 148, pp. 333–336, 2001.

[21] D. Das, 'A fuzzy multi-objective approach for network reconfiguration of distribution systems', IEEE Trans. Power Deliv., vol. 21, pp. 202–209, 2006.

[22] D. Das, 'Reconfiguration of distribution system using fuzzy multi-objective approach for network reconfiguration', Int. J. Elec. Power, vol. 28, pp. 331–338, 2006.
[23] J.S. Savier and D. Das 'Impact of network reconfiguration on loss allocation of radial distribution systems', IEEE Trans. Power Deliv., vol. 22, pp. 2473–2480, 2007.
[24] S. Ghosh and D. Das, 'Method for load flow solution of radial distribution networks', Proc. IEEE Genr. Transm. Distrib., pp. 641–648, 1999.
[25] L. Zadeh, 'Fuzzy sets', Inf. Contr., vol. 8, pp. 338–353, 1965.

Appendix 1

Table A1 Line and load data of example 5.1

Branch number	Sending end node	Receiving end node	R (Ω)	X (Ω)	PL (kW)	QL (kVar)
1	1	2	1.097	1.074	100.0	90.0
2	2	3	1.463	1.432	60.0	40.0
3	3	4	0.731	0.716	150.0	130.0
4	4	5	0.366	0.358	75.0	50.0
5	5	6	1.828	1.790	15.0	9.0
6	6	7	1.097	1.074	18.0	14.0
7	7	8	0.731	0.716	13.00	10.00
8	8	9	0.731	0.716	16.00	11.00
9	4	10	1.080	0.734	20.00	10.00
10	10	11	1.620	1.101	16.00	9.00
11	11	12	1.080	0.734	50.00	40.00
12	12	13	1.350	.9170	105.00	90.00
13	13	14	0.810	0.550	25.00	15.0
14	14	15	1.944	1.321	40.0	25.0
15	7	16	1.080	0.734	100.00	60.00
16	16	17	1.620	1.101	40.00	30.00
17	1	18	1.097	1.074	60.00	30.00
18	18	19	0.366	0.358	40.0	25.0
19	19	20	1.463	1.432	15.00	9.0
20	20	21	0.914	0.895	13.00	7.00
21	21	22	0.804	0.787	30.00	20.00
22	22	23	1.133	1.110	90.0	50.0
23	23	24	0.475	0.465	50.0	30.0
24	19	25	2.214	1.505	60.0	40.0
25	25	26	1.620	1.110	100.0	80.0
26	26	27	1.080	0.734	80.0	65.0
27	27	28	0.540	0.367	100.0	60.0
28	28	29	0.540	0.367	100.0	55.0
29	29	30	1.080	0.734	120.0	70.0
30	30	31	1.080	0.734	105.0	70.0
31	70	32	0.366	0.358	80.0	50.0
32	32	33	0.731	0.716	60.00	40.00
33	33	34	0.731	0.716	13.00	8.00
34	34	35	0.804	0.787	16.0	9.0

Table A1 Line and load data of example 5.1 (continued)

Branch number	Sending end node	Receiving end node	R (Ω)	X (Ω)	PL (kW)	QL (kVar)
35	35	36	1.170	1.145	50.0	30.0
36	36	37	0.768	0.752	40.0	28.0
37	37	38	0.731	0.716	60.00	40.00
38	38	39	1.097	1.074	40.00	30.00
39	39	40	1.463	1.432	30.00	25.0
40	34	41	1.080	0.734	150.0	100.0
41	41	42	0.540	0.367	60.0	35.0
42	42	43	1.080	0.734	120.0	70.0
43	43	44	1.836	1.248	90.0	60.0
44	44	45	1.296	0.881	18.0	10.0
45	42	46	1.188	0.807	16.0	10.0
46	46	47	0.540	0.367	100.0	50.0
47	44	48	1.080	0.734	60.0	40.0
48	37	49	0.540	0.367	90.0	70.0
49	49	50	1.080	0.734	85.0	55.0
50	50	51	1.080	0.734	100.0	70.0
51	51	52	1.080	0.734	140.0	90.0
52	70	53	0.366	0.358	60.0	40.0
53	53	54	1.463	1.432	20.00	11.0
54	54	55	1.463	1.432	40.00	30.0
55	55	56	0.914	0.895	36.0	24.0
56	56	57	1.097	1.074	30.00	20.00
57	57	58	1.097	1.074	43.00	30.00
58	54	59	0.270	0.183	80.0	50.0
59	59	60	0.270	0.183	240.0	120.0
60	60	61	0.810	0.550	125.0	110.0
61	61	62	1.296	0.881	25.0	10.0
62	57	63	1.188	0.807	10.0	5.0
63	63	64	1.188	0.807	150.0	130.0
64	64	65	0.810	0.550	50.0	30.0
65	65	66	1.620	1.101	30.0	20.0
66	64	67	1.080	0.734	130.0	120.0
67	67	68	0.540	0.367	150.0	130.0
68	68	69	1.080	0.734	25.0	15.0
Tie branches						
69	9	52	0.908	0.726		
70	9	40	0.381	0.244		
71	15	48	0.681	0.544		
72	24	69	0.254	0.203		
73	31	66	0.254	0.203		
74	47	62	0.254	0.203		
75	40	45	0.454	0.363		
76	41	61	0.454	0.363		
77	23	29	0.454	0.363		
78	9	15	0.681	0.544		
79	15	69	0.454	0.363		

Other Data: Current carrying capacity of all tie branches are 234.0 A. The current carrying capacity of branches 1–8, 17–23, 31–39, and 52–57 is 270 A. For branches 9–16, 24–30, 40–51, and 58–68, it is 208 A. Voltage magnitude of both the substations = 1.0 p.u. Loads are at the receiving end nodes.

Appendix 2

Table A2 Load data of example 5.2

Nodes	PL (kW)	QL (kVAr)	Nodes	PL (kW)	QL (kVAr)
2	120.0	108.0	36	72.0	48.0
3	72.0	48.0	37	48.0	36.0
4	180.0	156.0	38	36.0	30.0
5	90.0	60.0	39	180.0	120.0
6	21.60	13.0	40	72.0	42.0
7	21.60	17.0	41	144.0	84.0
8	15.60	12.0	42	108.0	72.0
9	19.0	13.0	43	21.60	12.0
10	24.0	12.0	44	19.20	12.0
11	19.20	11.0	45	120.0	60.0
12	60.0	48.0	46	72.0	48.0
13	126.0	108.0	47	108.0	84.0
14	30.0	18.0	48	122.40	79.20
15	48.0	30.0	49	120.0	84.0
16	72.0	36.0	50	168.0	108.0
17	48.0	30.0	51	72.0	48.0
18	18.0	11.0	52	24.0	13.20
19	15.6	8.4	53	48.0	36.0
20	36.0	24.0	54	43.20	28.80
21	108.0	60.0	55	36.0	24.0
22	60.0	36.0	56	51.60	36.0
23	72.0	48.0	57	96.0	60.0
24	120.0	96.0	58	288.0	144.0
25	96.0	78.0	59	150.0	132.0
26	120.0	72.0	60	30.0	12.0
27	120.0	66.0	61	12.0	6.0
28	144.0	84.0	62	180.0	156.0
29	126.0	84.0	63	60.0	36.0
30	96.0	60.0	64	36.0	24.0
31	72.0	48.0	65	156.0	144.0
32	15.60	9.6	66	180.0	156.0
33	19.20	11.8	67	30.0	18.0
34	60.0	36.0	68	120.0	72.0
35	48.0	33.60	69	48.0	36.0

Table A3 Tie branch resistance and reactance of example 5.2

Tie-branches	Resistance (Ω)	Reactance (Ω)
22–67	0.381	0.2445
67–15	0.454	0.3630
21–27	0.254	0.2030
9–50	0.681	0.5445
29–64	0.681	0.5445
45–60	0.254	0.2030
43–38	0.254	0.2030
9–15	0.454	0.3630

Chapter 6

Service restoration in distribution system

Biswarup Das[1]

When a fault occurs in a feeder of a distribution system, usually the circuit breaker located at the substation end of the feeder operates to trip the faulty feeder. As a result, electricity supply to a large number of loads gets interrupted. Under this condition, the task of the system dispatch centre is to locate and isolate the fault and subsequently restore supply to as large an out-of-service area as possible by appropriate switching actions. Theoretically, all the lost loads should be re-connected to alternative feeders to restore supply to them. However, this operation often leads to infeasible operation of the distribution system in terms of low bus voltages and line overloads. A simple example described below illustrates this issue.

6.1 Basic aspect of service restoration

Consider a sample distribution system shown in Figure 6.1. In this figure, bus 1 is the substation bus[2]. There are three main feeder sections, namely (i) feeder 1 (1-2-3-4-5-6), (ii) feeder 2 (1-26-27-28-29-30-31-32-33-34-35) and (iii) feeder 3 (1-36-37-38-39-40). In between these three feeders, there are ten lateral feeders. These are: (a) lateral 1 (2-7-8-9-26), (b) lateral 2 (3-10-11-12-13-28), (c) lateral 3 (4-14-15-16-17-30), (d) lateral 4 (5-18-19-20-21-32), (e) lateral 5 (6-22-23-24-25-34), (f) lateral 6 (27-41-42-43-44-36), (g) lateral 7 (29-45-46-47-48-37), (h) lateral 8 (31-49-50-51-52-38), (i) lateral 9 (33-53-54-55-56-39) and (j) lateral 10 (35-57-58-59-60-40). Each section of any lateral feeder is equipped with a switch. For example, the section between bus 2 and bus 7 contains switch number 1[3]. As can be seen from Figure 6.1, there are totally 49 switches in the system. However, as the distribution system usually operates in radial topology, some of these switches are open. The initial configuration of the system is shown in Figure 6.2. As can be seen from this figure, the load points 7, 10, 11, 14, 18, 19, 20 22, 23 are supplied by feeder 1, feeder 2 supplies the load points 8, 9, 12, 13, 15, 16, 17, 21, 24, 25, 41, 45, 49, 53, 57 and

[1]Department of Electrical Engineering, Indian Institute of Technology Roorkee, Roorkee 247 667, Uttarakhand, India, email: biswafee@iitr.ac.in
[2] Buses are numbered with numerical values without any parentheses.
[3] The switches are shown with the symbol × and are numbered with numerical values within a parentheses.

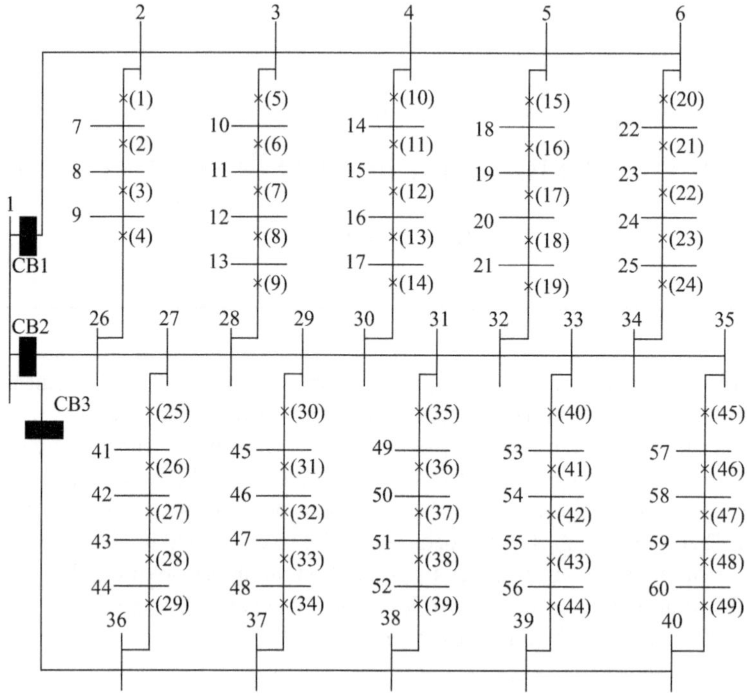

Figure 6.1 A sample distribution system

the load points 42, 43, 44, 46, 47, 48, 50 51, 52, 54, 55, 56, 58, 59, 60 are supplied by feeder 3. It is to be noted that all these three feeders are operated radially. The relevant data of this system are given in Table 6.1. In this system, the total real and reactive power load are 15,042 kW and 4997 kVAR, respectively.

Now let us assume that a fault occurs on feeder 1 and as a result, feeder 1 is tripped by the circuit breaker 1 (CB1). As a consequence, the load points supplied by feeder 1 loose supply. To restore supply to these load points, the most obvious way is to make the switches (2), (7), (11), (18) and (22) 'ON' such that these load points are now supplied by feeder 2. However, while transferring these load points to feeder 2, the switches (1), (5), (10), (15) and (20) are to be made 'OFF' so that when the fault on feeder 1 is repaired and feeder 1 is energised, feeder 2 and feeder 3 still operate in a radial fashion. When all loads of feeder 1 are shifted to feeder 2, the total real and reactive power load on feeder 2 becomes 11,106 kW and 3685 kVAR, respectively. Further, the total real and reactive power load on feeder 3 are 3936 kW and 1312 kVAR, respectively. Because of this heavy loading on feeder 2, the bus voltages in feeder 2 become quite low. The lowest voltage is 0.8898 p.u (at bus 22) which is lower than the lowest permissible limit (0.9 p.u). In fact, both at buses 22 and 23, the voltage magnitude becomes less than 0.9 p.u. Therefore, it is not feasible to shift all the loads to feeder 2 and operate the system within permissible voltage limits.

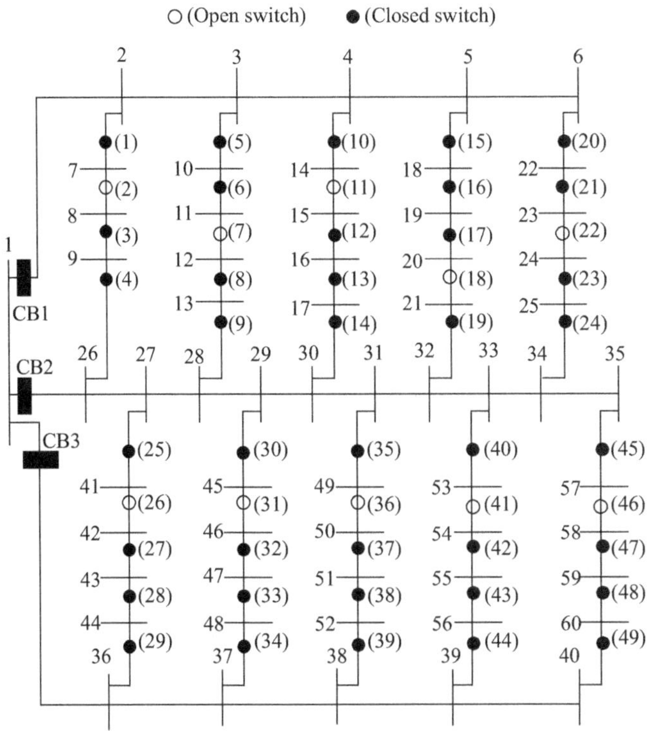

Figure 6.2 Initial configuration of the distribution system

Now, from Figure 6.2, it is observed that as there is no direct path between feeder 1 and feeder 3, it is also not possible to shift some of the lost loads (of feeder 1) to feeder 3. Therefore, the only alternative is to shift some of the original loads of feeder 2 to feeder 3 and subsequently shift all the lost loads of feeder 1 to feeder 2 by proper switching arrangement such that both the feeders operate in a radial fashion while maintaining all the node voltages within the prescribed limits. Now, as there are 44 switches which can be operated (switches (1), (5), (10), (15) and (20) are to be maintained in 'OFF' position), total number of possible combinations of switching operation is $2^{44} = 17,592.2$ billions (approximately). As it is computationally infeasible to check all of these combinations, some other method needs to be adopted to reduce the search space.

From the above discussion, it is observed that the objective of service restoration procedure is essentially the restoration of disconnected loads while maintaining the voltage constraints[4]. This objective can be translated into two objective functions, which should be satisfied simultaneously: (i) f_1: minimisation of unsupplied

[4] Other constraints are also considered; however, here only this constraint is considered for the sake of simplicity.

Table 6.1 System data for the three-feeder distribution system

Line data				Load at bus *j*		Line data				Load at bus *j*	
Bus *i*	Bus *j*	*R* (Ω)	*X* (Ω)	P_L (kW)	Q_L (kVAR)	Bus *i*	Bus *j*	*R* (Ω)	*X* (Ω)	P_L (kW)	Q_L (kVAR)
1	2	0.1833	0.194	0	0	2	3	0.0851	0.1487	0	0
3	4	0.3212	0.3693	0	0	4	5	0.3039	0.3451	0	0
5	6	0.176	0.1863	0	0	1	26	0.077	0.0815	0	0
26	27	0.055	0.0582	0	0	27	28	0.077	0.0815	0	0
28	29	0.1283	0.1358	0	0	29	30	0.2236	0.2367	0	0
30	31	0.11	0.1164	0	0	31	32	0.1833	0.194	0	0
32	33	0.154	0.163	0	0	33	34	0.3666	0.3881	0	0
34	35	0.1283	0.1358	0	0	1	36	0.11	0.1164	0	0
36	37	0.1833	0.194	0	0	37	38	0.154	0.163	0	0
38	39	0.3666	0.3881	0	0	39	40	0.1283	0.1358	0	0
2	7	0.022	0.0233	572	174	7	8	0.088	0.0931	936	312
8	9	0.2933	0.3105	189	63	9	26	0.2053	0.2173	0	0
3	10	0.2053	0.2173	336	112	10	11	0.1466	0.1552	657	219
11	12	0.8212	0.8693	783	261	12	13	0.3483	0.3686	729	243
13	28	0.1466	0.1552	0	0	4	14	0.1466	0.1552	477	159
14	15	0.1833	0.194	549	183	15	16	1.0851	1.1487	477	159
16	17	0.8505	0.9003	432	144	17	30	0.3519	0.3725	0	0
5	18	0.3519	0.3725	432	144	18	19	0.3813	0.4036	672	224
19	20	0.176	0.1863	495	165	20	21	0.0367	0.0388	207	69
21	32	0.088	0.0931	0	0	6	22	0.088	0.0931	522	174
22	23	0.2949	0.4293	391	131	23	24	0.7845	0.8305	312	104
24	25	0.6965	0.7373	549	183	25	34	0.2236	0.2367	0	0
27	41	0.0953	0.1009	386	129	41	42	0.6965	0.7373	146	49
42	43	0.154	0.163	103	34	43	44	0.1833	0.194	100	33
44	36	0.011	0.0116	0	0	29	45	0.011	0.0116	246	82
45	46	0.2053	0.2173	572	191	46	47	0.3666	0.3881	249	83
47	48	0.0073	0.0078	165	55	48	37	0.3666	0.3881	0	0
31	49	1.2831	1.3582	100	33	49	50	2.3462	2.4836	182	61
50	51	0.0367	0.0388	100	33	51	52	0.1466	0.1552	497	166
52	38	1.2611	1.3349	0	0	33	53	0.5866	0.6209	154	51
53	54	0.1833	0.194	199	66	54	55	0.5646	0.5976	507	169
55	56	0.6232	0.6597	150	50	56	39	0.7332	0.7761	0	0
35	57	0.6232	0.6597	503	167	57	58	0.7332	0.7761	351	117
58	59	0.4766	0.5045	484	161	59	60	0.6599	0.6985	131	44
60	40	0.8432	0.8926	0	0						

loads and (ii) f_2: minimisation of voltage deviation from the nominal voltage (1 p.u). Clearly, these two objective functions are contradictory to each other in the sense that when the amount of unsupplied load reduces (i.e. load connected to the system increases), the system voltages reduce (i.e. voltage deviation from the nominal voltage increases) and vice versa. Therefore, it is not possible to simultaneously minimise both these objective functions. As a result, the service restoration problem is essentially a multiobjective optimisation problem (MOOP).

To solve this MOOP while reducing the search space, several approaches have been developed in the literature. As the objective functions and constraints in a service restoration problem are generally non-differentiable, almost all of the reported methods are based on various types of non-mathematical optimisation techniques. These include heuristic methods [1–3], heuristic fuzzy reasoning technique [4], tree network-based approach [5, 6], artificial neural network and pattern recognition approach [7], ranking-based search technique [8], fuzzy evaluation of multi-criteria [9], combined fuzzy-genetic algorithm method [10], coloured Petri net [11], fuzzy cause-effect networks [12], G-net inference mechanism [13], expert system [14], quantitative evaluation model [15], meta-heuristic algorithms such as genetic algorithm [16], tabu search [16], reactive tabu search [16], simulated annealing [16], discrete differential mutation operator [17], evolutionary algorithm [18, 19], etc. Usually, the MOOP is converted to a single-objective optimisation problem (SOOP) by utilising suitable weighting factors as $f = \alpha f_1 + \beta f_2$, where α and β are the weighting factors [12, 16]. Subsequently, this SOOP is solved by using any suitable technique. However, in this method, the final solution depends on the choice of the weighting factors. As the choice of the weighting factors is quite subjective, the final solution obtained in this method can vary significantly depending on the choice of the weighting factors. Therefore, in the literature, methods for solving a MOOP without using any weighting factors have been developed. One of the most popular methods in this category is non-dominated sorting genetic algorithm II (NSGA-II) [20]. The application of NSGA-II for service restoration problem has been described in References 21 and 22. As NSGA-II does not require any weighting factors, the subjectivity of the obtained solutions is removed to a large extent. The basic steps of NSGA-II for service restoration are as follows [21, 22]:

Step 1: Create the initial population (P_0) of size N (i.e. the number of strings in a population) randomly. For the service restoration problem considered here, the length of each string (the number of bits in a string) is 49 (equal to the number of switches in this system). Here, the binary coding scheme is used and therefore, bit '1' and bit '0' represent the 'closed (ON)' and 'open (OFF)' switches, respectively. Also, while initialising the population, it is ensured that the faulty feeder is always isolated. For example, in Figure 6.2, if the fault occurs on feeder 1, the bit values corresponding to switches (1), (5), (10), (15) and (20) are always set to '0' before proceeding further. Also, if any string creates a loop in the system, a switch is randomly selected in the loop-forming path and made 'OFF' to maintain the radiality of the system. For example, in Figure 6.2, if the switches (30), (31), (32), (33) and (34) are all made 'ON' in a string, then any of these five switches is randomly selected and made 'OFF' before proceeding further. Also, set count $k = 0$.

Step 2: In this step, the population P_k is evaluated as follows. For each of the N generated strings in the population P_k, the amount of unsupplied load is calculated. Further, for each of these generated strings, AC load-flow study (by using backward/forward sweep method) of the corresponding network is carried out and subsequently, the maximum deviation of the bus voltages (from the nominal voltage) is calculated.

Step 3: The strings in the population P_k is sorted based on the level of non-domination. Each solution is assigned a rank based on its level of non-domination by other solutions, i.e. the solution(s) which is(are) not dominated by any other solution is(are) assigned a rank of 1, the solution(s) at the next level of non-domination is(are) assigned a rank of 2 and so on.

Step 4: In this step, an offspring population (Q_k) of size N is created by applying crowded tournament selection operator (CTSO), crossover and mutation operator on P_k as follows. Initially, two strings of P_k are chosen randomly and the string having better rank[5] is declared winner (winner 1). If both these two solutions have same rank, then the solution having higher crowding distance [20] is declared winner (winner 1). In the most rare case, if the crowding distances of both these chosen strings are also same, then the first chosen string is taken as the winner (winner 1). Further, two more strings of P_k are again chosen randomly (ensuring that any of the earlier two chosen strings are not repeated again) and from these two chosen strings, the second winner (winner 2) is selected following the same procedure as described above. Subsequently, single-point crossover operation is applied on these two winner strings to create two offspring strings and finally, mutation operator on these offspring strings is applied to enhance the diversity of the generated stings. Finally, necessary precautions (as described in Step 1) for isolating the faulty feeder and preventing the formation of any loop are undertaken before proceeding further.

By applying the above procedure repeatedly, finally the offspring population (Q_k) of size N is created.

Step 5: The offspring population Q_k is evaluated following the procedure described in Step 2. Subsequently, a combined population $R_k = P_k \bigcup Q_k$ is formed. Thus, the size of R_k is $2N$.

Step 6: In this step, non-dominated sorting is performed on R_k, by which R_k is divided into different fronts. The first front or best front (denoted as F_1) contains the strings of R_k which do not constraint dominate each other but constraint dominate all the other solutions of R_k. Among the remaining solutions (i.e. which are not in F_1), the solutions which do not constraint dominate each other but constraint dominate other solutions are kept in the second front (denoted as F_2). Similarly, among the strings which are not in the combined front $F = F_1 \bigcup F_2$, those strings are kept in the third front (denoted as F_3) which do not constraint dominate each other but constraint dominate other strings. This process is repeated till all the strings in R_k are assigned a front.

Step 7: In this step, the parent population P_{k+1} is created as follows. Initially, the solutions belonging to the set F_1 are considered. If size of F_1 is smaller than N, then all the solutions in F_1 are included in P_{k+1}. If after including all the solutions in F_1, the size of P_{k+1} is less than N, the solutions belonging to F_2 are included in P_{k+1}. If the size of P_{k+1} is still less than N, the solutions belonging to F_3 are included in P_{k+1}.

[5] Rank 1 is the best rank, followed by rank 2, rank 3, etc.

This process is repeated till the total number of solutions in P_{k+1} (let it be denoted as n) is greater than N. To make the size of P_{k+1} exactly equal to N, $(n - N)$ solutions from the last-included non-dominated front are discarded from P_{k+1}. To choose the solutions to be discarded, initially the solutions of the last-included non-dominated front are sorted according to their crowding distances and subsequently, the solutions having least $(n - N)$ crowding distances are discarded from P_{k+1}.

Step 8: If P_{k+1} comprises of strings only from the front F_1, the algorithm is considered to have converged and proceeds to Step 9. Otherwise, set $k = k + 1$ and go back to Step 2.

Step 9: In this step, the final solution from P_{k+1} is selected. Different approaches can be followed to select the final solution from P_{k+1}. One such approach is the use of 'max–min' method [23]. Another approach could be to consider the solution corresponding to least amount of unsupplied load. If the maximum voltage deviation (MVD) corresponding to this solution is within the acceptable limits, then this solution is finally chosen. Otherwise, from among the remaining solutions, choose the solution that restores maximum amount of load and simultaneously satisfies the voltage limit constraint.

The above procedure is illustrated with an example as described below. In this example, it is assumed that the population size (N) is 10.

Let us assume that a fault occurs on feeder 1. As a result, the circuit breaker 'CB1' trips to isolate feeder 1. To find a restoration schedule through NSGA-II, initially the population P_0 is created with size $N = 10$. To ensure that feeder 1 is always isolated in all the generated switch patterns, switches (1), (5), (10), (15) and (20) are always made 'OFF' as discussed in Step 1 above. In the next step, the 10 generated bit strings are evaluated by computing load-flow solution for each network corresponding to the strings and subsequently, the values of shedded load (SL) and the MVD are calculated for each of these strings. These results are shown in Table 6.2.

From the above results, the domination (**DOM**) matrix [24] is constructed which is shown in (6.1). The size of this matrix is $(N \times N)$ (in this case 10×10). For constructing this matrix, initially all the elements of this matrix are initialised to zero. Subsequently, if the ith bit string dominates the jth bit string, then the element (i, j) is set to unity. For example, from Table 6.2, it can be seen that both the LS and MVD values of string 2 are less than the LS and MVD values of string 5 and thus, string 2

Table 6.2 Results of evaluation of initial population

String no.	SL (kW)	MVD (p.u)	String no.	SL (kW)	MVD (p.u)
1	10,980	0.0322	2	11,906	0.0122
3	10,933	0.0526	4	9944	0.0400
5	12,374	0.0182	6	10,911	0.0439
7	12,503	0.0183	8	10,105	0.0358
9	10,249	0.0549	10	10,330	0.0638

dominates string 5. As a result, the element (2, 5) of the **DOM** matrix is set to unity. Similarly, all the other non-zero elements are set.

$$\mathbf{DOM} = \begin{bmatrix} 0 & 0 & 0 & 0 & 0 & 0 & 0 & 0 & 0 & 0 \\ 0 & 0 & 0 & 0 & 1 & 0 & 1 & 0 & 0 & 0 \\ 0 & 0 & 0 & 0 & 0 & 0 & 0 & 0 & 0 & 0 \\ 0 & 0 & 1 & 0 & 0 & 1 & 0 & 0 & 1 & 1 \\ 0 & 0 & 0 & 0 & 0 & 0 & 1 & 0 & 0 & 0 \\ 0 & 0 & 1 & 0 & 0 & 0 & 0 & 0 & 0 & 0 \\ 0 & 0 & 0 & 0 & 0 & 0 & 0 & 0 & 0 & 0 \\ 0 & 0 & 1 & 0 & 0 & 1 & 0 & 0 & 1 & 1 \\ 0 & 0 & 0 & 0 & 0 & 0 & 0 & 0 & 0 & 1 \\ 0 & 0 & 0 & 0 & 0 & 0 & 0 & 0 & 0 & 0 \end{bmatrix} \tag{6.1}$$

From the entries of the domination matrix, the domination inter-relationship among the strings can be determined. The domination matrix has a property that the kth column indicates by which strings the kth string is dominated while the jth row indicates the strings which are dominated by the jth string [24]. For example, the third column of the **DOM** matrix indicates that the third string is dominated by strings 4, 6 and 8, which is also evident in Table 6.2. Similarly, second row of this matrix shows that the second string dominates strings 5 and 7, which is again confirmed by the entries of Table 6.2.

Now, from (6.1), it is observed that the strings 1, 2, 4 and 8 are not dominated by any other strings and therefore, these strings are assigned rank 1. Further, string 5 is dominated by only one string and therefore, this string is assigned rank 2. Similarly, each of the strings 6, 7 and 9 is dominated by two other strings and therefore, these strings are assigned a rank 3. Finally, strings 3 and 10 are dominated by three other strings and therefore, these two strings are assigned a rank 4. Further, for all the strings, the crowding distances corresponding to each objective function are calculated following the procedure given in Reference 20. In this procedure, for each objective function, the strings are initially sorted in an ascending order. Subsequently, the strings corresponding to the lowest and highest objective function value are assigned a high value of crowding distance in an increasing order. In this chapter, this high value has been chosen as 10. It is to be noted that this value is not at all related to the number of population, N. Instead of 10, any other value such as 50, 100, 1000, etc. can also be chosen. Subsequently, for each string i in the sorted list, the crowding distance is calculated as $cwd(i) = \dfrac{f_{i+1} - f_{i-1}}{f^{\max} - f^{\min}}$, where f_{i+1} and f_{i-1} denotes the objective function value of the $(i+1)$th and $(i-1)$th string (in the sorted list), respectively, while $f^{\max}$ and $f^{\min}$ denote the maximum and minimum value of the objective function, respectively. After calculating the crowding distance of each string corresponding to each objective function value, the final crowding distance value (for each string) is computed as the sum of the crowding distances corresponding to both these objective functions. The rank, crowding distance for shedded load (CWSL), crowding distance for MVD (CWVD) and the final crowding distance (FCWD) for

Table 6.3 Ranks and crowding distances of initial population

String no.	Rank	CWSL	CWVD	FCWD	String no.	Rank	CWSL	CWVD	FCWD
1	1	10.0	0.193	10.193	2	3	0.0388	0.4520	0.4908
3	1	0.1226	0.2503	0.3729	4	1	10.0	10.0	20.0
5	4	0.0119	10.0	10.0119	6	2	0.0241	0.1498	0.1738
7	2	0.3243	0.1269	0.4512	8	1	0.4207	0.2488	0.6695
9	4	0.3666	0.4156	0.7821	10	4	0.2304	0.0251	0.2555

Table 6.4 Results of evaluation of the combined population

String no.	SL (kW)	MVD (p.u)	String no.	SL (kW)	MVD (p.u)
1	10,980	0.0322	2	11,906	0.0122
3	10,933	0.0526	4	9944	0.0400
5	12,374	0.0182	6	10,911	0.0439
7	12,503	0.0183	8	10,105	0.0358
9	10,249	0.0549	10	10,330	0.0638
11	10,980	0.0322	12	13,414	0.0098
13	12,874	0.0114	14	9137	0.0355
15	13,138	0.0183	16	10,911	0.0439
17	9944	0.0400	18	12,374	0.0182
19	13,428	0.0084	20	11,838	0.0152

each string are shown in Table 6.3. Further details of calculating the crowding distance can be found in Reference 20.

After the rank and crowding distances of the parent population are calculated, the offspring population is created as described in Step 4 above. Subsequently, the offspring population is evaluated and the values of SL and MVD are calculated for each of these strings in the offspring population. The values of SL and MVD for the combined population (formed by following the procedure described in Step 5) are shown in Table 6.4.

From the results of Table 6.4, the rank and crowding distances (CWSL, CWVD and FCWD) of all the strings belonging to the combined population are calculated and are shown in Table 6.5.

Following Step 7, the parent population for the next generation is subsequently created. From Table 6.5, it is observed that strings 2, 12, 13, 14, 19 and 20 have rank 1. As the total number of strings having rank 1 is 6, these strings are included in the parent population of the next generation. Further, Table 6.5 also shows that strings 1, 8 and 11 have rank 2. As the total number of strings having either rank 1 or rank 2 is $(6 + 3 = 9)$, these three strings (having rank 2) are also included in the parent population of the next generation. As one more string still needs to be chosen, the

Table 6.5 Ranks and crowding distances of combined population

String no.	Rank	CWSL	CWVD	FCWD	String no.	Rank	CWSL	CWVD	FCWD
1	2	0.0110	0.2501	0.2610	2	1	0.1249	0.0692	0.1941
3	7	0.0161	0.1994	0.2155	4	3	0.1881	0.0752	0.2632
5	4	0.1091	0.0549	0.1639	6	6	0.1354	0.0711	0.2065
7	5	0.1165	0.0018	0.1183	8	2	0.0711	0.0799	0.1510
9	5	0.0524	0.2016	0.2540	10	6	0.1543	10.0	10.1543
11	2	0.2000	0.0602	0.2602	12	1	0.0676	0.0537	0.1212
13	1	0.1480	0.0425	0.1905	14	1	10.0	0.0649	10.0649
15	7	0.1258	0.2501	0.3759	16	6	0.0051	0.1577	0.1628
17	3	0.0375	0.0711	0.1086	18	4	0.0301	0.0018	0.0319
19	1	10.0	10.0	20.0	20	1	0.2158	0.1099	0.3257

strings having rank 3 in Table 6.5 are considered. As can be seen from this table, both strings 4 and 17 have rank 3. As the total number of strings having either rank 1, 2 or 3 is $(6+3+2=11)$, out of strings 4 and 17, only one string needs to be chosen. Now, from Table 6.5, it is seen that string 4 has a higher value of FCWD as compared to that of string 17. As a result, string 4 is finally included in the parent population of the next generation. Thus, the strings which are finally selected for inclusion into the parent population of the next generation are: strings 2, 12, 13, 14, 19, 20, 1, 8, 11 and 4. As the parent population contains strings having different ranks, the algorithm proceeds further and goes back to Step 2.

For finding acceptable results with this method, the algorithm must be run several times with varying number of N. In this study, for a fault in feeder 1, one such acceptable result has been found for $N = 30$. With this choice of N, the final configuration suggested by the algorithm is shown in Figure 6.3. From this figure, it is observed that feeder 1 is isolated completely. Further, to maintain the system voltages within the permissible limit, the algorithm suggests the load points 10, 14, 18, 22, 23, 42, 43, 44, 55 and 57 to be cut-off, thereby making the total amount of load unsupplied to be equal to 3517 kW. The minimum voltage under this condition is 0.9345 p.u. Note that when feeder 1 is isolated by opening CB1, load points 7, 10, 11, 14, 18, 19, 20, 22 and 23 get de-energised resulting in a total of 4554 kW of load to remain unsupplied. Therefore, the suggested solution restores more than 1 MW of load.

It is to be noted that, by no means, the above solution suggested by the algorithm can be considered to be the most optimum one. It is quite possible that better solutions (as compared to the solution discussed above) may indeed exist. However, in order to ensure that the most optimum solution and/or the other better solutions are always obtained, all the 17,592.2 billion possible switch combinations need to be investigated. Even if it is assumed that investigation of one switch combination takes just 1 s

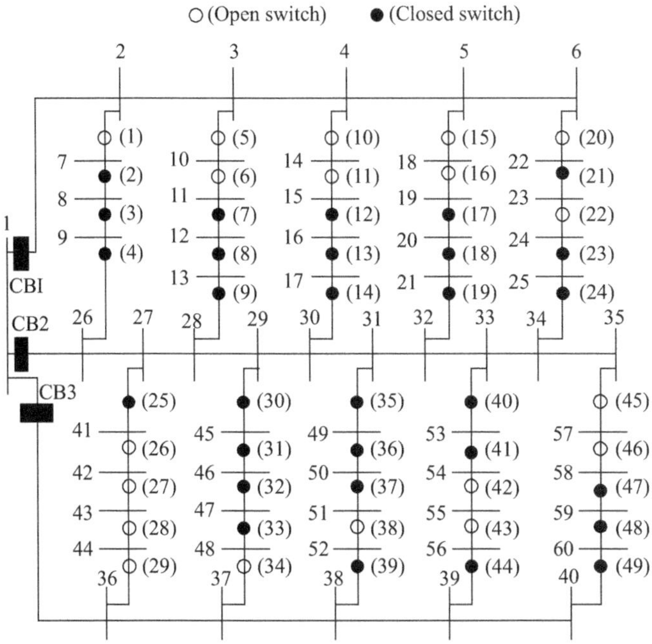

Figure 6.3 Final configuration of the distribution system

(on the currently available desktop or laptop computers), the total time required for investigating all these 17,592.2 billion combinations would be more than 5.5 Lakh years!! Therefore, when the service restoration schedule needs to be determined within next few hours, clearly the option of investigating all the possible options is infeasible. Under this circumstances, the only feasible option is to execute the above algorithm (or any other suitable meta-heuristic algorithm) several times with varying parameters and choose the best solution out of all the solutions obtained in different runs. Also, if access to any high-performance computing (HPC) platform is available, then the HPC platform can be utilised either for investigating all the possible combinations or for generating a large number of feasible solutions by executing the meta-heuristic algorithm many more number of times before selecting the final solution.

Now, in this chapter, for the purpose of illustration, only two objective functions have been considered. However, for realistic solution of service restoration problem, apart from the above two objective functions, several other aspects should also be considered. These are [21, 22]: (a) The required number of switching for restoring the loads should be as minimum as possible. This is due to the fact that increased number of switching operations reduces the life expectancy of the switches. Further, as the number of switching increases, the total time required by the field personnel to complete the switching actions also increases, which in turn, increases the time required for restoring the supply to the out-of-service loads. (b) No component (feeders, transformers, etc.) should be overloaded. (c) The radial structure of the network should be maintained. (d) The power losses in the system should be as minimum as possible. (e) Priority of the loads should be taken into account while finalising the service restoration schedule, i.e. high-priority loads should be given electricity supply before any other lower priority loads. Further application of NSGA-II for solving the restoration problem with all these objective functions and constraints is described in References 21 and 22.

6.2 Conclusion

In this chapter, the basic aspects of service restoration in a distribution system are discussed. The procedure for the application of a meta-heuristic algorithm (NSGA-II) for solving the service restoration problem is also discussed in detail with an illustrative example.

References

[1] Y.-Y. Hsu, H.M. Huang, H.C. Kuo, *et al.*, "Distribution system service restoration using a heuristic search approach", *IEEE Transactions on Power Delivery*, vol. 7, no. 2, pp. 734–740, April 1992.

[2] D. Shirmohammadi, "Service restoration in distribution networks via network reconfiguration", *IEEE Transactions on Power Delivery*, vol. 7, no. 2, pp. 952–958, April 1992.

[3] M. Gholami, J. Moshtagh and L. Rashidi, "Service restoration for unbalanced distribution networks using a combination two heuristic methods", *International Journal of Electrical Power and Energy Systems*, vol. 67, pp. 222–229, 2015.

[4] Y.-Y. Hsu and H.-C. Kuo, "A heuristic based fuzzy reasoning approach for distribution system service restoration", *IEEE Transactions on Power Delivery*, vol. 9, no. 2, pp. 948–953, April 1994.

[5] N.D.R. Sarma, V.C. Prasad, K.S. Prakasa Rao and V. Sankar, "A new network reconfiguration technique for service restoration in distribution networks", *IEEE Transactions on Power Delivery*, vol. 9, no. 4, pp. 1936–1942, October 1994.

[6] N.D.R. Sarma, S. Ghosh, K.S. Prakasa Rao and M. Srinivas, "Real time service restoration in distribution networks – a practical approach", *IEEE Transactions on Power Delivery*, vol. 9, no. 4, pp. 2064–2070, October 1994.

[7] Y.Y. Hsu and H.M. Huang, "Distribution system service restoration using the artificial neural network approach and pattern recognition method", *IEE Proceedings – Generation, Transmission and Distribution*, vol. 142, no. 3, pp. 251–256, May 1995.

[8] K.N. Miu, H.D. Chiang, B. Yuan and G. Darling, "Fast service restoration for large-scale distribution systems with priority customers and constraints",

IEEE Transactions on Power System, vol. 13, no. 3, pp. 789–795, August 1998.

[9] S.-J. Lee, S.-I. Lim and B.-S. Ahn, "Service restoration of primary distribution systems based on fuzzy evaluation of multi-criteria", *IEEE Transactions on Power Systems*, vol. 13, no. 3, pp. 1156–1163, August 1998.

[10] Y.-T. Hsiao and C.-Y. Chien, "Enhancement of restoration service in distribution systems using a combination fuzzy-GA method", *IEEE Transactions on Power Systems*, vol. 15, no. 4, pp. 1394–1400, November 2000.

[11] C.-S. Chen, C.-H. Lin and H.-Y. Tsai, "A rule-based expert system with colored Petri net models for distribution system service restoration", *IEEE Transactions on Power Systems*, vol. 17, no. 4, pp. 1073–1080, November 2002.

[12] C.-M. Huang, "Multiobjective service restoration of distribution systems using fuzzy cause-effect networks", *IEEE Transactions on Power Systems*, vol. 18, no. 2, pp. 867–874, May 2003.

[13] Y.-L. Ke, "Distribution feeder reconfiguration for load balancing and service restoration by using G-nets inference mechanism", *IEEE Transactions on Power Delivery*, vol. 19, no. 3, pp. 1426–1433, July 2004.

[14] M.-S. Tsai, "Development of an object-oriented service restoration expert system with load variations", *IEEE Transactions on Power Systems*, vol. 23, no. 1, pp. 219–225, February 2008.

[15] W.-H. Chen, "Quantitative decision-making model for distribution system restoration", *IEEE Transactions on Power Systems*, vol. 25, no. 1, pp. 313–321, February 2010.

[16] S. Toune, H. Fudo, T. Genji, Y. Fukuyama and Y. Nakanishi, "Comparative study of modern heuristic algorithms to service restoration in distribution systems", *IEEE Transactions on Power Delivery*, vol. 17, no. 1, pp. 173–181, January 2002.

[17] D.S. Sanches, J.B.A. London Jr., A.C.B. Delbem, *et al.*, "Multiobjective evolutionary algorithm with a discrete differential mutation operator developed for service restoration in distribution systems", *International Journal of Electrical Power and Energy Systems*, vol. 62, pp. 700–711, 2014.

[18] J. Stepien and S. Filipiak, "Application of the evolutionary algorithm with memory at the population level for restoration service of electric power distribution networks", *International Journal of Electrical Power and Energy Systems*, vol. 63, pp. 695–704, 2014.

[19] D.S. Sanches, J.B.A. London Jr. and A.C.B. Delbem, "Multi-objective evolutionary algorithm for single and multiple fault service restoration in large-scale distribution systems", *Electric Power Systems Research*, 110, pp. 144–153, 2014.

[20] K. Deb, A. Pratap, S. Agarwal and T. Meyarivan, "A fast and elitist multiobjective genetic algorithm: NSGA-II", *IEEE Transactions on Evolutionary Computation*, vol. 6, no. 2, pp. 182–197, April 2002.

[21] Y. Kumar, B. Das and J. Sharma, "Service restoration in distribution system using non-dominated sorting genetic algorithm", *Electric Power Systems Research*, 76, 768–777, 2006.

[22] Y. Kumar, B. Das and J. Sharma, "Multiobjective, multiconstraint service restoration of electric power distribution system with priority customers", *IEEE Transactions on Power Delivery*, vol. 23, no. 1, pp. 261–270, January 2008.

[23] M.R. Haghifam, H. Falaghi and O.P. Malik, "Risk-based distributed generation placement", *IET Generation, Transmission & Distribution*, vol. 2, no. 2, pp. 252–260, 2008.

[24] A. Alfari, "Multidisciplinary system design optimization (MSDO): multiobjective optimization (I), lecture 14". Available from http://ocw.mit.edu/courses/engineering-systems-division/esd-77-multidisciplinary-system-design-optimization-spring-2010/lecture-notes/MITESD-77S10-lec14.pdf.

Chapter 7
Feeder reconfiguration for load balancing

Adisa A. Jimoh[1]

7.1 Introduction

The electrical network system has a mandate to ensure that an adequate supply is available to meet the estimated load of the consumers in both the near and more distant future. With the increase in electric power consumption and the advancement in the distribution system technology, the future challenge is to further decrease the system technical loss, and improve the performance. In order to avoid excessive voltage drop and minimize loss, it may be economical to install apparatus to balance or partially balance the loads, even within the context of smart grid system in addition to employing demand response, demand side management, distribution generation, decentralized renewable energy, and micro grids. To meet this challenge, automation of the power distribution system needs to be widely adopted. The distribution automation can be defined as an integrated system concept [1, 2]. It includes control, monitoring, and decision to alter any kind of loads. The automatic distribution system provides directions for automatic reclosing of the switches and remote monitoring of the loads contributing towards phase balancing.

It is believed that the technology to achieve an automatic load balancing lends itself readily for the implementation of different types of algorithms for automatically rearranging the connection of consumers on the low-voltage end of a feeder for optimal performance [2]. All switches and circuit breakers involved in the controlled networks are equipped with facilities for remote operation. The control interface equipment must withstand extreme climatic conditions. Also, control equipment at each location must have a dependable power source. To cope with the complexity of the electricity distribution system, the latest information technology, communication, and power electronics equipment in distribution technologies are needed to be employed.

This chapter looks at automatic reconfiguration of a distribution low-voltage feeder by rearranging the load distribution such that the phase imbalance is always zero, or at the barest minimum achievable, but not more than the level allowed by regulation. The system is easily expandable to incorporate other distribution networks and feeder reconfigurations, including those through feeder bifurcation and the tie

[1] Department of Electrical Engineering, Tshwane University of Technology, Pretoria, South Africa

and sectionalizing switches controls. Problem formulation and solutions, and the methodology to realize this will be discussed. The chapter will conclude with a brief note on the practical feasibility and economic considerations of the system.

The phase voltage and current unbalances are major factors leading to extra losses, communication interference, equipment overloading, and malfunctioning of the protective relay which consequently results into degradation of service quality and operation efficiency [3]. Phase unbalance is also manifested in increased complex power unbalance, increased power loss, enhanced voltage drop, and increased neutral current.

In most of the cases, the phase voltage and current unbalances can be greatly improved by suitably arranging the connection phases between the distribution transformers and a primary feeder. It is also possible to reduce the phase current unbalances in every feeder segment by means of changing the connection phases [4]. The phase voltage unbalances along a primary feeder can also be improved in common cases by system reconfiguration, which involves the rearrangement of loads or transfer of load from heavily loaded area to the less loaded area.

7.2 Structure of the automated low-voltage distributed system

The low-voltage distribution system is formed from feeders that usually are three-phase, four-wire system with a radial or open-loop structure. To improve the system phase voltage and current unbalances, the connection between the specific feeder and the distribution transformer should be suitably arranged. The domestic loads are connected, as in most cases, in a single-phase. Figure 7.1(a) shows a typical radial distribution system with l primary feeders $F_{p1}, F_{p2}, \ldots, F_{pl}$. From each primary feeder, n secondary feeders are emanated to which consumer loads are connected [2]. Each secondary feeder, $F\mathrm{s}_1$, F_{s2}, …, F_{sk}, is three-phase, four-wire system that accommodates three-phase and single-phase consumer loads. The central computing unit (CCU) receives information from the loads connected to the secondary feeders under surveillance about their phase connection, consumption, and other pertinent required data, processes this information through the algorithm resident in it, and finally transmits control signal to the load selector switches (LSSs). Let us assume each feeder contains $n \times m$ domestic loads.

As shown in Figure 7.1(b) supposing each of the n secondary feeders has m number of loads connected to it and each load can be connected through the switch selector to only one of the three phases. The status of each of the loads connected to an unbalanced feeder is sent through a communication interface to a load re-arrangement/balancing algorithm in the CCU. In turn the algorithm returns, after processing, a re-arrangement status, through the same communication interface, that should be implemented by the LSS to yield a balanced feeder arrangement.

The transfer of the load from one phase to another is done via the three bilateral thyristors/TRIACs, Figure 7.2, such that at any moment in time only one switch is ON to prevent phase short-circuit. This is the main set of constrains of the system:

$$sw_{kj1} + sw_{kj2} + sw_{kj3} - 1 = 0 \tag{7.1}$$

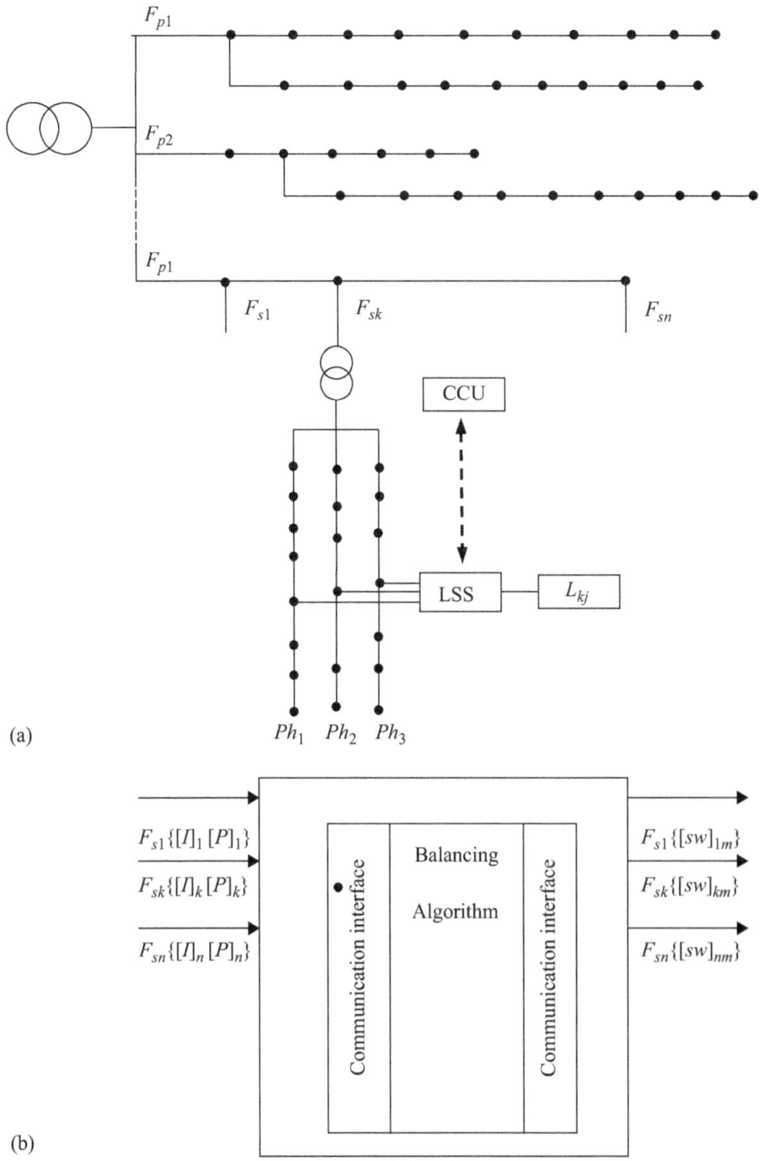

Figure 7.1 General representation of automated distributed system. © Elsevier 2011, Reprinted, with permission, from Reference 2

where sw_{kj1}, sw_{kj2}, and sw_{kj3} are switches that can only be 1 or 0 (conducting or not).

The status of the gate matrix is provided from the main controller. Actual transition from one phase to another is done at zero-crossing to avoid voltage spikes. This in return will introduce a dip of maximum 17 ms (see Figure 7.3) which is acceptable by the power quality standards.

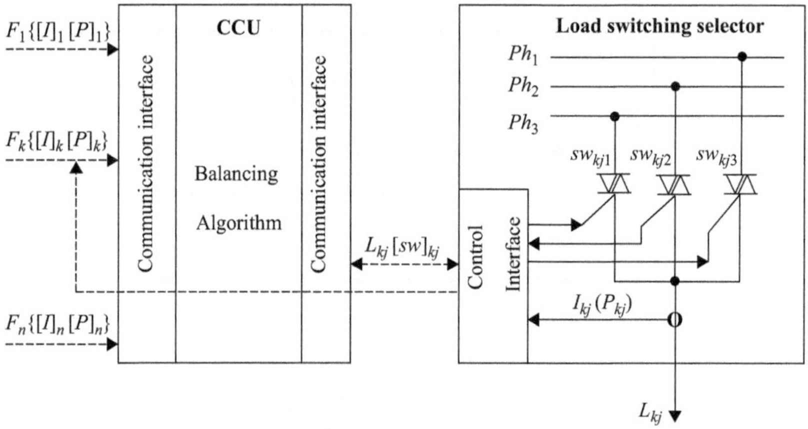

Figure 7.2 General description. © Elsevier 2011, Reprinted, with permission, from Reference 2

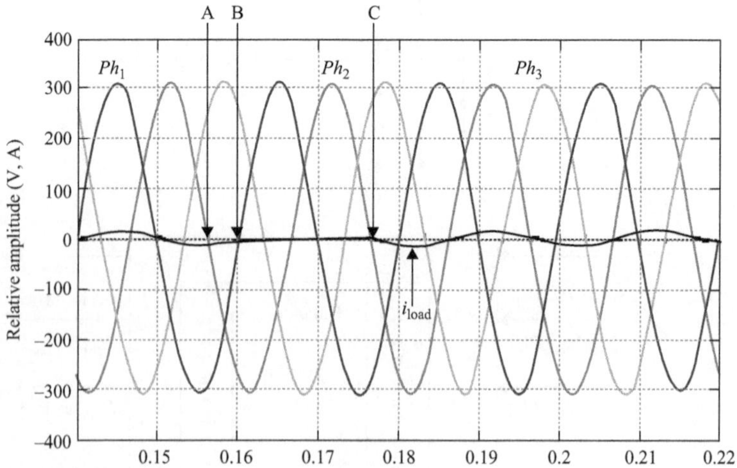

Figure 7.3 Load transition from one phase to another. © Elsevier 2011, Reprinted, with permission, from Reference 2

For example, the decision to reconfigure is taken at moment A in Figure 7.3. The zero-crossing switches OFF the load current (i_{load}) from phase 1 at moment B and switches ON again on phase 2 at moment C.

7.3 Problem formulation

The problem to be solved is formulated and solution is structured into an algorithm in the CCU. In formulating this problem, the redistribution of loads among the phases must be such that certain predefined objectives are satisfied. The objective functions that can be considered could be:

- **Total complex power unbalanced**

In a three-phase load system, the complex powers for a typical feeder i are expressed as $\bar{s}_i^j$ ($j = 1, 2, 3$ or a, b, c) for the loading of each phase [5]. The unbalance of these three complex powers s_i^{ul} can be evaluated as follows:

$$s_i^{ul} = \sqrt{(1/3)\sum_{j=a}^{c}\left|\bar{s}_i^j - \bar{s}_i^{id}\right|^2} \tag{7.2}$$

where $\bar{s}_i^{id}$ represents the ideal per phase loading and:

$$\bar{s}_i^{id} = (1/3)\sum_{j=a}^{c}\bar{s}_i^j \tag{7.3}$$

Thus an evaluation of $\bar{s}_i^{ul} = 0$ indicates that the complex power of ith feeder is balanced. The total complex power unbalance can be evaluated from:

$$s_T^{ul} = \sum_{i=1}^{n}\bar{s}_i^{ul} \tag{7.4}$$

where n is the number of feeders in the system under study. The term s_T^{ul} is used to evaluate the complex power unbalance of a feeder and the value of this describes the load unbalance situation at every feeder.

Equation (7.2) can be simplified, for the real power only, to obtain the following as the main objective function considered:

$$AU/ph = \left(\left|P_{ph1} - P_{ph2}\right| + \left|P_{ph2} - P_{ph3}\right| + \left|P_{ph3} - P_{ph1}\right|\right)/3 \tag{7.5}$$

where P_{ph1}, P_{ph2}, and P_{ph3} are the loads (power) drawn from the phases 1, 2, and 3, respectively. AU is the average power unbalance to be minimized.

- **Phasor current unbalance relationship**

In a three-phase unbalanced system, the objective could be to minimize the difference of the amplitude of the phase currents ($\hat{I}_{phik}$):

$$\text{Minimize } \boldsymbol{J} = \begin{vmatrix} I_{Ph1k} - I_{Ph2k} \\ I_{Ph1k} - I_{Ph3k} \\ I_{Ph2k} - I_{Ph3k} \end{vmatrix}$$

$$= \sqrt{\left(\left(I_{ph1k} - I_{ph2k}\right)^2 + \left(I_{ph1k} - I_{ph3k}\right)^2 + \left(I_{ph2k} - I_{ph3k}\right)^2\right)} \tag{7.6}$$

where I_{ph1k}, I_{ph2k}, and I_{ph3k} represent the currents (phasors) for each of the phases (1, 2, and 3) at an arbitrary point of connection k.

- **Neutral current of an unbalanced circuit**

The current at the neutral point of a balanced three-phase system should be 0. Therefore, the objective in balancing an unbalanced system is to minimize the neutral current as follows:

$$\text{Minimize } J_n = \left[I_N^2 - \left(\left(I_{ph1k} - I_{ph2k}\right)^2 + \left(I_{ph1k} - I_{ph3k}\right)^2 + \left(I_{ph2k} - I_{ph3k}\right)^2\right)\right]^2 \quad (7.7)$$

where I_N is the neutral current.

- **Zero and negative sequence currents of an unbalanced circuit**

For a balanced three-phase circuit, the zero and negative sequence components are zero. Therefore, the objective is to minimize the negative and zero sequence currents:

$$\text{Minimize } J_{0n} = \lambda \hat{I}^0_{phs} + \lambda \hat{I}^-_{phs} \quad (7.8)$$

For the real implementation of a control system, the following elements are necessary:

- A measurement system for measuring the loads.
- A communication system for sending the load data from each node to the control center.
- A communication system for sending the control signals to the switch breaker.
- The control cannot start if the above-described components and system are not properly installed and operating in healthy condition.

- **Minimization of power loss**

The minimization of the total real power losses arising from feeders can be calculated as follows [6]:

$$\text{Min } f_1(\overline{X}) = \sum_{i=1}^{N_l} r_i \frac{p_i^2 + q_i^2}{v_i^2} \quad (7.9)$$

where r_i, p_i, and q_i represent the resistance, real power, and reactive power of branch i, respectively, and v_i is the voltage on bus i. N_l is the total number of branches; $\overline{X}$ denotes the switch state vector, and $f_1(\overline{X})$ denotes the power lost from the system in state $\overline{X}$. A low value of $f_1(\overline{X})$ indicates economic operation of network.

- **Ensuring reliability of service**

From the operator's perspective, service reliability in distribution systems refers to the ability to support unexpected increasing loads and to supply loads following faults (restoration). A simple index to assess the service reliability is the capacity margin of a feeder and a transformer [6].

(a) *Capacity margin of feeders:* The capacity margin of feeders is calculated as follows [6]:

$$\mathrm{Min}\, f_2(\overline{X}) = 1 - \min_i \left\{ \frac{I_{iRate} - I_{iLoad}}{I_{iRate}} \right\}, \quad i = 1, 2, \ldots, N_l \tag{7.10}$$

where N_l is the total number of branches; and I_{iLoad} and I_{iRate} are the actual loading current and the rated current of branch i, respectively. The function $f_2(\overline{X})$ represents the relative value of the margin between the rated and the actual current among the feeders. A lower $f_2(\overline{X})$ indicates a greater current reserve in the feeders, implying that the considered system is more secure.

(b) *Capacity margin of transformers:* The capacity margin of transformers is calculated as follows [6]:

$$\mathrm{Min}\, f_3(\overline{X}) = 1 - \min_i \left\{ \frac{S_{iRate} - S_{iLoad}}{S_{iRate}} \right\}, \quad i = 1, 2, \ldots, N_l \tag{7.11}$$

where N_l is the total number of the transformers, and S_{iLoad} and S_{iRate} are the actual and rated megavolt amperes (MVA) loadings of transformer i, respectively. The function $f_3(\overline{X})$ specifies the relative value of the margin between rated and the actual MVA loading of the supply transformers. Lower $f_3(\overline{X})$ values indicate a larger reserve capacity in the supply transformers, implying that the operating state has higher service reliability.

Selecting a specific index for ensuring reliability of service is utility-dependent and would not alter the basic formulation presented in Reference 6. The above simple form of reliability is selected just for purposes of illustration.

- **Minimizing the deviation of bus voltage [6]**

Bus voltage is one of the most significant security and service quality indices, which can be described as follows:

$$\mathrm{Min}\, f_4(\overline{X}) = \max |V_i - V_{Rate}|, \quad i = 1, 2, \ldots, N_b \tag{7.12}$$

where N_b is the total number of buses; V_i and V_{Rate} are the actual and rated voltage on bus i and $f_4(\overline{X})$ represents the maximal deviation of the bus voltage in the system of interest. Lower $f_4(\overline{X})$ values indicate a higher quality voltage profile and better security of the considered system.

- **Minimizing switching operations [6]**

Minimizing the number of switching operations can be denoted as follows:

$$\mathrm{Min}\, f_5\left(\overline{X}\right) = \sum_{i=1}^{N_S} |S_i - S_{0i}| \tag{7.13}$$

Figure 7.4 Example of automated distribution feeder

where N_s represents the total number of switches; S_i and S_{0i} are the new and original states of switch i, respectively; and $f_5(\overline{X})$ represents the number of switching operations under state $\overline{X}$. A lower $f_5(\overline{X})$ value implies that less time is needed during the network reconfiguration process.

7.4 Solution

The solution objective for a feeder is to obtain a set of rearrangement of the connected loads at each node (or consumer point) such that the objective function is minimized. This is an optimization problem, which depending on the goal of the exercise could be formulated as either a single or multi-objective optimization problem [6–8]. While a single-objective optimization problem minimizes one of the objective functions in Section 7.3, subject to certain specified constraints, in the multi-objective optimization the goal is to obtain a pareto-optimal solution for optimizing a weighted combination of more than one objective function in Section 7.3, subject to certain specified constraints.

A number of solutions have been presented in the literature.

7.4.1 Heuristic method

Heuristic methods refer to experience-based techniques for problem solving, learning, and discovery that find a solution which is not guaranteed to be optimal, but good enough for a given set of goals [9]. Where the exhaustive search is impractical, heuristic methods are used to speed up the process of finding a satisfactory solution via appropriate shortcuts to ease the cognitive load of making a decision. Examples of this method include using a rule of thumb, an educated guess, an intuitive judgment, stereotyping, profiling, or common sense [9].

7.4.1.1 Description

For the phase and load balancing, a heuristic method was proposed in References 10, 11 on the sample distribution system shown in Figure 7.4 which consists of n loads, each having three switches to the three phases. Following from (7.1), the logic of load connection should be that: for each load, only one switch should be closed, other two should remain open, i.e., each load should be connected to only one of the three phases. The load currents are referred here by the term "load". The following initial assumptions are considered for the proposed method:

(i) $n/3$ is an integer.
(ii) The loads should be considered equally distributed per phase, i.e., an integer $n/3$ loads to be connected per phase.

The problem, therefore, is to find the optimum three sets of $n/3$ loads, with minimum differences among the individual sums of loads of the three sets. To achieve this, first we calculate the ideal phase balance current value I_{ideal}, which is equal to the one-third of the sum of the all n load currents:

$$I_{ideal} = \frac{1}{3}\sum_{j=1}^{n} I_{L_j} \tag{7.14}$$

In (7.14), I_{Lj} denotes the current of jth load. In the second step, optimally select three sets of currents for the three-phase currents I_{ph}, each set comprising of $n/3$ load currents $\{I_j, j = 1, \ldots, n/3\}$

$$I_{Load} = \{I_{L_j}, j = 1, \ldots., n\} \tag{7.15}$$

$$I_{ph} = \{I_j, j = 1, \ldots., n/3\} \quad \text{where} \quad I_j \in I_{Load} \tag{7.16}$$

Difference between the individual sum of these sets and the I_{ideal} should be *minimum*, ideally 0 for the perfect phase balance. So, three sets of $\{I_j, j = 1, \ldots, n/3\}$ have to be found, subject to the constraint:

$$\min\left|\sum_{j=1}^{n/3} I_j - I_{ideal}\right|, \quad \text{where} \quad I_j \in I_{Load} \tag{7.17}$$

This heuristic method can be implemented by structuring it to an algorithm and a subroutine as follows. The implementation takes as input the sequence of n load currents. It returns as output the sequence of the switch closing for each load, i.e., integer 1, 2, or 3 for each load, where 1, 2, 3 represent the switches for the respective phases as shown in Figure 7.4. Using the output switch-closing sequence and the load currents, we can calculate the three-phase currents and the differences between them, which indicate the quality of the phase balance. The implementation steps are depicted in the flow chart shown in Figure 7.5, where the left chart shows the main algorithm, and right chart shows a subroutine.

7.4.1.2 Main algorithm

Here as follows is the main algorithm for the implementation of the heuristic method:

- The n load currents are considered as vector.
- The output vector of the switching sequences is initialized for each load, which is also a vector of n elements.

- Then the quantity I_{ideal} is computed using (7.14).
- Check all the n loads to find the *first* set of $n/3$ load currents, to be connected to phase 1 such that phase 1 current (I_{ph1}) is closest to I_{ideal}. This is done by the subroutine "Calculate set of $n/3$".
- The switches used for I_{ph1} are updated by marking these switches as "1".
- Then remaining loads are checked to find the second set of $n/3$ load currents, to be connected to phase 2 such that phase 2 current (I_{ph2}) is closest to I_{ideal}. This is also done by the subroutine "Calculate set of $n/3$".
- The switches used for I_{ph2} are updated by marking these switches as 2.
- After finding the sequences for I_{ph1} and I_{ph2}, the rest of the load currents will be allocated to I_{ph3}.
- The switches used for I_{ph3} will be updated by marking these switches as 3.
- Using the output switching sequences of 1, 2, 3 for I_{ph1}, I_{ph2}, and I_{ph3} and the input load currents, the phase currents I_{ph1}, I_{ph2}, and I_{ph3} are computed. For example, I_{ph1} is calculated by adding all the $n/3$ load currents corresponding to the switches marked 1.
- Then the differences between I_{ph1}, I_{ph2}, and I_{ph3} are calculated which ideally should be zero. It indicates the quality of the outcome of the phase balancing procedure.
- The program returns:
  - (i) the switches used for each phase;
  - (ii) the phase currents I_{ph1}, I_{ph2}, and I_{ph3};
  - (iii) the differences between the phase currents.

7.4.1.3 Subroutine

The subroutine "Calculate set of $n/3$" used to choose the output sequences for I_{ph1} and I_{ph2} is presented; the sequential steps are:

- For I_{ph1}, we start with the n load currents.
- Mark the first element as 1.
- Assuming n is reasonably finite, iterate over $n - 1$ load currents for every possible combinations of the set of $(n/3 - 1)$ load currents. The elements in the sets are placed position independently, i.e., $\{1, 2, 3, \ldots, (n/3 - 1)\}$ is same as $\{2, 1, \ldots, (n/3 - 1), 3\}$.
- For each possible set, the difference parameter (ε) is calculated:

$$\varepsilon = I_{ideal} - \sum \text{set of } (n/3 - 1) \text{ currents} - \text{first current} \tag{7.18}$$

- Choose the set with the *minimum* value of ε as the optimum balance set.
- We return the set for the I_{ph1}.
- For I_{ph2}, start with the $(n - n/3)$ load currents.
- We mark the first element as 2.

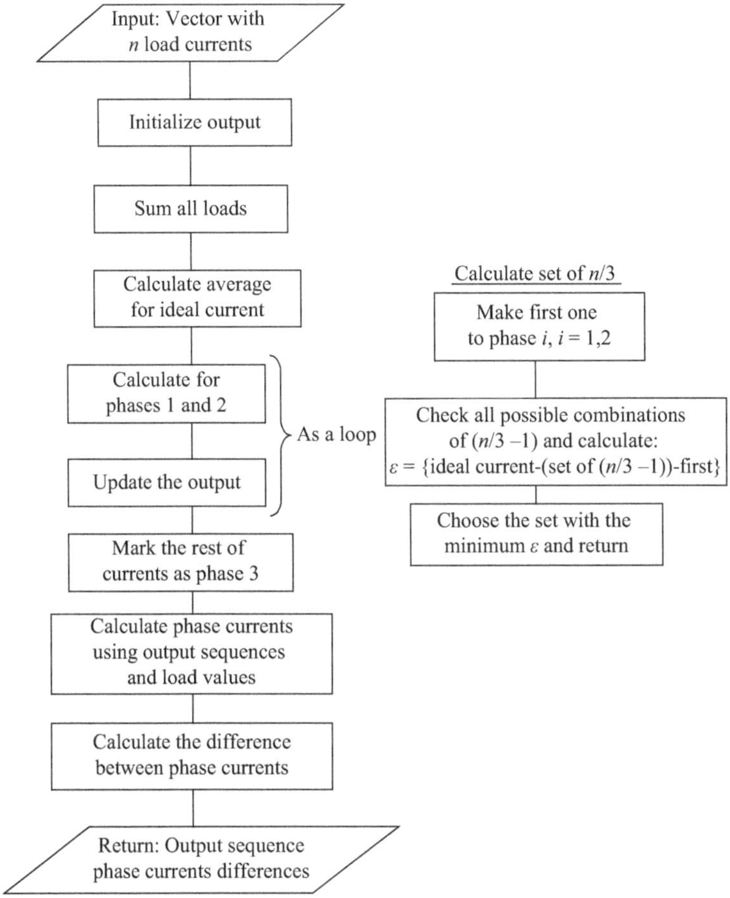

Figure 7.5 Flow chart of the implementation of the heuristic method for load balancing

- Iterate over $(n - n/3 - 1)$ load currents for every possible combinations of the set of $(n/3 - 1)$ load currents. The elements in the sets are placed position independently, i.e., $\{1, 2, 3, \ldots, (n/3 - 1)\}$ is same as $\{2, 1, \ldots, (n/3 - 1), 3\}$.
- For each possible set, the difference parameter (ε) is calculated with (7.18).
- Choose the set with the *minimum* value of ε as the optimum balance set.
- Return the set for the I_{ph2}.

This method is demonstrated through a good example in Reference 10.

7.4.2 Newton–Raphson-based solution method

In this optimization process, the neutral current gives immediate information about unbalancing of the phase currents/powers. Hence, the neutral currents will be used into the Newton–Raphson optimization process. An example of the Newton–Raphson-based method can be found in Reference 12.

Given a distribution system as shown in Figure 7.4, for a feeder j in a network with three phases with a known structure, the problem consists of finding a condition of balancing. In any connecting point, the mathematical model can be expressed as:

$$I_{ph1k} = \sum_{i=1}^{3} sw_{kji} I_k + I_{ph1(k+1)} \tag{7.19}$$

$$I_{ph2k} = \sum_{i=1}^{3} sw_{kji} I_k + I_{ph2(k+1)} \tag{7.20}$$

$$I_{ph3k} = \sum_{i=1}^{3} sw_{kji} I_k + I_{ph3(k+1)} \tag{7.21}$$

where I_{ph1k}, I_{ph2k}, and I_{ph3k} represent the currents (phasors) per phase (1, 2, and 3) after the k – an arbitrary point of connection; $sw_{kj1}, \ldots, sw_{kj3}$ are different switches (the value of "1" means the switch is ON and "0" means it is OFF). $I_{ph1(k+1)}$, $I_{ph2(k+1)}$, and $I_{ph3(k+1)}$ represent the currents (phasors) per phase (1, 2, and 3) at point "$k+1$", which is situated immediately after the point "k" (conventionally, the first point is at the starting of the feeder). Balancing the phase current of any feeder, the neutral current should be minimized and consequently the power losses associated is diminished. Therefore, with the phase current/power unbalance being the primary objective the second objective is as in (7.6), which minimize the difference of the amplitude of the phase currents ($\hat{I}_{phik}$), and the third objective function is to minimize the neutral current, equation (7.7).

The fourth objective functions will be to minimize the negative and zero sequence currents. According to Fortescque [13–14] the three unbalanced phasors of a three-phase system can be resolved into three balanced systems phasors known as positive, negative, and zero sequence components. The relation between the symmetrical components and the phase current is:

$$I_{phs} = A^{-1} I_{phk} \tag{7.22}$$

where A represents the main matrix and the symmetrical component operator is given by $a = 1/60°$. Then (7.22) becomes:

$$\begin{bmatrix} I^0_{Phs} \\ I^+_{phs} \\ I^-_{phs} \end{bmatrix} = \frac{1}{3} \begin{bmatrix} 1 & 1 & 1 \\ 1 & a & a^2 \\ 1 & a^2 & a \end{bmatrix} \begin{bmatrix} I_{ph1k} \\ I_{ph2k} \\ I_{ph3k} \end{bmatrix} \tag{7.23}$$

Writing as three separate equations, where $I^0_{phs}, I^+_{phs}, I^-_{phs}$ are, respectively, the zero sequence current, the positive sequence current, and the negative sequence currents.

$$I^0_{phs} = (I_{ph1k} + I_{ph2k} + I_{ph3k})/3 \tag{7.24}$$

$$I^+_{phs} = (I_{ph1k} + aI_{ph2k} + a^2 I_{ph3k})/3 \tag{7.25}$$

$$I^-_{phs} = (I_{ph1k} + a^2 I_{ph2k} + aI_{ph3k})/3 \tag{7.26}$$

Applying the Lagrange multiplier for the constraint to balance the three-phase system results in:

$$\lambda I_{phs}^{0} = \lambda(I_{ph1k} + I_{ph2k} + I_{ph3k})/3 \tag{7.27}$$

$$\lambda I_{phs}^{+} = \lambda(I_{ph1k} + aI_{ph2k} + a^2 I_{ph3k})/3 \tag{7.28}$$

$$\lambda I_{phs}^{-} = \lambda\left(I_{ph1k} + a^2 I_{ph2k} + aI_{ph3k}\right)/3 \tag{7.29}$$

To have a balanced distribution feeder, the equations presented as constraints (7.27) and (7.29) will be equal to zero. The combination of (7.6), (7.7), (7.27), and (7.29) can be integrated as a multi-objective optimization problem as follows:

$$\text{Minimize: } J_{oj} = J + J_n + \lambda I_{phs}^{0} + \lambda I_{phs}^{-} \tag{7.30}$$

To solve the minimization problem, the gradient of the least square objective function J_{oj} as defined in (7.30) can be expressed in terms of x and set equal to zero:

$$J_x = \left[\frac{\partial J_{oj}}{\partial sw_{k1i}}, \frac{\partial J_{oj}}{\partial sw_{k2i}}, \frac{\partial J_{oj}}{\partial sw_{k3i}}\right] = 0 \tag{7.31}$$

where $x = [sw_{ki}, sw_{ki}, sw_{ki}]$ is the vector of the switching matrix.

Equation (7.31) is a system of non-linear equations. To solve the system of non-linear equations, these equations are linearized around the current solution x_k by using a truncated Taylor series expansion:

$$J_{xx}(x_k)\Delta_{xk} + J_x(x_k) = 0 \tag{7.32}$$

where $J_{xx}(x_k)$ is the 3×3 Hessian matrix, containing the second-order derivatives of the objective function J_{oj} evaluated at point x_k, and $J_x(x_k)$ is the gradient of J evaluated at point x_k. The correction vector Δx_k can then be calculated by solving the following system of linear equations:

$$J_{xx}(x_k)\Delta_{xk} = -J_x(x_k) \tag{7.33}$$

Initially, an arbitrary value is given to the parameter vector x_k and then an iterative procedure is used to obtain a better value of the parameter vector:

$$x_{k+1} = x_k + \Delta_{xk} \tag{7.34}$$

7.4.3 Fuzzy logic-based load balancing

This section describes the application of fuzzy logic (FL) to the solution of this problem. This has been presented in Reference 2. Here, the objective functions are formulated as fuzzy sets since they are imprecise. FL has two different meanings. In a narrow sense, FL is a logical system, which is an extension of multi-valued logic

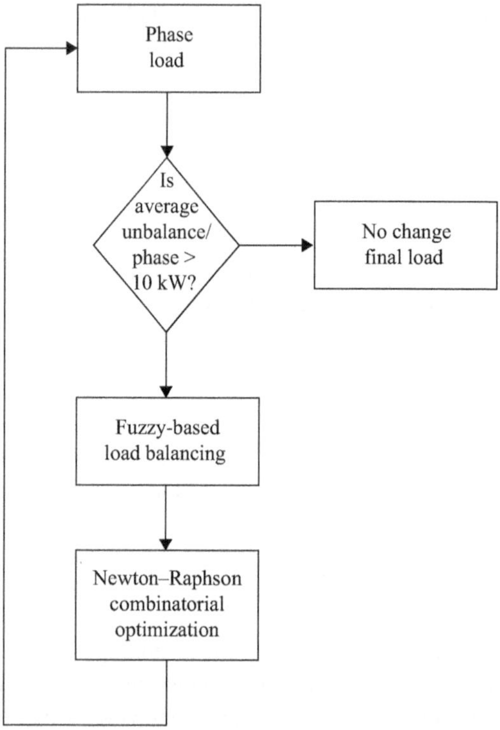

Figure 7.6 Architecture for the combined FL and Newton–Raphson load-balancing system

[15–17]. However, in a wider sense FL is almost synonymous with the theory of fuzzy sets, a theory which relates to classes of objects with imprecise boundaries in which membership is a matter of degree. Even in its more narrow definition, FL differs both in concept and substance from traditional multi-valued logical systems.

A FL-based load-balancing technique can be used along with the Newton–Raphson method to solve the objective function as a combinatorial problem. The architecture of the algorithm is as shown in Figure 7.6. The input is the total phase load (for each of the three phases). The average unbalance (AU) per phase, calculated according to (7.5), is checked against a specified threshold of say 10 kW. If the AU per phase is below 10 kW, it can be assumed that the system is more or less balanced and discard any further load balancing. Otherwise, it goes for the FL-based load balancing. This 10 kW threshold is arbitrarily/conveniently chosen in order to reduce the rate of system reconfiguration.

The output from the fuzzy-based load-balancing step is the load-change values for each phase. A *negative* value indicates that the specific phase has surplus load and should *release* that amount of load, while a *positive* value indicates that the specific phase is less loaded and should receive that amount of load.

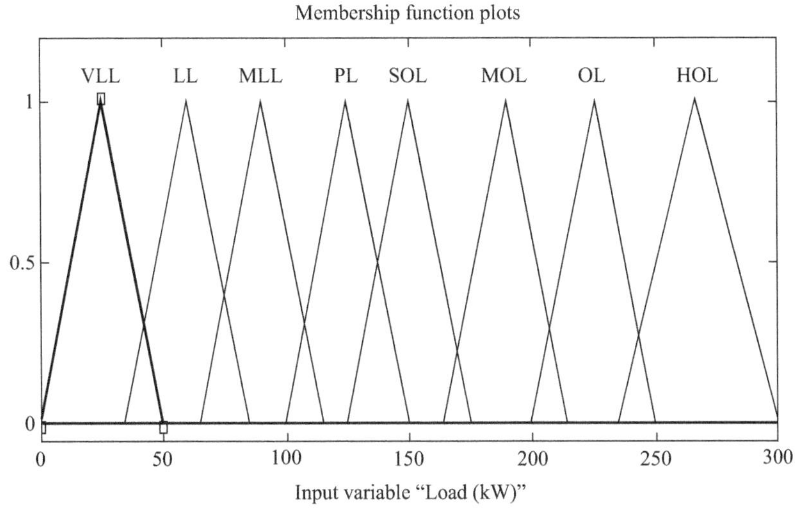

Figure 7.7 Fuzzy membership functions for the input variable

This load-change configuration obtained here is the starting point for a more refining method (in this case Newton–Raphson; in some other cases this could be any of the expert system or evolutionary mathematical-based methods), which tries to optimally shift the specific number of load points. However, sometimes the expert system may not be able to execute the exact amount of load change as directed by the fuzzy step. This is because the actual load points for any phase might not result in an optimum combination which sums up to the exact change value indicated by the fuzzy step. So, the best possible change from the expert system is implemented, and iteratively check the system unbalance until the AU below 10 kW is achieved.

7.4.3.1 Fuzzy controller: input and output

To design the fuzzy controller, first there are specified the input and the output variables. For the load-balancing purpose, which is to obtain a set of rearrangement of the connected loads at each node (or consumer point) such that the objective function is minimized, it is chosen the input as "Load", i.e., the total phase load (kW) for each of the three phases, and the output as "Change", i.e., the change of load (kW, positive or negative) to be made for each phase. For the input variable, Figure 7.7 and Table 7.1 show the fuzzy nomenclature and the respective triangular fuzzy membership functions [15]. And for the output variable, Table 7.2 shows the fuzzy nomenclature and Figure 7.8 shows the corresponding triangular fuzzy membership functions [15, 16].

7.4.3.2 Fuzzy rules and surface

Next, the IF-THEN fuzzy rule set [15, 16] governing the input and the output variables is described in Table 7.3.

Table 7.1 Fuzzy nomenclature for the input variable

S. no.	Input (load) description	Fuzzy nomenclature	kW range
1	Very less loaded	VLL	0–50
2	Less loaded	LL	35–85
3	Medium less loaded	MLL	65–115
4	Perfectly loaded	PL	100–150
5	Slightly overloaded	SOL	125–175
6	Medium overloaded	MOL	165–215
7	Overloaded	OL	200–250
8	Heavily overloaded	HOL	235–300

Table 7.2 Fuzzy nomenclature for the output variable

S. no.	Output (change) description	Fuzzy nomenclature	kW range
1	High subtraction	HS	−150 to −85
2	Subtraction	S	−100 to −50
3	Medium subtraction	MS	−65 to −15
4	Slight subtraction	SS	−50 −25
5	Perfect addition	PA	0–50
6	Medium addition	MA	35–85
7	Large addition	LA	65–115
8	Very large addition	VLA	100–150

Corresponding to the fuzzy input, output variables and the associated rule set, the fuzzy surface [15] is shown in Figure 7.9, depicting the non-linear relationship between the input and the output variables. A detailed description of an application of this method can be found in Reference 2.

7.4.4 Neural network-based method

Neural network is applied to solve the network and feeder reconfiguration problem. This has already been well documented by Salazar et al. [18] as well as by Kim et al. in Reference 19.

The method uses the neural network to control the switch-closing sequence of each load for the minimum power loss which will lead to the optimal phase balance. The inputs to the neural network are the unbalanced load currents and the outputs are the switch-closing sequences for each load.

The input layer of the network has N input neurons, N being the number of unbalanced load currents to be controlled.

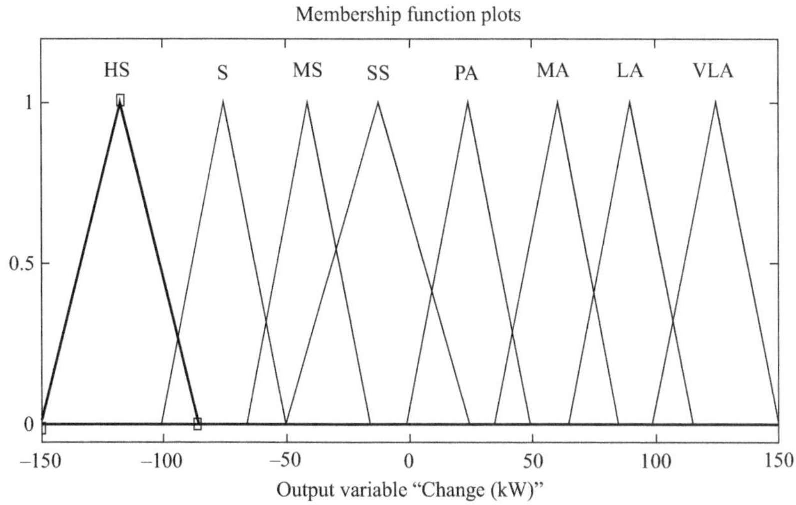

Figure 7.8 Fuzzy membership functions for the output variable

Table 7.3 Fuzzy rules for the input and the output variables

Rule no.	Rule description
1	If load is VLL then change is VLA
2	If load is LL then change is LA
3	If load is MLL then change is MA
4	If load is PL then change is PA
5	If load is SOL then change is SS
6	If load is MOL then change is MS
7	If load is OL then change is S
8	If load is HOL then change is HS

For a particular feeder, the following column vector has been assumed as the input:

$$C = [I_{L1}, \ \ldots, \ I_{LN}]^T \tag{7.35}$$

The output of the network is in the range $\{1, 2, 3\}$ for each load, i.e., which switch (to the specific phase) should be ON for that specific load and moment in time.

7.4.4.1 Neural network structure

The radial basis network [20] has been used for this application. Experimentations with the back propagation and the radial basis network indicated faster training and better convergence for the latter. Radial basis networks may require more neurons than

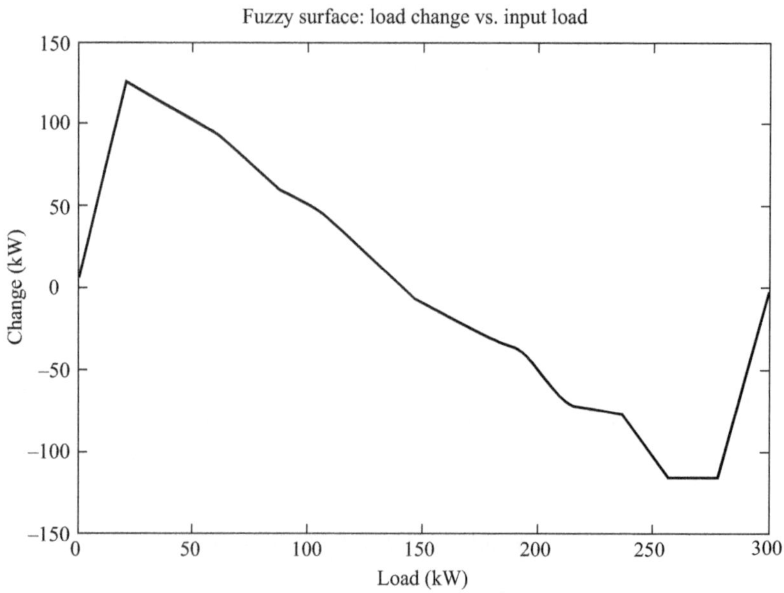

Figure 7.9 Non-linear relationship between the input and the output variables

the standard feed-forward back propagation networks, but often they can be designed in a fraction of the time needed to train the standard feed-forward networks. They work best when many training vectors are available [21]. MATLAB® neural network toolbox [22] has been used for the implementation. As a result of repeated simulations studies with different kinds of radial basis networks, the generalized regression neural network (GRNN) [22] has been found to produce the best result; a GRNN is often used for function approximation. It has a radial basis layer and a special linear layer.

7.4.4.2 Network training

The neural network was trained using a set of 500 real historical data for 15 randomly selected houses in a South African city. The real data set consisted of unbalanced load data that include average load current values per house in a specific locality of the city for the different times of each day in a month.

For illustration purpose, we will assume that there are two loads (L_1 and L_2) connected per phase, and these loads are equally distributed per phase, the problem then is to find the optimum three sets of two loads, with *minimum* differences among the individual sums of the three sets. To achieve this, the ideal phase balance current value I_{ideal} is first calculated, which is equal to the one-third of the sum of the all 15 load currents I_L:

$$I_{ideal} = \frac{1}{3}\sum_{j=1}^{N} I_{L_j} \tag{7.36}$$

In the second step, we optimally select our three sets of currents for the three-phase currents I_{ph}, each set comprising $\left(\frac{N}{3}\right)$ load currents $\{I_j, j = 1, 2, \ldots, \left(\frac{N}{3}\right)\}$.

$$I_{Load} = \{I_{L_j}, j = 1, \ldots, N\} \tag{7.37}$$

$$I_{ph} = \{I_j, j = 1, \ldots, n/3\} \text{ where } I_j \in I_{Load} \tag{7.38}$$

The difference between the individual sum of these sets and the I_{ideal} should be *minimum*, ideally 0 for the perfect phase balance. So, it is needed to find three sets of $\{I_j, j = 1, \ldots, n/3\}$, subject to the constraint:

$$\min\left|\left(\sum_{j=1}^{n/3} I_j\right) - I_{ideal}\right|, \text{ where } I_j \in I_{Load} \tag{7.39}$$

Following this, the output switching sequences are obtained as the target data set for training the networks. The balanced phase currents I_{ph1}, I_{ph2}, and I_{ph3} have been computed using the output switching sequences and the input load currents. For example, I_{ph1} is calculated by adding the two load currents corresponding to the switches marked "1". Then the differences between I_{ph1}, I_{ph2}, and I_{ph3} have been computed, which ideally should be zero. The differences indicate the quality of the phase balance [23].

The above-mentioned neural network is then trained using the real and simulated unbalanced load as the input vector, and the output switching sequences as the target vector. Then, the network is tested with different unbalanced load data set. The output was the optimal switching sequences of {1, 2, 3} for the three phases as explained above. Using the similar procedure as explained above, the balanced phase currents have been computed and the differences between the phase currents (ΔI) and the results indicate the quality of the balance.

$$\Delta I_{max} = \max\{||I_{ph1}| - |I_{ph2}||, \; ||I_{ph2}| - |I_{ph3}||, \; ||I_{ph3}| - |I_{ph1}||\} \tag{7.40}$$

7.4.5 Adding decaying self-feedback continuous Hopfield neural network method

Fei et al. [24] developed an improvement over the traditional continuous Hopfield neural network (CHNN) by adding an extra self-feedback to every neuron of CHNN, thus the energy of CHNN will not always decrease, but can increase or be maintained. The energy increasing may lead to avoiding the local optimal values. This improvement, which then results in adding decaying self-feedback continuous Hopfield neural network (ADSCHNN), becomes useful for solving a class of combinatorial optimization problems (COPs) of which the load balancing problem belongs. Most COPs are nondeterministic polynomial-hard and difficult to solve [25]. The neural networks have become powerful tools to solve the COP [26].

7.5 Application to load-balancing problem

7.5.1 Problem analysis and energy function construction

From the description in Section 7.3, it is expressed that the load-balancing problem is solved in terms of minimum real power loss. In a three-phase, four-wire system, a branch of a feeder (7.9) becomes:

$$\begin{aligned}\sum_{i=1}^{3} r_i \frac{P_i^2 + Q_i^2}{|V_i|^2} &= \sum_{i=1}^{3} r_i \frac{|V_i|^2|I_i|^2\cos^2\varphi + |V_i|^2|I_i|^2\sin^2\varphi}{|V_i|^2} \\ &= \sum_{i=1}^{3} r_i|I_i|^2 = r_1|I_1|^2 + r_2|I_2|^2 + r_3|I_3|^2 \end{aligned} \tag{7.41}$$

In general, each phase has the same internal resistance r which is constant. Therefore, we have:

$$\sum_{i=1}^{3} r_i \frac{P_i^2 + Q_i^2}{|V_i|^2} = r\left(|I_1|^2 + |I_2|^2 + |I_3|^2\right) \tag{7.42}$$

constraining to $|I_1| + |I_2| + |I_3| = C$, where C can be complex or real constant depending on the load. To minimize the total real power losses means:

$$\begin{aligned}&\min\left(|I_1|^2 + |I_2|^2 + |I_3|^2\right) \\ &\text{subject to } \ |I_1| + |I_2| + |I_3| = C\end{aligned} \tag{7.43}$$

We will use the method of Lagrange multipliers to solve (7.43). Create the non-constrained function as:

$$\begin{aligned}L(|I_1|, |I_2|, |I_3|, \lambda) = {} & |I_1|^2 + |I_2|^2 + |I_3|^2 \\ & + \lambda(|I_1| + |I_2| + |I_3| - C)\end{aligned} \tag{7.44}$$

The gradient for this new function is:

$$\begin{aligned}\frac{\partial L}{\partial |I_1|} &= 2|I_1| + \lambda_1 = 0 \\ \frac{\partial L}{\partial |I_2|} &= 2|I_2| + \lambda_1 = 0 \\ \frac{\partial L}{\partial |I_3|} &= 2|I_3| + \lambda_1 = 0 \\ \frac{\partial L}{\partial \lambda_1} &= |I_1| + |I_2| + |I_3| - C = 0\end{aligned} \tag{7.45}$$

From (7.45), we have $|I_1| = |I_2| = |I_3| = \frac{1}{3}C$. Therefore, when $|I_1| = |I_2| = |I_3| = \frac{1}{3}C$, the total real power losses are minimal. If the loads are pure resistive, the minimum power losses are achieved when $P_1 = P_2 = P_3 = \frac{P}{3}$, where P_i $(i = 1, 2, 3)$ is the real power per phase and P is the sum of three-phase real powers. So, we can solve the load-balancing problem by distributing equally the load current or power to three phases, according to the load property.

7.5.2 Energy function construction for the ADSCHNN

From the above analysis, we know that the load-balancing problem means all the loads are distributed to three phases equally, with minimum differences among the individual sums of three phases. For this purpose, either load power or load current can be used. So, there is an ideal phase balance of load $Load_{ideal}$, which is equal to the one-third of the sum of all the loads:

$$Load_{ideal} = \frac{1}{3}\sum_{j=1}^{n} Load(j) \tag{7.46}$$

where n is the number of all the loads. The load balancing is complete, if the sum of every phase loads satisfies:

$$\sum_{i=1}^{m} Load_{phase}(i) = Load_{ideal} \tag{7.47}$$

where m is the number of load points which are connected to one phase. Therefore, in a three-phase four-wire system, we have load balancing when:

$$Load_{phase1} = Load_{phase2} = Load_{phase3} \tag{7.48}$$

In order to solve the load-balancing problem, the solution of load-balancing problem is mapped to the ADSCHNN. So, a transposition matrix with size $(3 \times n)$ is needed to show the configuration of all the loads. The component of the transposition matrix is either 1 or 0. The transposition matrix also indicates the neuron output. At the same time, the energy function for load-balancing problem is constructed considering under restrictions:

(1) The transposition matrix has only one "1" component in the one column.
(2) The sum of the all elements of transposition matrix is n.
(3) The difference among the individual sums of all the loads in three phases is minimum, which is the total line losses are minimum.

(1) Point (1) above means each load is connected to only one feeder. (2) Point (2) above indicates the number of closed switches equals the number of loads connected. (3) Point (3) above is the objective function. From (1), (2), and (3), we construct the

energy function as (7.49), where *Load* is a matrix with size $(1 \times n)$ containing all the loads. A, B, C are the coupling parameters corresponding to the constraints and the objective function.

$$E = \frac{A}{2}\sum_{j=1}^{n}\sum_{i=1}^{3}\sum_{\substack{l \neq i \\ l=1}}^{3} x_{ij}x_{lj} + \frac{B}{2}\left(\sum_{i=1}^{3}\sum_{j=1}^{n} x_{ij} - n\right)^2 + \frac{C}{2}\sum_{i=1}^{3}\left(\sum_{j=1}^{n} x_{ij}Load(j) - Load_{ideal}\right)^2 \tag{7.49}$$

The first two terms in (7.49) correspond to (1) and (2). It is, if (1) and (2) are satisfied at the same time, that the first two terms of (7.49) are equal to zero, otherwise they are not zero. So the two terms are the constrained terms. When the constrained terms are equal to zero, every one load only belongs to one phase. The third term is objective function. $x_{ij} = 1$ denotes that load j is connected to Ph_i, while $x_{ij} = 0$ denotes that load j is not connected to Ph_i. If the difference among the individual sums of all the loads current in three phases is the smallest, the objective function is minimum.

From (7.49), we have:

$$\frac{\partial E}{\partial x_{ij}} = A\sum_{\substack{l \neq i \\ l=1}}^{3} x_{lj} + B\left(\sum_{i=1}^{3}\sum_{j=1}^{n} x_{ij} - n\right) + C\left(\sum_{j=1}^{n} x_{ij}Load(j) - Load_{ideal}\right) \tag{7.50}$$

Putting (7.50) into the following equations:

$$\begin{cases} y_i(k+1) = \alpha y_i(k) + \lambda\left(-\dfrac{\partial E'}{\partial x_i}\right) + z_i(k)x_i(k) \\ x_i(k) = \dfrac{1}{1 + e^{-y_i(k)/\varepsilon}} \\ z_i(k+1) = (1-\beta)z_i(k) \end{cases} \tag{7.51}$$

we get the discrete dynamics of the ADSCHNN for the load-balancing problem as follows:

$$\begin{cases} y_{ij}(k+1) = a y_{ij}(k) + \lambda\left[-A\sum_{\substack{l \neq i \\ l=1}}^{3} x_{lj} - B\left(\sum_{i=1}^{3}\sum_{j=1}^{n} x_{ij} - n\right) - C\left(\sum_{j=1}^{n} x_{ij}Load(j) - Load_{ideal}\right)\right] + z_{ij}(k)\, x_{ij}(k) \\ x_{ij}(k) = \dfrac{1}{1 + e^{-y_{ij}(k)/\varepsilon}} \\ z_{ij}(k+1) = (1-\beta)z_{ij}(k) \end{cases} \tag{7.52}$$

For an example, the ADSCHNN is used to optimize the practical field data used in Reference 10, where the loads were addressed in terms of currents. The parameters for the ADSCHNN are set as:

$$\begin{aligned} &\alpha = 1, \quad \lambda = 0.25, \quad A = 0.1, \quad B = 0.1, \\ &C = 0.35, \quad z_{ij}(0) = 0.13, \quad \beta = 0.001 \end{aligned} \tag{7.53}$$

Initial conditions of the ADSCHNN are $y_{ij} = 1$ $(i = 1, 2, 3;\ j = 1, 2, \ldots, n)$. The results are shown in Table 7.4, where "1" means the respective load is connected to ph_1, "2" to ph_2, "3" to ph_3, ΔI_{ph-max} is the maximum difference of the phase currents, which ideally should be zero if there is totally balanced.

7.5.3 Particle swarm optimization

Particle swarm optimization (PSO) is an intelligent algorithm developed by Kennedy and Eberhart in 1995. The motivation of the PSO algorithm was social and has been applied to optimal flow and power loss minimization [27–29].

The position (which is the current in this particular case) of the jth particle at time t is a D-dimensional will be denoted as follows:

$$I_j(t) = (I_{j,1} I_{j2}, \ldots, I_{j,D}) \in S \tag{7.54}$$

where S is a feature space.

The velocity of this particle at time t is also a D-dimensional vector denoted by:

$$V_j(t) = (V_{j,1},\ V_{j,2}, \ldots, V_{j,D}) \in S \tag{7.55}$$

The best previous position of the jth particle at time t is a point in S, which is denoted by:

$$I_{b,j} = (I_{b,j,1},\ I_{b,j,2}, \ldots, I_{b,j,D}) \in S \tag{7.56}$$

The global best position ever attained among all particles is a point in S which is denoted by:

$$I_{gb} = (I_{gb,1}, I_{gb,2}, \ldots, I_{gb,n}) \in S \tag{7.57}$$

The new velocity and the new position (current in this case) are given, respectively, by (7.55) and (7.57).

$$V_j^{t+1} = wV_j^t + C_1 r_1 (I_{b,i} - I_j^t) + C_2 r_2 (I_{gb} - I_j^t) \tag{7.58}$$

$$I_j^{t+1} = I_j^t + I_j^{t+1} \tag{7.59}$$

where w is the inertia weight; C_1 and C_2 are acceleration coefficients; r_1 and r_2 are two separately generated uniformly distributed random numbers in the range [0, 1] added in the model to introduce stochastic nature.

The solution objective for a feeder is to obtain a set of rearrangement of the connected loads at each node (or consumer point) such that the objective function is minimized. This is a non-linear problem that will involve a number of trial and errors.

Table 7.4 Results for neural network, heuristic, and ADSCHN methods

Current	**Unbalanced**		**Balanced**		
		Switch	**Neural network**	**Heuristic**	**ADSCHNN**
I_1 (A)	40.16	1	1	1	1
I_2 (A)	92.61	2	2	1	2
I_3 (A)	90.77	3	3	2	3
I_4 (A)	40.61	1	1	3	1
I_5 (A)	88.47	2	3	3	1
I_6 (A)	5.73	3	1	1	3
I_7 (A)	34.93	1	3	3	2
I_8 (A)	80.50	2	1	2	3
I_9 (A)	0.97	3	2	2	2
I_{10} (A)	13.75	1	1	1	2
I_{11} (A)	20.07	2	3	2	2
I_{12} (A)	19.67	3	2	2	1
I_{13} (A)	59.77	1	3	1	2
I_{14} (A)	26.94	2	2	3	1
I_{15} (A)	19.68	3	2	3	3
I_{16} (A)	1.51	1	1	1	3
I_{17} (A)	73.93	2	2	2	3
I_{18} (A)	44.06	3	3	3	1
I_{19} (A)	92.24	1	1	2	2
I_{20} (A)	46.13	2	1	1	1
I_{21} (A)	41.44	3	2	2	3
I_{22} (A)	83.77	1	3	3	1
I_{23} (A)	51.99	2	1	1	3
I_{24} (A)	20.06	3	3	2	2
I_{25} (A)	66.54	1	2	3	2
I_{26} (A)	82.97	2	2	1	3
I_{27} (A)	1.94	3	3	3	1
I_{28} (A)	67.44	1	1	3	1
I_{29} (A)	37.56	2	1	2	2
I_{30} (A)	82.34	3	1	1	2
I_{31} (A)	94.06	1	1	1	3
I_{32} (A)	22.88	2	2	2	1
I_{33} (A)	60.07	3	1	1	1
I_{34} (A)	48.11	1	3	3	2
I_{35} (A)	88.23	2	1	3	1
I_{36} (A)	75.44	3	1	3	3
I_{37} (A)	45.19	1	2	2	2
I_{38} (A)	1.83	2	3	3	2
I_{39} (A)	81.31	3	2	3	2
I_{40} (A)	60.92	1	1	1	3
I_{41} (A)	78.40	2	3	2	1
I_{42} (A)	91.25	3	2	1	3
I_{43} (A)	73.08	1	2	2	2
I_{44} (A)	17.45	2	2	2	1
I_{45} (A)	44.02	3	1	1	1
I_{ph1} (A)	822.1	–	746.1	761.4	770.24
I_{ph2} (A)	809.9	–	778.5	786.1	770.36
I_{ph3} (A)	678.8	–	788.1	771.4	770.19
$I_{ph\text{-}max}$ (A)	143.3	–	42	24.7	0.17

A method is used to sense the relative loading of the phases, and another method is used to edge towards the minimized objective

$$J_n = \left(\sum_{l_1=1}^{N} I_{ph1} - \sum_{l_2}^{N} I_{ph2} \right)^2 + \left(\sum_{l_1}^{N} I_{ph1} - \sum_{l_3}^{N} I_{ph3} \right)^2 + \left(\sum_{l_2}^{N} I_{Ph2} - \sum_{l_3}^{N} I_{ph3} \right)^2 \quad (7.60)$$

The conditions taken into consideration are:

(1) the system loss must be minimized;
(2) the voltage magnitude of each node must be within permissible limit $|V_{\min}| \leq |V_i| \leq |V_{\max}|$;
(3) current capacity of each branch $|I_i| \leq I_j$, max.

By changing the switches on distribution feeders, some load currents can be transferred from heavily loaded feeders or transformers to relatively less heavily loaded feeders or transformers. In this way, the loads on transformers and feeders will become more balanced and the risk of overloads can be reduced, which can be represented by

$$|I_{ti}| \leq I_{t,i}^{\max}, \quad i = 1, 2, \ldots, N_t$$

where $|I_{ti}|$ and $I_{t,i}^{\max}$ are the current amplitude and maximum current of the ith transformer, respectively. N_t is the number of feeders connected to the given transformer. For algorithms and further understanding, refer References 30, 31.

7.6 Practical feasibility and economic considerations

The technologies for the implementation of the scheme in Section 7.1 are already in existence. They comprise, basically, the following, which nowadays are common feature in telecontrol and smart grid systems:

- metering sensors
- wireless data communication, such as Wi-Fi
- programmable single-chip microcomputer or microcontroller
- actuators
- static power electronics switches with soft switching.

These technologies are well proven for applicability within the described scheme proposed here; even though they are continuously being further developed for ruggedness, robustness, and reliability, which are considered the major challenges of the practical implementation of the idea.

Several literatures [2–5,10,11] have elucidated the technical and economic consequences of phase unbalance in the low-voltage side of the distribution networks:

- real power loss
- voltage drop
- communication interference

- equipment overloading and malfunctioning
- malfunction of the protective relay
- overstressing cables and transformers
- poor service quality and operation.

All these are undesirable as they lower the overall energy and service efficiency, and have serious economic and financial consequences to the utility.

Currently, most utilities minimize this problem of unbalance in the low-voltage feeders by changing the connection phase of some feeders manually after some field measurement and analysis. This manual correction normally will require the use of additional manpower, service interruption, and ensuring additional safety measures. Although in some cases this process can improve the phase current balance, this strategy is more time consuming, requires supply interruption, unsafe and only last for a while before the process is repeated again [32].

The schemes described in this chapter eliminate manual correction of phase unbalance, thus eliminating their limitations, including additional cost overheads and technical shortcomings. Furthermore, the technical problems listed above and their economic and financial consequences are addressed as well. Since these schemes do not entail any running costs, the capital cost of putting these schemes in place is easily paid off within short period of time by the savings in the correction of the technical consequences and overhead costs involved with the manual solution.

7.7 Conclusions

Automatic reconfiguration of a distribution low-voltage feeder by rearranging the load distribution such that the phase imbalance is always zero, or at the barest minimum achievable, but not more than the level allowed by regulation has been discussed in this chapter. The system is easily expandable to incorporate other distribution networks and feeder reconfigurations, including those through feeder bifurcation and the tie and sectionalizing switches controls.

The formulation of the optimization problem has been presented and possible solution methods have also been discussed. A brief insight into the practical feasibility and economic considerations of the system concludes the presentation.

References

[1] D.C. Walters and G.B. Schele, "Generic algorithm solution of economic dispatch with valve point loading," *IEEE Transactions on Power Systems*, Vol. 8, No. 3, pp. 1325–1331, 1991.

[2] M.W. Siti, A.A. Jimoh, and D.V. Nicolae, "Distribution network phase load balancing as a combinatorial optimization problem using fuzzy logic and Newton–Raphson," *Electric Power System Research*, Vol. 81, pp. 1079–1087, 2011, doi: 10.1016/jepsr.2010.12.06.

[3] M.E. Baran and W. Kelly, "State estimation for real time monitoring of distribution systems," *IEEE Transactions on Power Systems*, Vol. 9, No. 3, pp. 1601–1609, 1994.

[4] S. Cinvalar, J.J. Grainger, H. Yin, and S.S.H. Lee, "Distribution feeder reconfiguration for loss reduction," *IEEE Transactions on Power Delivery*, Vol. 3, No. 3, pp. 1217–1223, 1988.

[5] T. Chen and J. Cherng, "Optimal phase arrangement of distribution connected to a primary feeder for system unbalance improvement and loss reduction using genetic algorithms," *IEEE Transactions on Power Systems,* Vol. 15, No. 3, pp. 994–999, 2000.

[6] Y.-T. Hsiao, "Multiobjective evolution programming method for feeder reconfiguration," *IEEE Transactions on Power Systems*, Vol. 19, No. 1, pp. 594–599, February 2004.

[7] T. Niknam, E.A. Farsani and M. Nayeripour, "An efficient multi-objective modified shuffled frog leaping algorithm for distribution feeder reconfiguration problem," *European Transactions on Electrical Power*, Vol. 21, No. 1, pp. 721–739, January 2011.

[8] T. Niknam, "An efficient hybrid evolutionary algorithm based on PSO and HBMO algorithms for multi-objective distribution feeder reconfiguration," *Energy Conversion and Management,* Vol. 50, pp. 2074–2082, 2009.

[9] http://en.wikipedia.org/wiki/Heuristic

[10] M.W. Siti, D.V. Nicolae, A.A. Jimoh, and A. Ukil, "Reconfiguration and load balancing in the LV and MV distribution networks for optimal performance," Paper TPWRD-00418-2006, *IEEE Transactions on Power Delivery*, Vol. 22, No. 4, pp. 2534–2540, October 2007.

[11] M. Siti, A. Jimoh and D. Nicolae, "Automatic load balancing in distribution feeder through reconfiguration," *IECON 05*, Raleigh, North Carolina, USA, 2005.

[12] M.W. Siti, D.V. Nicolae, J.A. Jordaan, and A.A. Jimoh: "Distribution feeder phase balancing using Newton–Raphson algorithm-based controlled active filter", Neural Information Processing, in *Lecture Notes in Computer Science*, ISBN: 978-3-540-69159-4, Springer, Berlin, ISSN: 0302-9743 (Print) 1611-3349 (Online), Vol. 4985/2008, pp. 713–720, June 2008.

[13] C.L. Fortescue, "Method of Symmetrical Co-Ordinates Applied to the Solution of Polyphase Networks", *Transactions of the American Institute of Electrical Engineers (Trans. AIEE)*, pt. II, Vol. 37, no. 2, pp. 1027–1140, 1918.

[14] G.C. Paap, "Symmetrical components in the time domain and their application to power network calculations", *IEEE Transactions on Power Systems*, Vol. 15, no. 2, pp 522–528, 2000.

[15] B. Kosko, *Fuzzy Engineering*, Prentice-Hall, Englewood Cliffs, NJ, 1999.

[16] MATLAB® Documentation – Fuzzy Logic Toolbox, Version 6.5.0.180913a Release 13, The Mathworks Inc., Natick, MA, 1999.

[17] M. Sugeno, *Industrial Applications of Fuzzy Control,* Elsevier Science, Amsterdam, 1985.

[18] H. Salazar, R. Gallego and R. Romero, "Artificial neural networks and clustering techniques applied in the reconfiguration of distribution systems," *IEEE Transactions on Power Delivery*, Vol. 21, No. 3, pp. 1735–1742, July 2006.

[19] H. Kim, Y. Ko and K.-H. Jung, "Artificial neural-network based feeder reconfiguration for loss reduction in distribution systems," *IEEE Transaction on Power Delivery*, Vol. 8, No. 3, pp. 1356–1366, July 1993.

[20] C.C. Liu, S. J. Lee, and K. Vu, "Loss minimization of distribution feeders: optimality and algorithms," *IEEE Transactions on Power Delivery*, Vol. 4, No. 2, April 1989.

[21] C.B. Alexandre, A. Carlos and Newton, "Main chain representation for evolutionary algorithms applied to distribution system reconfiguration," *IEEE Transactions on Power Systems*, Vol. 20, No. 1, pp. 425–436, February 2005.

[22] A. Ukil, W. Siti, and J. Jordaan, "Feeder load balancing using neural network," *Lecture Notes in Computer Science*, Vol. 3972, pp. 1311–1316, 2006.

[23] S. Civanlar and J.J. Grainger. "Distribution feeder reconfiguration for loss reduction," *IEEE Transactions on PWRD-3*, Vol. 3, issue 3, 1998, pp. 1217–1223.

[24] C. Fei, G Qi, and A. Jimoh, "Adding decaying self-feedback continuous Hopfield neural network and its application to TSP," *International Review on Modelling and Simulations*, Vol. 3, No. 3, pp. 272–282, June 2010.

[25] T. Kwok and K.A. Smith, "Experimental analysis of chaotic neural network models for combinatorial optimization under a unifying framework," *Neural Networks*, Vol. 13, pp. 731–744, 2000.

[26] R.L. Wang, S.S. Guo, S. Fukuta, P. Zhang, and K. Okazak, "An improved internal dynamics in binary neural network for solving optimization problems," *International Conference on Neural Networks and Brain, 2005. ICNN B '05*, Beijing, China, 2005, pp. 1364–1366.

[27] J. Kennedy, "Small worlds and mega-minds: effects of neighborhood topology on particle swarm performance," in *Proceeding of the 1999 Congress on Evolutionary Computation, 1999. CEC '99*, Washington, DC, Jul. 1999, vol. 3, pp. 1931–1938.

[28] J. Kennedy and R. Mendes, "Neighborhood topologies in fully-informed and best-of-neighborhood particle swarms," in *Proceedings of the 2003 IEEE International Workshop on Soft Computing in Industrial Applications, 2003. SMCia/03*, Washington, DC, Jun. 2003, pp. 45–50.

[29] H. Fan and Y. Shi, "Study on V_{max} of particle swarm optimization," in *Proceedings of Workshop on Particle Swarm Optimization*, Purdue School of Engineering and Technology, Indianapolis, IN, April 2001.

[30] M.R. Alrashidi and M. El-Hawary, "A survey of particle swarm optimization applications in power system operations," *Electric Power Components and Systems*, Vol. 34, No. 12, pp. 1349–1357, December 2006.

[31] T.M. Khalil and A.V. Gorpinich, "Reconfiguration for loss reduction of distribution systems using selective particle swarm optimization," *International Journal of Multidisciplinary Sciences and Engineering*, Vol. 3, No. 6, pp. 16–21, June 2012.

[32] O.M. Popoola, A.A. Jimoh, and D.V. Nicolae, "On-line remote and automatic switching of consumers' connection for optimal performance of a distribution feeder," in *The Institute of Electrical Electronic Engineering Conference (IEEE AFRICON 2007)*, Windhoek, Namibia, 26–28, pp. 1–6, September 2007.

Chapter 8

Volt/VAR control in distribution systems

Ilya Roytelman[1] *and Jose Medina Palomo*[1]

8.1 Voltage standards

The main objective of the electrical distribution system is to provide reliable high-quality power supply. Voltage is the key parameter defining the quality of electrical service. The majority of customers (small and medium size residential, commercial, and agricultural) do not have means to regulate voltages. The customer voltage is regulated by distribution utilities. The customer voltage changes during the day, staying within the limits defined by standards. An example of the typical 24-hour voltage at the customer service delivery point is shown in Figure 8.1.

Voltage standards are different from country to country. Voltage limits are defined for normal conditions and non-normal conditions (emergency, after outage). Under normal condition, the voltage is allowed to deviate ±5% from nominal voltage. Under non-normal conditions, the minimum voltage limit decreases by an additional 3–5% and maximum voltage limit stays at the same or is increased by 1%. Non-normal voltages are allowed only for a period of time specified in the standards.

The American National Standard Institute (ANSI) standard C84.1 1995 [1] defines two customer voltages: Service Voltage (supplier connection to consumer) and Utilization Voltage (equipment connection point–customer outlet). The standard defines normal conditions as Range A and non-normal conditions as Range B. Voltages in Range B are below and above range A. The duration of Range B limits is not defined directly, but corrective measures should be undertaken within a reasonable time after they occur.

All ANSI values are defined not in percentage, but in per unit values on a base voltage of 120 V. For both Range A and Range B the nominal utilization voltage is 115 V. Range A voltage limits:

- Maximum utilization and service voltage: 126 V
- Minimum service voltage: 114 V
- Minimum utilization voltage: 110 V

[1] Open Access Technology International, Inc., Minneapolis, MN, USA

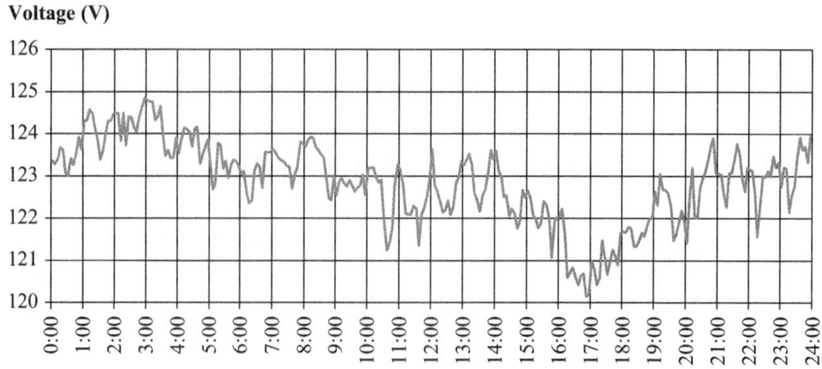

Figure 8.1 Typical customer voltage during 24 hours

Range B voltage limits:

a) Maximum utilization and service voltage: 127 V
b) Minimum service voltage: 110 V
c) Minimum utilization voltage: 107 V

Comparing voltage magnitudes for normal and non-normal conditions, it can be observed that high non-normal voltage limit is only slightly above the high normal limit (1 V), but non-normal minimum voltage limit is 4 V below the low normal limit. The standard reflects the fact that voltage above limit may damage utilization equipment or trigger over voltage protection equipments. Low voltage, up to a certain limit, affects only the performance of the equipment.

In addition to the voltage magnitude range, the ANSI standard provides recommendations on voltage unbalance at three phase buses measured at the customer service point under no-load conditions (limited to 3%). Voltage unbalance is defined as:

$$\text{Voltage unbalance} = (\text{maximum deviation from average voltage})/(\text{average voltage}) \times 100\% \tag{8.1}$$

The standard on voltage unbalance exists only for American distribution systems which are structurally asymmetrical with unbalance loading both at the primary and secondary distribution sides. European distribution systems are symmetrical and balanced at the medium voltage level, but asymmetrical and unbalanced at the low voltage distribution level. The fact that the Voltage Standards for American and European systems are different also indicates that these systems have different means (equipment) for voltage control.

8.2 Volt/VAR control methods

Standards require that voltages at the customer buses (service delivery points and utilization points) are maintained inside the required limits. This is accomplished

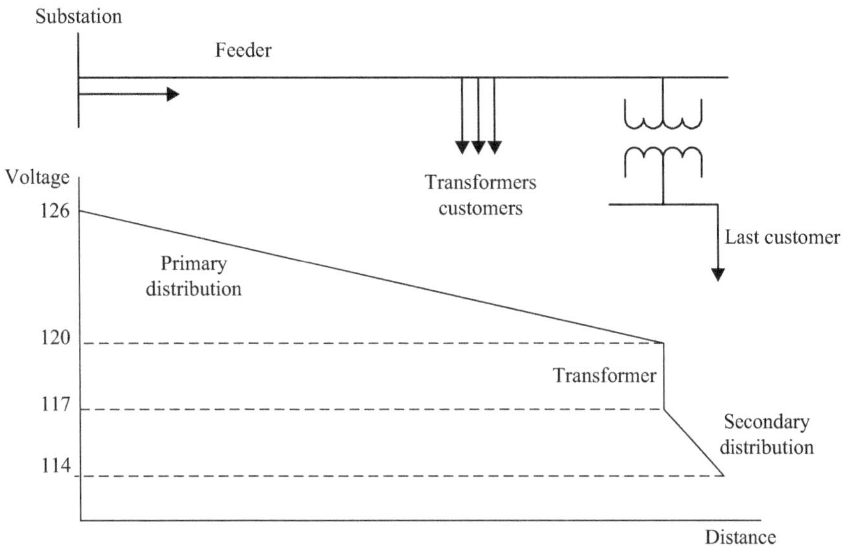

Figure 8.2 Voltage profile along the feeder

first of all through the proper design of the distribution systems. Wires, transformers, and other equipment are selected based on the maximum allowed voltage drops. For an American system, the peak loading voltage drop in the secondary distribution is limited to 2 V, service voltage drop to 1 V, and distribution transformer voltage drop to 3 V. Therefore, under ANSI range A for a minimum service voltage of 114 V, the voltage at the primary of the distribution transformer cannot be lower than 120 V. The highest feeder voltage in range A is limited to 126 V for two reasons. The first one is that a customer could be connected directly to this point. And the second one is that the primary voltage of a distribution transformer is limited by the excitation of the transformer, which limits the voltage at the primary to a value of nominal voltage plus 5%. Therefore, the voltage along the primary distribution feeder under normal conditions (Range A) should be between 126 and 120 V (Figure 8.2).

In traditional distribution systems, there are two types of equipment that affect voltages: transformers (autotransformers), and capacitors. For an ideal two-winding transformer (zero resistance, no magnetic flux leakage), the ratio of the primary (P) to the secondary (S) voltage is equal to the ratio of the number of turns in the primary (W_p) winding to the number of turns in secondary winding (W_s). Distribution system consumers are typically connected to the secondary winding of a distribution transformer and current flows from the primary side to the secondary side. Therefore, secondary voltage (V_s) may be presented as below:

$$V_s = V_P \frac{W_s}{W_P} \tag{8.2}$$

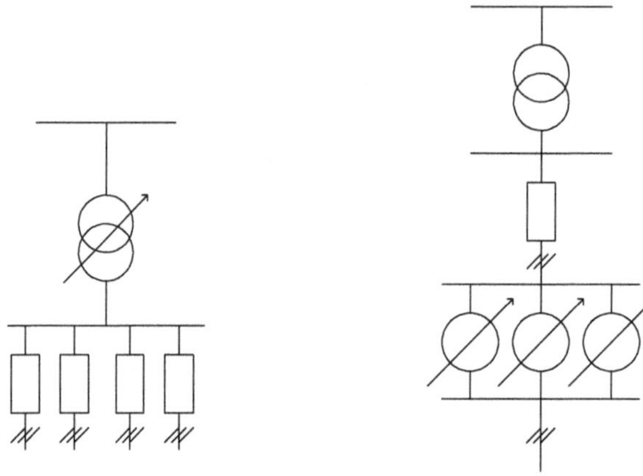

(a) (b)

Figure 8.3 Substation transformer with LTC (a) and feeder step voltage regulator (b)

As it follows from (8.2), for a given primary voltage (V_P), it is possible to get desired secondary voltage V_s by changing the ratio of number of turns. For example, in order to increase the secondary voltage, the number of turns in the secondary winding should be increased or the number of turns in the primary winding should be decreased.

Almost all power system transformers and autotransformers have changeable turns ratio implemented through tap changing mechanism. There are two types of tap changing mechanisms:

- Load Tap Changer (LTC): This allows switching under transformer normal working conditions;
- De-energized (No-Load) Tap Changer: In this case, switching can be done only for de-energized equipment (the transformer should be fully disconnected from the grid).

LTC mechanism is much more complicated and expensive than No-LTC. In the distribution systems, LTCs are installed on substation transformers and on step voltage regulators (SVRs), which are autotransformers (Figure 8.3). Substation transformer LTC regulates the voltage at the feeder head busbar which can have several feeders connected. The substation transformer is core type three phase with gang-operated LTC (all three phases have the same tap positions).

SVR [2,3] typically regulates the voltage on a single feeder or part of the feeder. The SVR may be single phase or three phases. Single-phase SVR regulates the voltage per phase. Single-phase SVRs are often installed at the feeder heads and are connected externally to form a three-phase regulator. Typical connections of single-phase SVRs are three regulators in grounded wye or closed delta, or two regulators in an open wye or open delta.

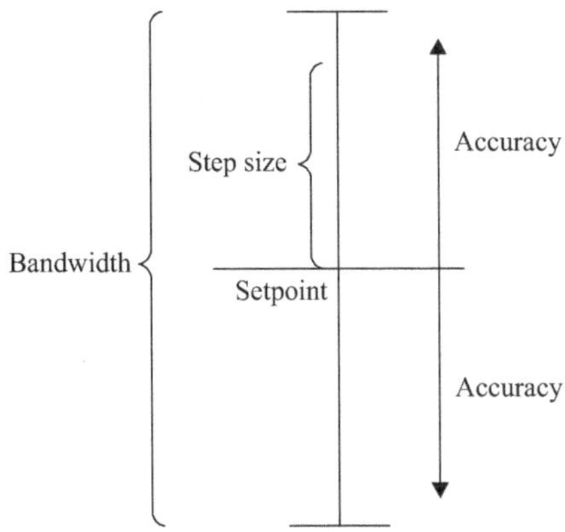

Figure 8.4 Transformer bandwidth, step size, and accuracy

Three-phase SVR is gang operated so that the taps on all windings are changed simultaneously. They can be connected in a three-phase wye or closed delta.

LTCs on transformer or SVR have the typically same regulation range and number of steps: ±10% with 16 steps up and 16 steps down. It makes step size equal to 0.625% or 0.75 V on 120-V base.

The main advantage of LTC is flexibility of voltage regulation: taps can be changed at any moment when adjusting voltage is needed in response to variable loading. LTC mechanisms are equipped with local automatic controllers which continuously monitor (measure) electrical parameters (voltage, current, and phase angle) at the LTC location and correspondingly adjust the tap positions.

Voltage can only be regulated downstream (as judged by the current (I)) the transformer/autotransformer. Because of this, the local controller has a functionality which automatically determines the direction of the flow (typically using active power direction, P_{flow}), i.e., forward or reverse. In case of reverse flow the LTC is automatically blocked or switched to regulate voltage on the opposite bus.

The simplest local controller can be set to maintain a constant voltage, which is independent of the loading condition and voltage at the primary side:

$$|V_{meas} - V_{setting}| < \delta \tag{8.3}$$

where δ is the accuracy of the voltage regulator (half a bandwidth is the allowed deviation from the set point in either direction).

The accuracy cannot be bigger than the tap changer step size (0.75 V). Modern voltage regulators operate with a bandwidth defined by the user. Typical bandwidth is 2–3 V, which makes the accuracy equal to 1–1.5 V (Figure 8.4).

Figure 8.5 shows the daily voltages per phase on the SVRs regulated three-phase bus where the controller voltage setting for each phase is 125 V and bandwidth is 2 V.

As it can be seen from Figure 8.5, phase voltages change between 126 V (setting plus half bandwidth, 125 V + 1 V) and 124 V (setting minus half bandwidth, 125 V − 1 V). They are close to the voltage setting but not necessarily close to each other.

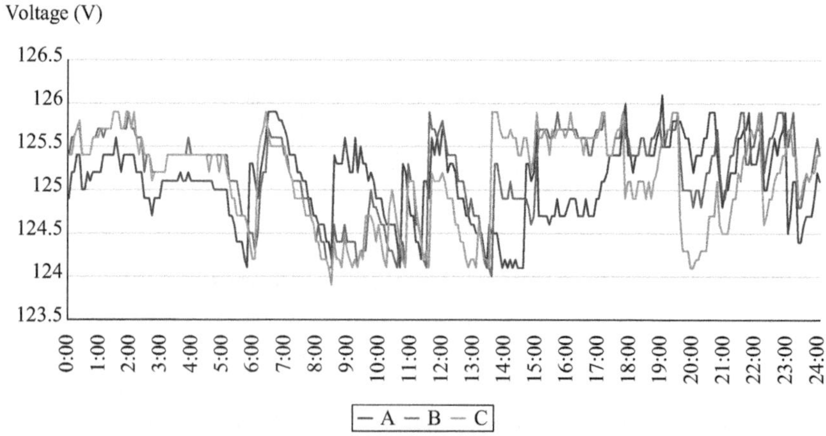

Figure 8.5 Example of bus voltages over a 24-hour period, regulated per phase: controller setting is 125 V and bandwidth 2 V

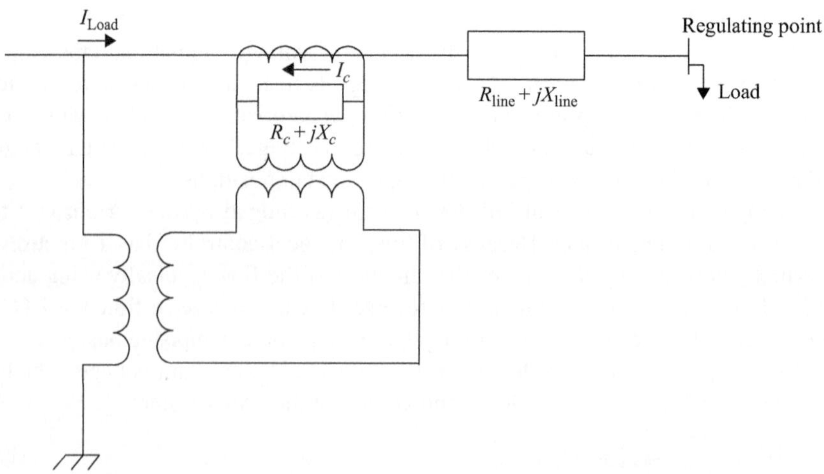

Figure 8.6 Step voltage regulator line drop compensator circuit

A general approach to voltage regulation is to keep LTC transformer/SVR output terminal voltage (target bus) higher when loading is higher and lower when loading is lower. The simplest local controller that achieves this goal sets the voltages depending on the time of day. Modern controllers keep voltage constant not at the output terminal but at the target point located downstream somewhere on the feeder (typically load center). It is accomplished through the line drop compensator which is part of the local controller (Figure 8.6).

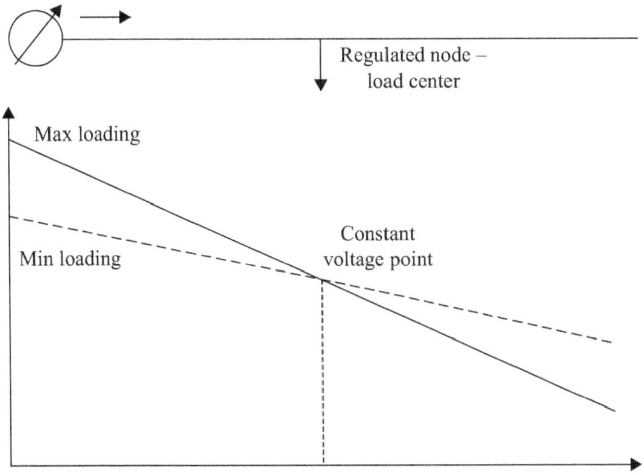

Figure 8.7 Feeder voltage profiles for maximum/minimum loading under regulator with line drop compensation—target bus voltage is constant

The line drop compensator simulates the voltage drop of the distribution line from the voltage regulator till the point where it is desired to maintain the voltage constant (Figure 8.7). In order to accomplish this, the controller changes the output voltage taking into account this voltage drop:

$$V_{\text{set}} = V_{\text{base}} + (R_c + j \times X_c) \times (I_{C_{\text{act}}} + j \times I_{C_{\text{react}}}) \tag{8.4}$$

where, V_{set} is the voltage setting; V_{base} is the no load voltage setting; R_c and X_c are the compensator resistance and reactance; and $I_{C_{\text{act}}}$ and $I_{C_{\text{react}}}$ are the active and reactive component of the current through the compensator.

The voltage increase due to compensator circuit is equal (in per unit) to the voltage drop from the regulator output terminal through the distribution line to the target point where load is connected:

$$(R_c + j \times X_c) \times (I_{\text{act}} + j \times I_{\text{react}}) = (R_{\text{line}} + j \times X_{\text{line}}) \times (I_{\text{act}} + j \times I_{\text{react}}) \tag{8.5}$$

where R_{line} and X_{line} are the resistance and reactance of the line; I_{act} and I_{react} are the active and reactive component of the load current.

If the compensator circuit impedance is zero, the target point is the output at the regulator terminal. It should be pointed out that the value of the compensator impedance in per unit is equal to the actual impedance of the line between the regulator and the target point only if there are no loads/laterals in between. The presence of loads/laterals makes the accurate calculation of compensator impedance both laborious and complicated. It requires knowledge of the feeder topology, the electrical parameters, the typical peak, average and minimal loading, and it leads to multiple power flow calculations. To simplify this calculation process, manufacture companies

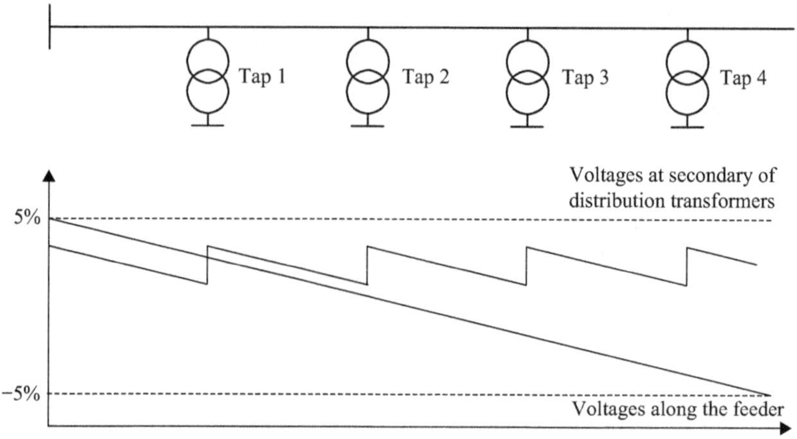

Figure 8.8 Voltage profiles along the feeder and at the distribution transformer secondary

usually provide a simplified guide on how to determine an approximate value of the compensator impedance. This compensation impedance in the local controller is not set in ohms, but in volts.

Regulator voltage settings are subject to high and low limits (typically 126 and 120 V). These limits are enforced by local controllers. The local controllers prevent LTCs from making voltages too high or too low.

Distribution transformers (240/120 V for American systems and 380/220 V for European systems) are typically equipped with No-LTCs (4 steps, 2.5% each or 2 steps, 5% each). No-LTCs are used to compensate the static part of the primary feeder voltage drop and the drop through the distribution transformer itself. The diagram shown in Figure 8.8 illustrates how tap positions are set to steps 1–4 depending on the voltage drop. It makes transformer secondary voltages practically the same (changing in 2.5% steps) independently from electrical distance to supply bus.

No LTCs are traditionally used in European distribution systems with big size distribution transformers (250–1000 kVA), where the primary grid topology does not change frequently. The tap positions are calculated offline as part of seasonal planning based on the primary grid topology and on the substation bus voltage regulation rules. No-LTCs are used to compensate voltage drops with the condition that under minimum loading the transformer secondary voltage should not violate the high voltage limit.

In American systems, distribution transformers are typically smaller (10–150 KVA) and do not always have No-Load Tap Changers and even if they have, they are mainly kept in the neutral position. Only a few transformers per feeder may have a different tap position usually to respond to customer complains. In general, there is a lack of flexibility in using No-Load Tap Changers (taps cannot be adjusted to changing load conditions easily) which significantly limits their application. Feeders where service restoration can be accomplished through backfeed reconfiguration

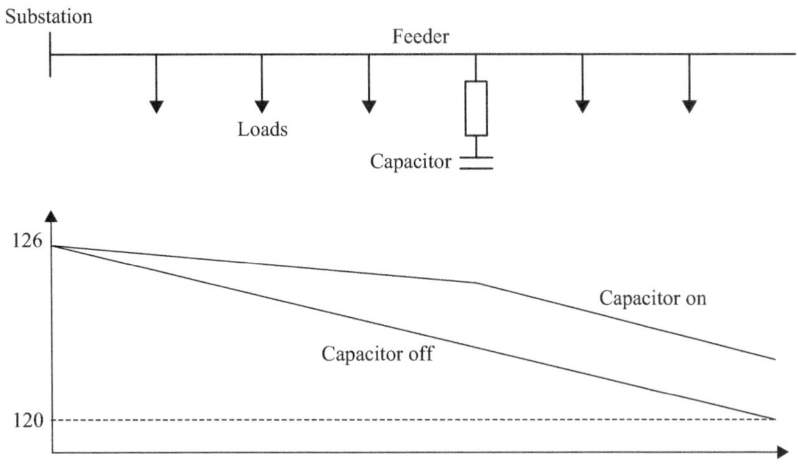

Figure 8.9 Voltage profiles along the feeder with capacitor ON and OFF

(reconnecting the end of the feeder to a different feeder) practically always keep No-Load Tap Changers in neutral positions.

Another type of equipment widely used for voltage control in the distribution grid is shunt capacitors. Under stable voltage, these capacitors generate a predictable amount of reactive power which is consumed by medium- and small-size customers who do not have their own means to generate reactive power. Bigger commercial and industrial customers are required to keep power factor close to unity using their own compensation (reactive power generating) devices, so utility shunt capacitors are not needed. Shunt capacitors are used in overhead lines where reactive power losses can reach 20% of consumed reactive power, not in underground cables which generate reactive power.

Shunt capacitors do not directly change voltage but can decrease voltage drop by compensating reactive power and reducing the total current from the substation to the capacitor location. A reduced current leads to a smaller voltage drop and therefore increases the voltage along the feeder (Figure 8.9). Smaller shunt capacitors are usually connected to the feeder at all times and may either be three-phase or single-phase capacitors. Bigger capacitors are switchable and typically are three-phase capacitors. They are connected in either wye or delta.

Switchable shunt capacitors are equipped with local controllers which are different from LTC controllers. They do not consider accuracy: control actions are limited to switching on/off the capacitors based on a minimum/maximum value of a given parameter. For example, shunt capacitors operated on a time trigger are switched ON at a certain time and switched OFF at a different time later in the day. Shunt capacitors triggered on voltage are switched ON when the voltage is below a minimum value and switched OFF if the voltage is above a maximum value. Furthermore, the current status of these capacitors depends on their recent history. For instance, if they reach a certain voltage moving from a lower value, they will be kept ON, whereas at the same voltage, they will be kept OFF if they reach the value moving from a higher value. As an example, let us assume that the minimum voltage setting is 112 V and the maximum voltage setting is 126 V. During the normal daily operations the voltage drops to 112 V and the capacitor is switched ON. The voltage now rises to 126 V later in the day. At this point the capacitor is switched OFF. Looking now at a voltage between 126 V and 112 V, e.g., 120 V, when this value of 120 V is reached from 112 V, the capacitor is switched ON. However, if the same value of 120 V is reached from 126 V, the capacitor will be switched OFF.

Switched capacitors in the same operational area are typically de-energized during the minimum consumption time period (from midnight till early morning) as they are not needed. This daily de-energization also serves a dual purpose: it resets all capacitors to the same status simplifying the control strategies.

Modern capacitor controllers support the following triggering signals for their operation: time, temperature, voltage, reactive power, and current. Typically a few triggering signals are used simultaneously with a priority order. For example, temperature and voltage may be used together, where voltage has the override capability.

8.3 Volt/VAR regulation by local controllers

Local automatic controllers are an integral part of modern distribution systems ensuring voltage quality under changing power demand and network topology. They control the tap positions of the LTC transformers and SVRs and the statuses of switched shunt capacitors. LTC local controllers change the tap positions in such a way that the measured voltage remains within a half bandwidth of the voltage setting. For shunt capacitors the capacitors are switched ON/OFF if some parameter is above or below a certain value. For example, a capacitor will be switched ON if the voltage is below a lower limit and it will be switched OFF if the voltage is above a higher limit.

The simplest local controllers operate on nonelectrical measurements, which are correlated with the level of electrical consumption: time of the day and air temperature (heat index).

The presence of a variety of local controllers, connected to the same grid and responding to the same voltages and currents according to their own internal logic, results not simply in a set of automatic devices but also in a locally distributed non-centralized voltage and reactive power control system. The key feature of this control system is an obvious need for coordination between different electrically connected controlling devices to prevent them from oscillating and blocking each other. For example, without coordination the upstream and downstream voltage regulators can respond to the same signal for voltage change, but the downstream device can move tap positions in the opposite direction later responding to upstream regulator change.

In more detail, let us consider a feeder with two SVRs: substation regulator at the feeder head and line regulator in the middle of the feeder (Figure 8.10).

Under the condition that feeder is mainly supplied from substation (no distributed energy resource/distributed generation (DG) creating reverse flow), each voltage regulator controls the voltages downstream from the regulated bus up to the next regulator

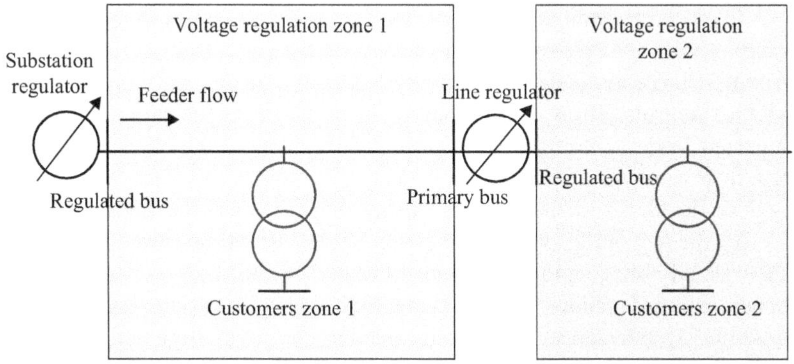

Figure 8.10 Feeder with two step voltage regulators and two voltage regulation zones

or the end of the feeder. From this observation comes the term voltage regulation zone (VRZ)—the part of the radially connected distribution grid, located downstream from the voltage regulator up to another voltage regulator or the end of the feeder, where this regulator sets voltages practically independent from any other voltage setting device.

In the absence of centralized Volt/VAR control (VVC), voltage regulators of the adjacent VRZs are coordinated by introducing time delays in the controllers. The objective of this delay is to prevent oscillation of the control actions (line regulator moves taps in one direction and immediately after in the opposite direction). The most common logic used for assigning delays to the controllers is that the shortest time delay (quick response) should be set on controllers that are closest to the supply substations.

Typically, substation regulators have the shortest time delay, whereas line regulators have longer delays. The response time of a substation regulator (controller time setting + time delay of the executable mechanism) should be smaller than the time delay setting of a line regulator.

In case of switched capacitors, coordination is not needed if capacitor controllers are responding to non-electrical parameters such as time or temperature. In any other case, the capacitor controllers are coordinated with LTC controllers, they have the longest time delay. Capacitors connected to the same feeders are mutually coordinated too (the downstream capacitor has a shorter time delay than the upstream capacitor).

8.4 Centralized VVC

The complexity of voltage and reactive power control in modern distribution systems limits the capabilities of local automatic controllers explained in the earlier section. The settings and parameters of the local controllers are calculated mainly using offline studies. Because of the diversity of the actual operating conditions resulting from the changes in power demand and network configuration, it is not feasible for such an approach to provide effective settings for all conditions.

Additionally, coordination problems exist with conventional voltage and VAR (Volt/VAR) controls. To avoid oscillations in the VVC in the same feeder (group of connected feeders), compromises are typically made in the choice of controller settings, their bandwidth, and their time delays. These compromises reduce the efficiency of the localized VVC functions. Also, using only local measurements without considering the rest of the system, the global Volt/VAR optimization is unattainable.

Centralized voltage and VAR control became one of the most desirable distribution automation functions starting in the 1980s [4], with varying levels of interest through time [5]. It began as centralized capacitor control management which is still widely used separately from voltage control function. The Cooper's capacitor control can serve as an example of such system that became popular in the industry [6].

One of the main reasons behind centralized capacitor control (VAR dispatch) systems is the requirement from transmission systems to keep the power factor at the feeder heads (the point at which the distribution feeder is connected to the transmission system) and/or substation transformers close to unity. Supervisory control and data acquisition (SCADA) systems covering distribution systems supply substations including feeder heads make the feeder head power factor visible to both transmission and distribution systems operators. The development of low-cost radio–based one-way communication link to switched capacitors triggered the implementation of centralized capacitor controlled systems. Nowadays it is not uncommon to have two-way wireless communications to the capacitors.

In parallel, centralized voltage control, independent from capacitor control, was being developed under SCADA operations too. The voltage control function uses SCADA capability to control LTC taps, and communication links to measure voltage at the feeder ends where voltage was expected to be the lowest.

The development of advanced distribution system SCADAs, which cover not only supply substations but also remote telemetry units (RTUs)/intelligent electronic devices along the feeders, leads to the creation of distribution management systems (DMS), closely integrated with SCADA. One of the DMS functions is optimal VVC. This real-time control function can be rule based, but it can be based on other DMS network applications such as Topology Processing, Distribution State Estimator, and Distribution Power Flow (Section 8.7).

Centralized VVC function allows the following capabilities to become feasible [7]:

- Voltage and capacitor controllers respond in coordination to changing system conditions (time-dependent loading and grid topology due to feeder reconfiguration).
- VVC can have different objectives depending on requirements and conditions.
- The goal of the VVC is to be optimal at the system level not just at the local level.

The primary objective of any VVC is to keep the distribution grid voltage within the permissible limits as prescribed in the standards and, to a lesser degree, to meet loading constraints. The limits of power factor at the feeder heads and/or substation transformers are considered as additional constraints.

As long as the primary objective is satisfied, secondary objectives may be selected (e.g., minimization of power demand or power loss). From an operational stand point, it is desirable to attain the objective in the least possible number of control actions (steps) starting from the current status. The simplest optimization objective is to remove the constraint violations within the system.

All secondary objectives use the physical characteristics of the load to change the consumed power (energy) by adjusting the voltage. Power system analysis typically describes customer load using the ZIP model which represents every load as a combination of constant power, constant impedance, and constant current. For example, any residential load can be represented as a combination of all three components—constant impedance (lighting), constant current (electronic devices), and constant power (electrical motors):

$$P = \left(\frac{V^2}{V_{nom}^2} Z_P + \frac{V}{V_{nom}} I_P + P_P \right) P_{nom} \tag{8.6}$$

where $Z_P + I_P + P_P = 1$, P_{nom} is active power consumption under nominal voltage V_{nom}, V is actual voltage.

Reactive power consumption can be described by a similar equation as (8.6) but with different values for ZIP factors:

$$Q = \left(\frac{V^2}{V_{nom}^2} Z_Q + \frac{V}{V_{nom}} I_Q + P_Q \right) Q_{nom} \tag{8.7}$$

where $Z_Q + I_Q + P_Q = 1$, and I_Q is often negative. The reactive power load model lacks physical meaning. Even constant P consumers such as synchronous (inductive) motors have Q consumption dependent on voltage.

The dependences of P and Q on voltage are approximately shown in Figure 8.11: monotonous direct line for P and parabola for Q [8,9]. The lowest point of the parabola representing Q corresponds to voltage (0.7–0.85) per unit. If the voltage drops below this value the load starts to increase the consumption of reactive power which can cause voltage collapse. However, distribution systems operate in the limits defined by the voltage standard: not more than +5% and −10%. In this range, both P and Q have monotonous dependence on voltage with Q having a steeper dependence. This fact is important for understanding and predicting VVC actions for different objectives.

The most typical VVC objectives are:

- Power demand minimization: The demand is measured at the feeder heads or at the transformers in the supply substation. Actually, it is the sum of customer demand (loads) and power losses. As it follows from (8.6) and (8.7) for loads close to constant Z and I, this objective leads to voltage reduction. However, for loads close to constant P, current decreases with voltage increase and demand reduction is achieved only through power loss minimization. The objective function can

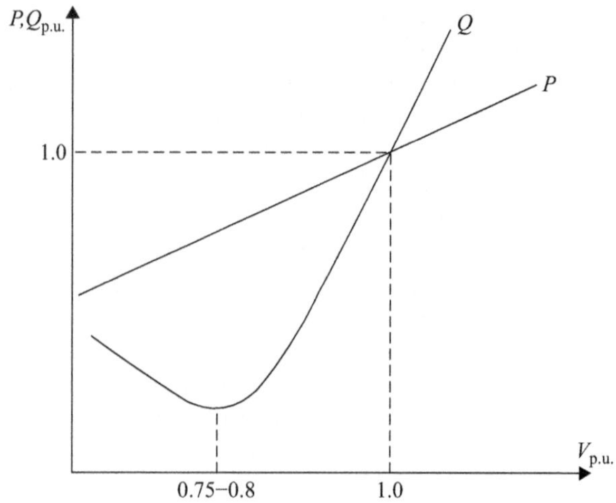

Figure 8.11 Typical load active (P) and reactive (Q) dependence on voltage

be expressed as the sum of voltage-dependent demand and losses which are also voltage dependent.

$$P_{Dem} = \sum P_i\,(V) + \Delta P\,(V) \tag{8.8}$$

- Power loss minimization: sum of copper and no-load losses in lines, transformers, and capacitors:

$$\Delta P = \sum R_{ij} \frac{P_{ij}^2 + Q_{ij}^2}{V_i^2} + \sum G_j V_j^2 \tag{8.9}$$

Loss minimization in distribution systems is usually accomplished by voltage reduction. It is different in transmission systems where it is achieved by increasing the voltage. The reason behind this discrepancy is the dependence on voltage of distribution loads as opposed to transmission loads which are considered as constant power loads.

Power loss minimization seems to be the most trivial text book objective. However, its practical implementation is very challenging. Power losses in the distribution system are not trivial to measure as most lines do not have adequate measuring devices. They can be calculated for some parts of the grid (using power flow) and estimated for the rest. This calculation has never been considered reliable, and it cannot be easily verified.

What can be verified is energy loss—difference between energy through the supply substation and the energy sold to the customers. Substation active power flow is practically measured by the SCADA system instantaneously. Sometimes, there are energy measurements available at the supply substations which measure the energy once in every few minutes. The critical item is the periodicity of the customer meter readings of consumed energy. Advanced metering infrastructure (AMI) provides interval energy measurements in intervals of minutes. This just allows the calculation of a ratio of energy sold to energy bought over a period of time. This ratio is also used to estimate power losses [10].

For all practical purposes, power loss minimization can be achieved through a different objective, by ensuring unity power factor. Keeping power factor along the whole feeder close to unity leads to the elimination of reactive power flow and related active power losses. Switchable capacitors along the feeder can accomplish this objective reasonably well and significantly decrease active power losses. An additional reduction can be accomplished by adding coordinated voltage control: increasing voltages for constant power loads or decreasing voltages for constant admittance loads.

There are other objectives used by centralized VVC in some relatively rare cases.

- Maximization of generated reactive power (transmission VAR support): In this case, the reactive power flow from the distribution to the transmission system is maximized by switching on distribution feeder capacitors. The reactive power at the primary side of substation transformers is typically directly measured. The generated reactive power comes from the capacitors which are considered to be negative for the purpose of calculation while the consumption (loads and losses) is considered positive as shown in (8.10):

$$Q_{gen} = abs\left[-\sum_{Caps} Q_j(V) + \sum_{Load} Q_i(V) + \Delta Q\right] \tag{8.10}$$

  This objective is used when transmission system has a deficit of reactive power. The challenge of this objective is probable voltage increase along the feeder because of capacitors, which is eliminated by reducing the voltage by using voltage regulators.
- Profit maximization: In this case, the difference between energy sales and primary cost of the energy (the cost of production or the price of purchased energy from another producer) is maximized:

$$\text{Profit} = \sum P_i^{\text{Load}} C_i^{\text{Load}} - \sum P_j^{\text{injected}} C_j^{\text{prime}} \tag{8.11}$$

Where P_i^{Load} and C_i^{Load} represent load and energy price (price of energy sold) for the ith customer; and P_j^{Injected} and C_j^{Prime} represent injected power and its cost at substation j.

In order to clarify the meaning of equation (8.11), let us assume that energy price is the same for all customers, and prime cost is the same for all substations. In this case, (8.11) can be rewritten as:

$$\text{Profit} = P^{\text{Load}}(C^{\text{Load}} - C^{\text{Prime}}) - P^{\text{Loss}} C^{\text{Prime}} \tag{8.12}$$

where $P^{\text{Loss}} = P^{\text{injected}} - P^{\text{Load}}$ is total power loss.

From (8.12), it follows that profit is determined by total load power, total power loss, customer energy price, and primary cost. Because both load and loss depend on voltage, revenue can be optimized by the VVC.

8.5 Centralized VVC modes of operation

Centralized VVC can be implemented in two different ways:

- VVC blocks local controllers and directly changes LTC tap positions and capacitor switch statuses.
- VVC changes settings of the local controllers, which, in turn, changes tap positions of LTC or execute capacitor switching.

The first approach—blocking local controllers—was widely used by VVC when local controllers did not allow changing settings remotely or were not reliable. Modern local controllers do not have any constraints to be operated remotely and the second approach is now widely used in the industry to control LTCs. However, capacitor switching is still often executed by VVC directly. Capacitor local controllers operated by time of day and temperature can easily be blocked; however, changing their local controller settings can cause them to fall out of synchronization with the rest of capacitors and it is not usually done.

It is always assumed that centralized VVC is running periodically responding to the changing demand. The periodicity is typically between 5 and 15 minutes. In addition, VVC can be executed manually or triggered automatically by an event (circuit breaker tripping, loading above threshold, etc.).

Centralized VVC can be executed in two different modes: advisory mode and closed loop mode. Both modes use real-time measurements, and their solutions can be implemented in the field. However, advisory mode requires user's review and approval before implementation. Closed loop executes control actions automatically. This change from advisory to closed loop should not be taken lightly. During review in advisory mode, the user's experience may lead to excluding low effective control actions and changing the order of execution. Replacing an experienced operator by an automated sequence of execution as is done in closed loop is an additional step included in the VVC closed loop solution.

Closed loop mode of VVC execution is actually the only mode used for everyday operations in the distribution systems. The operator may be included in VVC only in some very rare cases. The closed loop mode should satisfy all requirements that guarantee a reliable execution both in normal and abnormal conditions:

- No violations on the intermediate steps when a few control actions are executed consecutively as part of the same optimal solution (e.g., when an LTC moves voltage down, no violation should occur before a capacitor is switched ON).
- The solution is implemented only if objective function improvement is above a certain threshold; i.e., low effective control actions are not implemented.

- VVC reruns automatically if any control action has failed during execution (communication error, local devices failure, capacitor fuse is melted) and failed equipment is excluded from the rerun execution.
- Verification of the implemented control actions: directly through two-way communication and indirectly through monitoring change in the real-time measurements.
- Implementation of delay of VVC control actions in transient system conditions (change of the measured values occurs faster than expected).
- Heartbeat functionality for local controllers included in VVC: the controller is automatically switched to default setting if VVC does not send any command for a time period longer than a predefined value.

It should be pointed out that modern voltage regulator controllers have heartbeat as part of their functionality [6]. In addition, voltage regulator controllers have minimum and maximum voltage limits values. All these features enhance the reliability of closed loop execution.

VVC is never solved for the whole distribution system in one single step. It is solved per subsystem, where each subsystem is usually the electrically connected part of the grid supplied from the transmission system through the same substation transformer. Subsystems are typically radial but may be weakly meshed. A simplified flowchart of closed loop VVC is shown schematically in Figure 8.12.

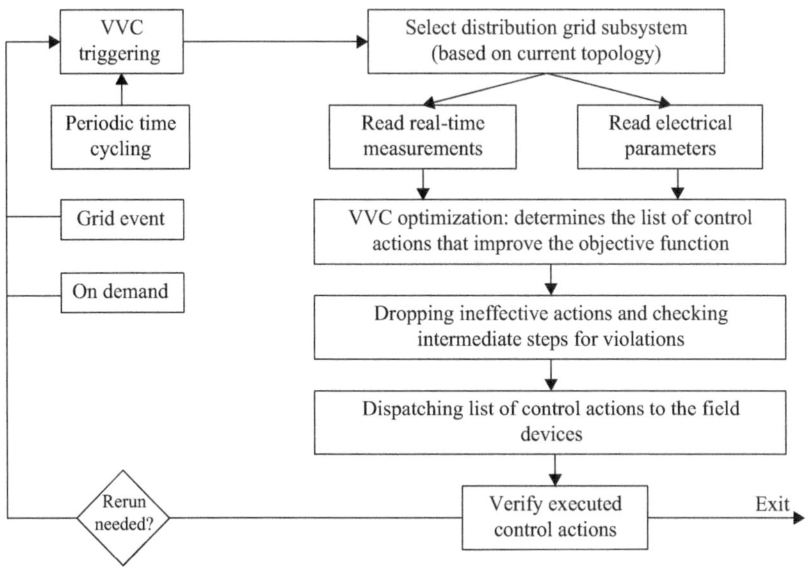

Figure 8.12 Closed loop Volt/VAR control simplified diagram

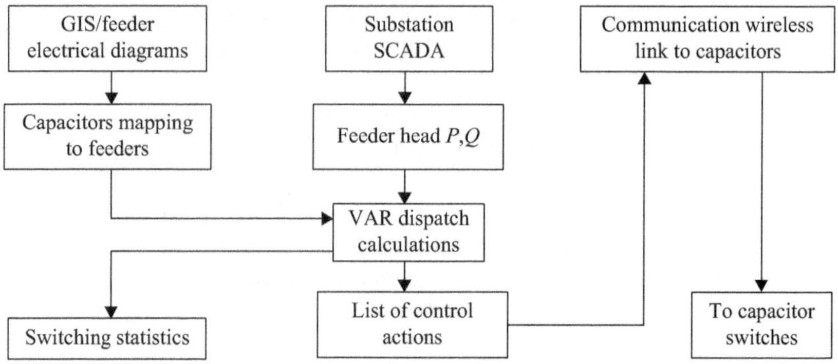

Figure 8.13 Rule-based VAR dispatch system—data flow

8.6 Rule-based voltage control and VAR dispatch

The most common implementations of centralized VVC systems (voltage control and VAR dispatch) are rule based. The control actions follow a set of predefined rules determined by engineering, planning, and analysis. The algorithms are simple and transparent for operations personnel to follow. Rule-based VVC is a good option in part of the distribution systems where voltages and reactive power can be measured and monitored by operators (e.g., values at feeder heads measured by SCADA). Their control actions are effective in extreme cases—peak and light loading.

The data flow of a VAR dispatch system, which uses SCADA measurements on the feeder heads and wireless (radio) link to control the capacitors, is shown on Figure 8.13. The mapping of capacitors to feeders and their sequence within the feeder is also required. The source for these data is typically from the geographical information system (GIS) or the feeder electrical diagrams if GIS is not available.

A typical set of control rules for VAR dispatch system may look as below:

- All capacitors are switched OFF at midnight (any time between 0 and 4 AM);
- Only one capacitor per feeder is switched ON or OFF in the same set of switching actions;
- Switching action success/failure is verified through the change in reactive power flow at the corresponding feeder head/phase;
- Capacitor current status is monitored, and the status and switch action are stored by the VAR dispatch system documenting the capacitor was switched ON/OFF with a timestamp indicating the time;
- Capacitors are switched ON/OFF based on the feeder head P and Q values and on power factor requirements.

Let us consider an example showing how capacitor switching rules can be implemented for a feeder with two switch capacitors (Figure 8.14). It is obvious that both

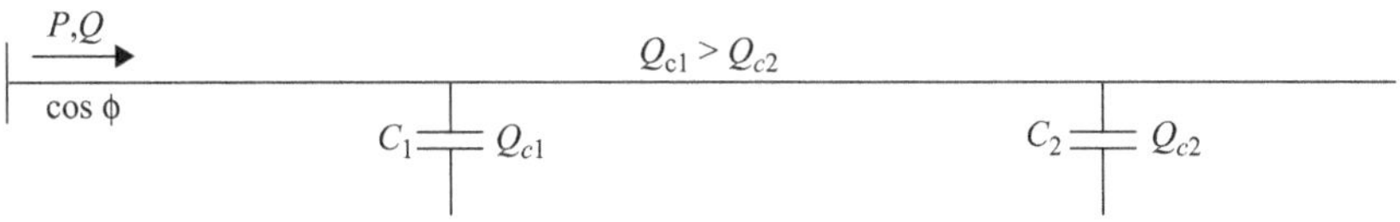

Figure 8.14 Feeder with two switchable capacitors

Table 8.1 Capacitor statuses depending on feeder loading

Number	Feeder loading and power factor	Capacitor 1	Capacitor 2
1	Peak loading, cos ϕ lagging < limit	ON	ON
2	Light loading, cos ϕ leading < limit, Q flow $< Q_{C1}$, $Q < Q_{C2}$	OFF	OFF
3	Medium loading, cos ϕ lagging < limit, $Q > Q_{C1}$, $Q < Q_{C1}+Q_{C2}$	ON	OFF
4	Medium loading, cos ϕ lagging < limit, $Q < Q_{C1}$; $Q > Q_{C2}$	OFF	ON

capacitors are ON during peak loading time, and both capacitors are OFF in light loading period (cases 1 and 2, Table 8.1).

Cases 3 and 4 of medium load in the feeders are common in the real world. Both cases require only one capacitor to be ON: capacitor 1 in case 3 and capacitor 2 in case 4. However, transitioning the capacitor controls between these cases is not trivial. During the capacitor transition from one state to the next, the capacitors inevitably will be both ON or both OFF which may by itself be the cause of voltage violations. In the case when both capacitors are ON, high-voltage violations may happen. In case when capacitors are OFF, low voltage violations are possible.

Centralized voltage control is not as widely used as centralized capacitor control. It is used only when the number of VRZs is more than one (e.g., substation transformer with LTC and SVR along the feeder), and there are communication links to the voltage sensors monitoring the maximum and minimum voltages along the feeder.

The practical need for integrated voltage and VAR control may be illustrated by the following example. Consider a feeder with a voltage regulator at the feeder head and a switched capacitor in the middle of the feeder (Figure 8.15). In the medium loading conditions, voltage along the feeder is inside the allowable limits, but the power factor at the feeder head is not. Switching the capacitors ON corrects the power factor issue but leads to high-voltage violation. The high-voltage violation can be prevented if the voltage regulator decreases its voltage setting.

As it can be seen from Figure 8.15, coordinated VVC makes the voltage at the two ends of the feeder closer to the nominal value by switching the capacitor ON and turning the regulator down.

Integrated voltage and VAR control can be rule based. However, most integrated VVC systems are power flow based and are integral part of SCADA-based DMS.

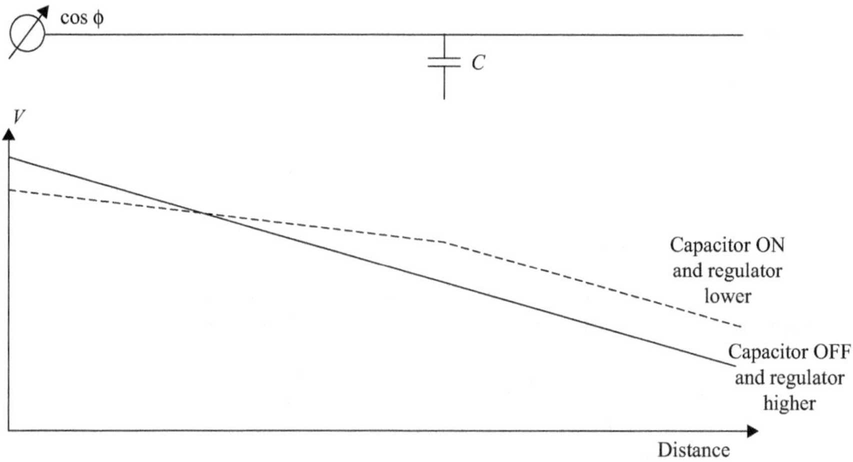

Figure 8.15 Switching capacitor requires voltage regulator down

8.7 Power flow–based VVC

VVC as a single function, integrated in SCADA/DMS, provides a mathematically accurate optimal solution. Based on Distribution System State Estimator (DSE) solution as an input and power flow as an internal tool used by the optimization engine to evaluate possible control strategies, the VVC allows a wide choice of objectives which can be optimized with high mathematical accuracy.

The optimal Volt/VAR solution, as any other optimization problem, may be found as the extremum of a specially constructed analytical function. In some implementations the control actions which are inherently discrete (reactive power generated by the capacitor bank and LTC voltage increments) are treated as continuous variables. The solution is then discretized at the optimal point after a continuous solution is found. This process can skew the solution especially when large bank sizes are involved and voltage constraints limit the LTC capabilities. That is why discrete optimization methods suitable for nonlinear objective functions and inequality constraints (multistep discrete programming search) are more typically used for VVC.

An additional problem comes from solving the distribution systems power flow (DSPF). Even when solving each subsystem independently, the DSPF deals with an electrical grid which is structurally asymmetrical having unbalanced load (forcing a three-phase power flow model) resulting in a system of a few thousand nonlinear equations. Reliable convergence of the DSPF iterative process while dealing with inaccurate actual data is accomplished by adding heuristic rules that are incorporated into the power flow algorithm. Because of this, the power flow equations are typically not integrated in any analytical model used for practical VVC optimization.

VVC integration in SCADA-based DMS, running DSE, provides the VVC following features which are difficult to achieve otherwise:

- Correct topological model based on SCADA switch statuses is verified by real-time measurements.
- Real-time measurements are verified for consistency and feasibility through the topological model and DSE.
- Distribution transformer loads are adjusted based on real-time measurements. These loads serve as inputs to power flow used for VVC.
- The voltage drop in the secondary distribution can be simulated based on transformer loading using power flow (in case of no measurement is available in the secondary distribution which was common before the advent of AMI).
- The global optimal solution includes both voltage and VAR controls.
- The execution of control actions can be monitored through SCADA functionality.
- VVC is triggered again if any action fails and a new solution is found excluding devices with failures.

In a typical VVC optimization algorithm the DSE solution is used as an input. Then, a power flow solution determines the values of all relevant variables together with an initial objective function (IOF) value. This solution is used as a starting point for optimization. The optimization engine determines the sets of control actions which may improve current objective function. Power flow is used to test the effectiveness of the control actions. When the optimization engine cannot find a better set of controls, the IOF is compared with the optimal value of the objective function at this point. If the difference is above a threshold, the final set of controls (list of optimal LTC settings and capacitor bank statuses) is prepared for execution by SCADA.

The most time-consuming part of such algorithms is solving the power flow. Even if the optimization is done per subsystem, consisting of a single feeder or a group of electrically connected feeders, the power flow model may easily include more than a few thousand nodes (one feeder may supply up to 500–700 distribution transformers).

The fact that helps to reduce the power flow execution time is that the network topology remains unchanged. Both radial (forward/backward sweep technique [3]) and meshed network power flow (nodal admittance matrix [11] or loop impedance matrix [12]) methods use preordering and matrix factorization which for an unchanged topology can be done only once and saved and reused. The VVC actions are taken into account by representing the effect of the control action by fictitious compensating injection currents instead of changing the impedance or admittance which would require refactorization of the system matrices.

In case of switched capacitors, the compensation current is directly calculated as $J_C = V(\Delta Y)$ where ΔY is a negative (positive) admittance of the additional capacitor banks which were switched ON (OFF).

In the case of LTC transformers, simulating the change in the transformer tap positions by using compensation currents are calculated as follows. Consider the transformer, connected between buses and K and M, with an initial tap ratio t and a physical admittance Y (Figure 8.16(a)). Admittances, which simulate no-load losses,

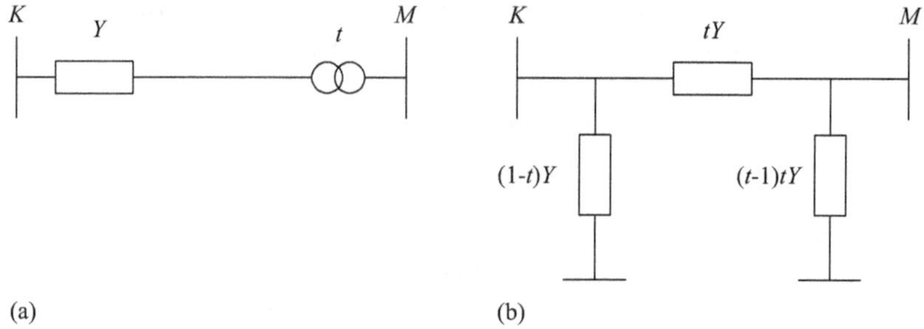

Figure 8.16 Transformer with tap ratio "t"(a) and its model indirectly representing "t" (b)

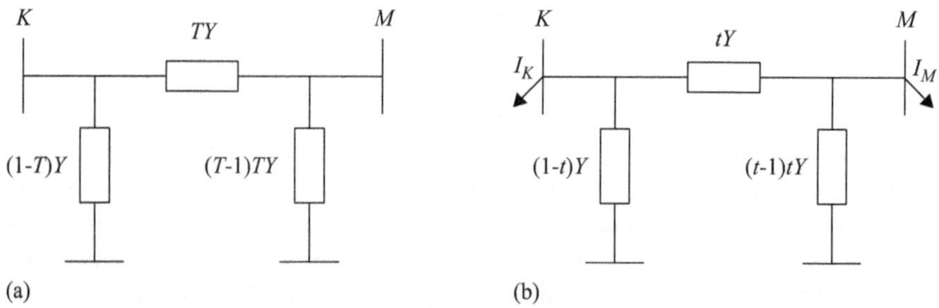

Figure 8.17 Figure 8.17: Transformer with new ratio T (a) and its equivalent (b) which schematically is the same as on Figure 16(b) but with fictitious currents making (b) electrically the same as (a)

are not shown, because they do not affect the model. This transformer is described by a Π model with indirect representation of the transformer tap ratio (Figure 8.16(b)) by its series and shunt admittances.

When the transformer tap ratio is changed from t to $T = t + \Delta\underline{t}$ (LTC moved to new position), the transformer model changes as shown in Figure 8.17(a). In this case, the corresponding matrix should be refactorized. Another way to simulate the tap positions change that does not require refactorization is to modify the model shown in Figure 8.16(b) by adding fictitious injection currents as is shown in Figure 8.17(b).

The fictitious injection currents I_K and I_M are determined by enforcing the condition that systems 8.17(a) and 8.17(b) are equivalent. In other words, under the same voltages V_K and V_M, currents in nodes K and M are equal in both systems:

$$V_K(1 - T)Y + (V_K - V_M)TY = V_K(1 - t)Y + (V_K - V_M)tY + I_K \qquad (8.13)$$

$$V_M(T - 1)TY + (V_M - V_K)TY = V_M(t - 1)tY + (V_M - V_K)tY + I_M \qquad (8.14)$$

Solving (8.13) for current I_K and (8.14) for current I_M leads to (8.15) and (8.16), respectively:

$$I_K = -V_M Y \Delta t \tag{8.15}$$

$$I_M = ((2t+\Delta t)V_M - V_K)\,\Delta t Y \tag{8.16}$$

8.8 Advanced metering infrastructure as voltage monitoring tool

AMI completely changed secondary distribution system—a part of the distribution system which has never had real-time measurements [13]. AMI provides real-time observability of the customer service delivery points: outage alarms, energy consumption, voltage measurements, and voltage violation alarms. It fully changed the quality of voltage control in distribution system. Until the implementation of AMI, most real-time VVC solutions estimated or simulated the secondary distribution system.

The main goal of advanced meters was to measure and record usage data at relatively short time intervals (hour or less), and to provide usage data to both consumers and energy companies at least once every 24 hours. Advanced meters can now also provide voltage information in the form of alarms and analog values. A voltage alarm is a discrete message (single bit) indicating that the meter voltage is below its low limit or above its high limit for a predefined period of time (30–120 seconds). The actual AMI voltage limits are set as part of the meter configuration. Both voltage limits and measurement interval values are meter configurable parameters which are typically set to be equal for groups of meters serving the same type of customers. In real time, only a discrete message is sent as the limits are violated.

More elaborated advanced meters can provide direct voltage measurements. Integrated in AMI, these meters may provide almost real-time voltage measurements at the expense of more bandwidth and infrastructure. Voltage measurements from AMI meters require more communication bandwidth than voltage alarms which can be transmitted in real time as digital signals taking up very limited bandwidth. Because of this, the utilities use a limited number of real-time voltage AMI meters. They are called bellwether meters. In spite of the fact that bellwether meters provide limited observability of the secondary distribution, they are very important in determining the trend in the voltage change.

Both bellwether AMI voltages and voltage alarms are integrated in the modern centralized VVC systems. As a result, AMI provides observability of the voltages in the secondary distribution system. Before AMI, secondary distribution voltages were calculated based on measurements taken from the primary distribution system. It should be emphasized that voltage standards are enforced at the service and utilization voltages which lacked visibility before AMI.

8.9 Voltage control and demand response

Demand response (DR) has a long history of being used by utilities for peak-load reduction and emergency control (overload relief through load control). Advances in communication (AMI), load metering and load control Home Area Network (HAN) and Building Energy Management Systems (B-EMS) have made DR more reliable and predictable. As a result DR is now included in the USA government energy policy: the US Congress, as part of the Energy Independence and Security Act (EISA), required the Federal Energy Regulatory Commission to develop a National Action Plan for DR [14].

Practical implementation of a coordinated DR for reducing the electricity consumption responding to operational requirement or market need is not a trivial task [15]. In addition to engineering constraints, it requires customer engagement. Among other tasks, it involves the creation of DR programs (agreement between DR aggregator and customers). Customers need to be incentivized to enroll in these programs, and participate in the programs when an event is called. In addition, customers that participate in DR are paid, and these payments increase the cost of DR.

Conservation voltage reduction (CVR, which is an established terminology but dispatchable voltage reduction (DVR), is more correct term), which is the reduction of power demand and energy consumption resulting from changing (typically decreasing) the feeder voltage [9], is a type of DR that does not require direct customer participation. The advantage of CVR is the possibility of a wide implementation of DR for the whole feeder/ substation, not just for a priory selected group of consumers, without customer involvement. CVR provides both demand (power) and energy reduction estimated to be approximately 0.5–4% of the load depending on the type of load in a specific feeder [9]. Operations personnel uses approximate load reduction estimates based on the ratio that a 1% voltage reduction produces a demand reduction of 1% and an energy reduction of 0.5%.

Dependence of consumer demand (load) on voltage has a complex characteristic (see Section 8.4 on power to voltage dependence), but from a CVR stand point, it may be divided into two classes: loads without a thermal cycle and loads with a thermal cycle [9]. Loads without a thermal cycle can be described through ZIP model. Thermal cycle loads have a varying duration of the cycle depending on the voltage value and in a simplified manner can be described as constant energy loads: lower voltage means less power demand but longer duration. Examples of loads with a thermal cycle are Heating, Ventilation, and Air Conditioning (HVAC).

The fact that voltage reduction decreases both demand and energy consumption for the majority of the loads with exception of HVAC and synchronous electrical motors is well known in the power industry. The CVR is not a new technology and has been effectively used since the mid-seventies.

From an algorithm stand point, CVR is similar to centralized VVC with the objective of demand minimization, but it is much simpler. In CVR the direction of voltage change is predefined. Preliminarily studies remove from CVR those feeders with predominately constant power loads where voltage reduction does not lead to demand reduction. The goal of CVR control is the biggest load reduction without violating voltage limits.

Switchable capacitor control is typically coordinated with CVR solution by keeping power factor at the feeder head and along the feeder close to unity but leading, not lagging. Leading power factor increases voltages at the feeder electrical ends creating room for general voltage reduction along the feeder (Figure 8.9).

AMI revolutionized the nature of the CVR. Available in almost real-time, measurements from the customer service delivery points make it possible to achieve deeper voltage reduction and higher demand/energy saving without customer voltage violations and using much simpler algorithmic solutions. A typical AMI-based CVR has a lot of emphasis on how to combine SCADA measurements and AMI measurements in one solution.

The stochastic nature of the small customer consumption, the wide variety of the distribution grid operating conditions, and the unbalanced phase loading, theoretically require having at least one bellwether meter at each distribution transformer, which is practically impossible. It is also impossible to find the meter points with the lowest voltage point under every loading/topological condition. That is why bellwether meters are combined with alarms from other AMI meters. The number of AMI meters reporting voltage alarms is typically high. Bellwether meters show trends and voltage alarms alert under voltage violations.

Bellwether and AMI Alarm meters are installed in each VRZ per phase. Voltage alarms received during CVR allow for corrective actions preventing actual voltage violations. The AMI meters sends an alarm signal identifying which alarm (high or low) was violated. The voltage alarm signal does not contain the voltage value.

Voltage alarms serve as indicators of the lowest/highest voltages in the distribution grid. The capability of the AMI communication to change the voltage alarm settings remotely allow to identify the "worst" grid areas without actually violating customer voltages. The electrical connection between the AMI meter and the corresponding distribution transformers and feeders is always visible to the operational personnel through GIS, Outage Management System (OMS), or DMS.

8.10 VVC performance

The performance of any centralized VVC system is of great practical interest for the distribution utility company (owner of the grid). Statistical reports showing the performance of the Volt/VAR function (referred to as VVC performance numbers) are needed for operational control (to estimate change in loading), for economic studies (to calculate benefits), for reducing power and energy losses among others.

The methodology for calculating the VVC effectiveness is not well established in spite of the fact that it is based on a simple idea [16]. The value of the objective function when VVC is running should be compared to the value of this objective function when VVC is not running under the same loading and topological conditions. The objective function may be measured directly or calculated based on real-time measurements. The performance is the difference or ratio between these two values.

Real-time VVC, which controls LTC taps and capacitor statuses, affects power system real-time conditions. Now, for a given time period, there can only be either of the two options—(a) VVC is being executed and (b) VVC is not being executed. If VVC is being executed (say), then we can measure the values of different parameters directly thereby directly calculating the value of the objective function. However, for the same time period, it is impossible to measure the values of the same parameters when VVC is not running (as these two events, VVC running and VVC not running, are mutually exclusive to each other). Therefore, for the same time period, the value of the objective function can only be indirectly estimated or calculated (through simulation studies) by making some assumptions on grid conditions (such as LTC tap positions and capacitor statuses) when the VVC is not running.

There are different approaches to resolve this ambiguity of collecting two sets of data when one set is measured and the second is simulated based on these measurements. For example, it is possible to use similar load pattern days: one set of measurements is taken on the day in which VVC is running and another set of measurements is taken on a day without VVC with the same topology and approximately the same loading condition. It is also possible to differentiate demand changes due to VVC actions from the demand change due to the natural time dependency. This can be achieved by using data from a subsystem where demand is well correlated with the subsystem under consideration, but VVC is not running. For example, if the substation has homogenous load structure, then another bus section of this substation can be used to monitor the natural time dependent changes of load levels to compare them to the load levels of the bus section where VVC was implemented.

Calculating the performance of CVR has its own methodology based on regular DR event performance calculations. Voltage reduction is a type of DR, which has a well-developed methodology for calculating performance. The reasons for this are both technical and economical. First of all, DR events run only for a limited time, whereas VVC can run during 24 hours, 7 days a week. Second, DR participants are often paid based on their performance during DR events.

The DR event performance is calculated based on customer baseline (CBL). The baseline is the amount of electricity which the customer would have consumed on this day and time under normal conditions (including no participation in the DR event). The actual customer demand and energy during a DR event is measured by AMI meters. DR performance (demand and energy reduction) is the difference between the baseline and actual consumption.

There are several common techniques for calculating CBL, e.g., day matching, statistical regression analysis, and rule based. In day-matching approach, historical data for the same day type but without a DR event is used to estimate consumption without DR event.

In the statistical approach, day types based on weekdays are introduced following the observation that power consumption has a periodicity that depends on the day of the week. For groups of small customers (residential, small commercial, etc.) or big consumers the consumption has a stable weekly periodicity. Baselines are calculated per day type.

The power industry traditionally assumes that according to their consumption pattern, the week day types may be divided into Monday, Tuesday through Thursday, Friday, and Saturday and Sunday. Holidays are usually represented as Sunday but can also be considered as a separate type. Each day of the week is often represented by an individual day type.

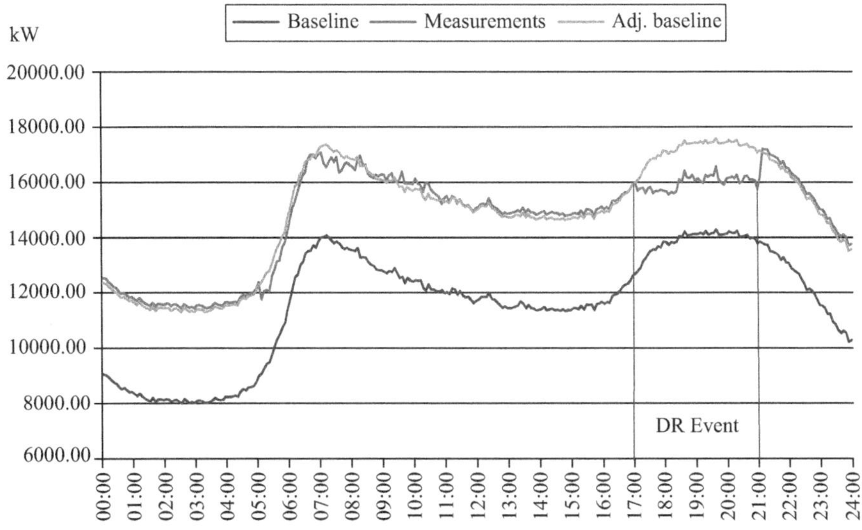

Figure 8.18 CVR event performance (measurements and baselines)

The statistical baseline of a specific day type is updated every time new measurements are available. The most common baseline calculations are based on linear regression (exponential smoothing):

$$\text{Baseline}(d,j) = \text{Baseline}(d,j) \times (1 - \alpha) + \text{Consumption}(d,j) \times \alpha \tag{8.17}$$

where α is a smoothing factor in the range of 0–1; Consumption (d,j) is measured energy consumed for day d, hour j.

Typical consumption patterns for the day (baselines) can be different from the actual measurements taken on the day of a CVR event. As a result, before using baselines for performance calculations the baselines are adjusted for the conditions of the particular day of the event. The adjustment consists in determining the difference between the baseline and the actual customer load before DR event started (morning adjustment) and adjusting (scalar or additive) the baseline values based on this difference (Figure 8.18). The difference between adjusted baseline (green) and actual consumption (red) is the achieved demand reduction used to calculate the performance of the event.

When computing baselines per feeder, there is an issue with load changes due to feeder reconfiguration. Depending on how often the topology changes, it may not be always possible to build a baseline per feeder or group of feeders (feeder head busbar) sharing a common voltage regulating device based on aggregating all the customer meter baselines. To solve this issue, sometimes, feeder baselines are built for normal topology only, where the same set of customer meters is supplied through the same feeder.

However, for an operational zone or for the distribution area of a small/medium utility, the supplied set of customer meters does not depend on the grid topology.

The same set of customers simply can be supplied through different electrical paths defined by the electrical topology, but the overall set remains constant. Because of this, the baselines for zones or entire utilities are meaningful and their voltage reduction performance can be calculated.

Substation transformers and feeder heads have measurements of active power (voltage, current, and power factor) typically coming from RTU through SCADA. Substations may have interval energy meters too. It means that the power/energy measurements required for baseline calculations are available.

8.11 Impact of distributed energy resources

Distribution systems are designed assuming that the supply substation is the primary source of electrical power in spite of the fact that small-capacity distributed generators have always been connected to the feeder. The recent increase in the capacity level of DG, penetrating some distribution systems, impacts both voltage regulation and capacitor switching. A detailed compilation of the system issues which may be encountered as DG penetration increases in distribution system is presented in [17]. The graph shown in Figure 8.19 presents a simplified illustration on how the presence of DG impacts the voltage profile along the feeder.

In case 1, the feeder is fully supplied from the substation and the SVR installed at the feeder head sets the voltage along the feeder. In case 2, the feeder is supplied both from the substation and from the DG. The active power flow at the SVR location is still fed from the substation (forward flow), and the SVR controls the voltage of the feeder until the flow is reversed and the loads are supplied from the DG. Downstream from this flow division point, the DG sets the feeder voltage.

In case 3, the DG supplies power for the whole feeder and changes flow direction through SVR. As soon as SVR flow is reversed, the SVR cannot regulate voltage at the same bus as before. Any transformer/autotransformer regulates voltage in the direction of the current, which is typically the same as direction of the active power flow. Most SVR local controllers monitor the direction of active power and automatically change the mode of operation from forward to reverse when the flow direction is changed. In the reverse mode, SVR can regulate voltage on the opposite bus (in case shown on Figure 8.19—the substation bus bar) or be blocked. The SVR, connected to the substation directly, should be blocked in the reverse mode because substation voltage is regulated by transmission system. As a result, feeder voltage profile is set by DG only.

If DG has the capability to regulate voltage and reactive power, the feeder may have a normal voltage profile through DG VVC system. Until now, only traditional type DG (synchronous generator) could regulate voltage and reactive power (through the DG's excitation system and also by generating or consuming reactive power) which allowed them to support the feeder power factor.

Renewable generation sources—wind, solar, batteries—have limited technical capabilities to regulate voltage and reactive power. Connection of the wind generators

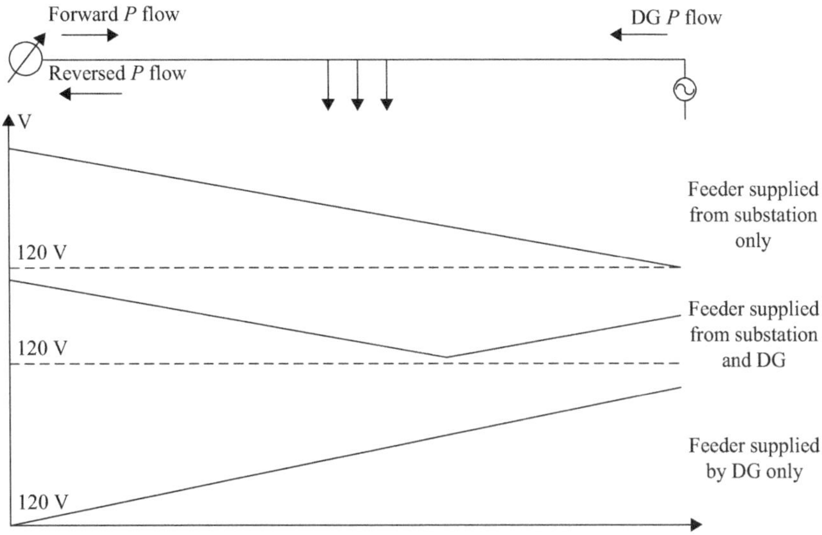

Figure 8.19 Voltage profile along the Feeder supplied from substation and DG

and photovoltaic (PV) panels to the AC grid, which is operated at constant frequency, is not a trivial task. Wind generators are induction (asynchronous) machines which consume reactive power. Capacitors are often used to compensate reactive power consumed by wind generators. Wind generators do not have constant speed of the turbine and therefore do not generate power at constant frequency. Typically, their power is converted in DC and supplied through an inverter, which transforms it to the required frequency AC. Solar PV panels produce direct current power. Any voltage or reactive power support is accomplished through the inverter.

Both wind and PV may be integrated into big generation farms having sophisticated inverter systems which can change voltage and both consume and generate reactive power. This solves the VVC problem. However, this type of generation farms is typically connected to the transmission network and not to the distribution systems.

Distribution systems are dealing with an increasing number of wind generators and PV panels, owned by small consumers (residential and commercial). They typically have a capacity of single-digit kilowatt, and are connected to the customer service delivery points (secondary distribution). These generation assets usually operate at unity power factor to maximize revenue and do not have any Volt/VAR regulation system. Utilities limit the percentage of solar and wind power per feeder/substation to avoid voltage violations.

In addition, small wind and solar generation resources are nondispatchable, as they produce electricity depending on wind and solar availability. Their peak production usually happens when consumption is low (day time for solar and night for wind) which may reverse the direction of the power through feeder and voltage regulators and create problems keeping voltages inside acceptable range.

8.12 About the Authors

The authors of this chapter are both PhD electrical engineers in power systems with about 65 years of combined industry experience. They started their careers in the development of SCADA based Energy Management Systems (EMS) which lead them to the Distribution Management Systems (DMS) and later to Smart Grid.

Both authors were educated in Europe and have practiced in the United States for more than half of their professional lives. It gives the authors a unique capability to know both North American and European type distribution systems. The diversity of distribution systems which does not exist in transmission can be felt across both its design and operations including voltage control and reactive power compensation.

In this chapter on voltage and reactive power (Volt/VAR) control in distribution electrical power systems the authors tried to use their practical knowledge and experience. They are not teaching, but discussing. At the same time, they provide precise and mathematically strict description of the VVC foundation. Their experience allowed them to emphasize on the main solutions used in real life rather than the more theoretical solutions with little industry application.

References

[1] American National Standard for Electric Power—Systems and Equipment Voltage Ratings (60) Hertz, ANSI C84.1-1995, National Electrical Manufacturers Association, Rosslyn, Virginia, 1996.
[2] IEEE Standard Requirements,. Terminology and Test Code for Step-Voltage and Induction Regulators, ANSI/IEEE C57.15-1996, Institute of Electrical and Electronic Engineers, New York, 1988.
[3] W.H. Kersting, Distribution System Modeling and Analysis, 2007, CRC Press, Baca Raton, Florida, pp. 421.
[4] N.S. Markushevich, Voltage and VAR Control in Automated Distribution Systems, Third International Symposium of Distribution Automation and Demand Side Management, Florida, USA, 1993, pp. 478–485.
[5] R.W. Uluski, VVC in the Smart Grid Era, Proceedings of IEEE Power and Energy Society General Meeting, 2010, pp. 1–7.
[6] http://www.cooperindustries.com/content/public/en/power_systems/products.
[7] I. Roytelman, B.K. Wee, R.L. Lugtu, Volt/VAR control algorithm for modern distribution management system, IEEE Trans. Power Syst., 10(3), pp. 1454–1459, 1995.
[8] Determining load characteristics for transient performance, EPRI EL-850, Volume 4, Project RP849-1, Final Report, March 1981.
[9] Evaluation of Conservation Voltage Reduction (CVR) on National Level, Pacific Northwest National laboratory report prepared for U.S. Department of Energy, 2010, pp. 106.
[10] C.A. Dortolina, R. Nadira, The loss that is unknown is no loss at all: a top-down/bottom-up approach for estimating distribution losses, IEEE Trans. Power Syst., 20(2), pp. 1119–1125.

[11] T.H. Chen, M.S. Chen, K.J. Hwang, P. Kotas, E.A. Chebli, Distribution system power flow analysis—a rigid approach, IEEE Trans. Power Deliv., 6(3), pp. 1146–1152, 1998.

[12] C.S. Cheng, D. Shirmohammadi, A three-phase power flow method for real-time distribution system analysis, IEEE Trans. Power Syst., 10(2), pp. 671–679, 1995.

[13] Assessment of Demand Response and Advanced Metering, Federal Energy Regulation Commission Staff Report, December 2012, p. 120.

[14] National Action Plan on Demand Response, the Federal Energy Regulation Commission Staff, June 17, 2010, 79 p.

[15] J. Medina, N. Muller, I. Roytelman, Demand response and distribution grid operations: opportunities and challenges, IEEE Trans. Smart Grid, 1(2), pp. 193–198, 2010.

[16] I. Roytelman, B.K. Wee, R.L. Lugtu, T.M. Kulas, T. Brossart, Pilot project to estimate the centralized Volt/VAR control effectiveness, IEEE Trans. Power Syst., 13(3), pp. 864–869, 1998.

[17] R.A. Walling, R. Saint, R.C. Dugan, J. Burke, L.A. Kojovic, Summary of distributed resources impact on power delivery systems, IEEE Trans. Power Deliv., 23(3), pp. 1636–1644, 2008.

Chapter 9
Fault location

Yuan Liao[1]

9.1 Introduction

Prompt and accurate location of faults in an electric power distribution system can help engineers quickly identify and repair faulted components, reduce outage time, speed up system restoration, and thus greatly improve system reliability [1–4]. Fault location has been an important function in the distribution management system (DMS). Fault location problem has been a long-standing research topic, and diverse methods have been proposed in the past depending on different assumptions and simplifications.

Most of the existing methods are impedance-based methods that utilize fundamental frequency voltages and currents, which are usually recorded at the local substation [4, 5]. Additional measurements obtained at other places on the distribution systems may also be harnessed. Methods may be applicable to radial networks or applicable to both radial and nonradial networks. Traveling wave theory has also been employed for fault location. Traveling wave method imposes higher demand on instrument transformers and recording devices. There are also approaches that exploit short-circuit simulation studies. In this method, a fault is imposed at a specified location in a simulation model of the power network, and simulation study is performed to obtain simulated quantities, which usually include voltages and currents. The simulated quantities are then compared with actually recorded quantities. Based on certain criterion, another fault location is attempted to obtain new simulated quantities. The steps are repeated until the simulated quantities match the recorded quantities. The quantities can be in time domain or in phasor domain depending on the accuracy of the instrument transformers and recording devices.

The difficulties of distribution system fault location are attributed to the characteristics of distribution systems including unbalances, existence of laterals, sparsity of monitoring devices, and uncertainty of network data. Recent developments in distribution systems have introduced new challenges for this problem. An increasing number of distributed generations (DGs) including solar and wind power sources emerge in distribution systems to harness more renewable energy resources. Power flow is no longer unidirectional as in traditional distribution systems, and becomes bidirectional depending on the location and outputs of DGs. On the other hand, new opportunities arise for solving this problem in a better way. Advanced metering infrastructure (AMI)

[1]Department of Electrical and Computer Engineering, University of Kentucky, Lexington, KY 40506, Email: yliao@engr.uky.edu

is being deployed in distribution grids, which will provide two-way communication between customers and utilities. Consumer load data at a desired time interval, say 15 minutes, will be available to the control room for improved operation, control, and protection purposes. More protection and monitoring devices including feeder terminal units (FTUs), digital fault recorders (DFRs), digital relays, power quality meters, reclosers, and phasor measurement units (PMUs) are being deployed in the distribution networks to make the network more resilient, reliable, and smarter. As a result, an increasing volume of data will become available for fault location applications. The data can be exploited to develop better load model and network model, help identify faulted segment, and obtain more accurate fault location estimate.

By taking advantage of the increasingly available data, this chapter puts forward a possible fault location solution that may solve some of the challenges.

9.2 Overall idea of the fault location system

Figure 9.1 depicts the overall structure of the fault location system. The measurement devices represent any device that can provide analog or status data, which includes

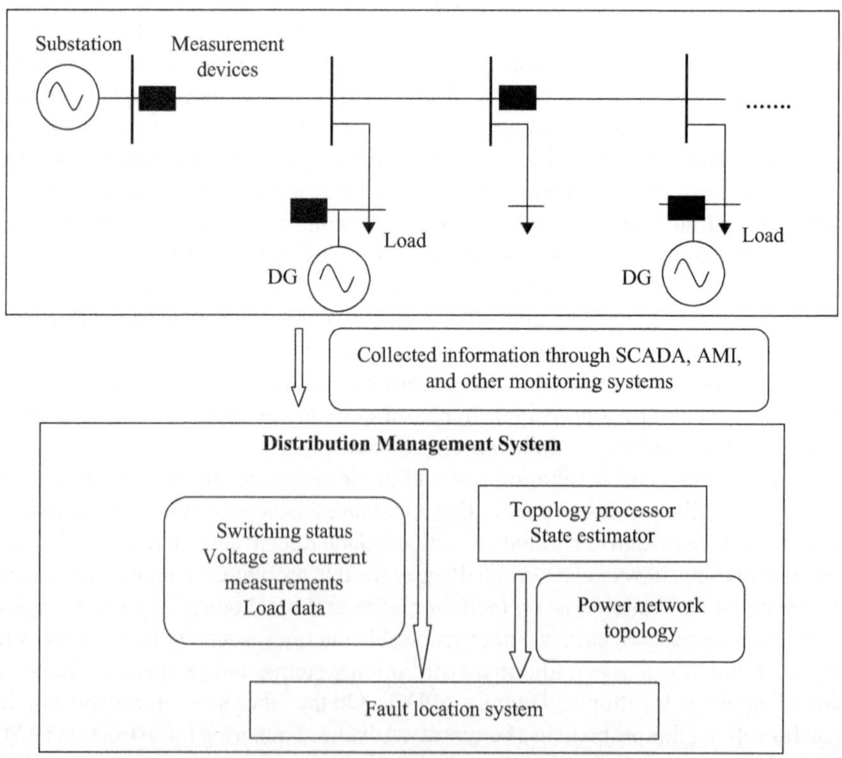

Figure 9.1 Overall structure of the fault location system

protection and control devices (e.g., digital relays, reclosers, and intelligent switches), FTUs, AMI, and general monitoring devices (e.g., DFRs, power quality meters, and PMUs). The supervisory control and data acquisition (SCADA) system periodically polls data from the power network or receives data when defined events such as faults occur. The DMS encompasses a set of functions for operating and controlling the system. Based on real-time network information, the topology processor/state estimator is utilized to provide an accurate model of the distribution network for enhanced fault location [6]. The fault location system may be integrated with a utility's SCADA/DMS and other monitoring systems; data exchanges and software interface definitions, etc., will need to be considered.

The fault location system first identifies the faulted section–based graph theory and fuzzy logic, and then pinpoints the fault location based on optimization theory [7]. The network is described by a directed graph or digraph, including vertices and edges with direction that connect them. The vertices represent the switching points, and the edges with specified direction represent the feeders between vertices. When a fault occurs, the current measurements from the vertices across the network will be sent through the communication system to the fault location system. An efficient graph theory–based approach is described to identify the faulted section. Once the faulted section is determined, a general fault location method is put forth to pinpoint the exact location of the fault. Optimization theory is used in order to make use of the most of available data.

Distribution systems are equipped with breakers for switching on or off circuits and breaking the fault currents, and sectionalizing switches for isolating and reconfiguring the feeders. Feeder automation devices such as FTU controls the status of the sectionalizing switches and detects fault current and reports the overcurrent status and direction of the fault current to the control center. Utilization the overcurrent status and fault current direction obtained from the power network provides a feasible way for speedy identification of the faulted section. Matrix based on graph theory establishes the network topology. Then the overcurrent statuses obtained by FTUs are processed to determine the faulted section [7]. For improved performance, current quantities instead of switching status, along with fuzzy logic, can be exploited. The basic principle is similar to differential protection: a section is considered faulted if the sum of all currents flowing into the section is greater than a threshold. In practice, various factors such as loading in this area and potential measurement errors need to be considered. Application of fuzzy logic helps to handle the uncertainties. The incidence matrix of the network as defined in reference 7 remains the same, which will be updated according to the real-time topology of the system obtained from SCADA/DMS. Instead of using a discrete value of 0 or 1 to signify normal current and overcurrent, the current quantities flowing through the monitored locations will be utilized to generate the fault vector. In fuzzification step, the actual current value is transformed to the fuzzy-matching degree based on defined fuzzy membership functions. The matching degree will constitute the fault vector. Then, the multiplication of the incidence matrix and fault vector is performed to generate the faulted section identification vector. Existence of an element with a value significantly greater than zero or a threshold will indicate a faulted section. This procedure can be applied to phase current or superimposed phase current of each phase. The method will be efficient and applicable to systems having multiple sources and simultaneous faults. Note that a membership function at each current measurement location needs to be defined considering that the load current will be different at each location.

The basic idea for fault location is presented as follows [4, 8]. A fictitious bus is added at the fault point. Then the transfer impedance between any bus and the fictitious bus, and the driving point impedance at the fault point can be obtained as a function of the fault location. Based on short-circuit analysis theory, the bus voltage and branch current at any location across the network can be expressed in terms of the according transfer impedance and driving point impedance. Consequently, the bus voltage and branch current can be derived as a function of the fault location. Thus, utilizing measured voltage and current at certain locations such as the local substation can readily determine the fault location.

To account for the intrinsic unbalances of the distribution system, the distribution system is represented in three-phase domain, and the phase-domain short circuit analysis technique is utilized to derive the fault location. Therefore, the presented methods naturally accommodate any unbalance in a system. The methods are also applicable to both radial and meshed networks with single or multiple sources including DGs. Fault location methods for single faults are described first, followed by methods for simultaneous multiple faults.

9.3 Fault location method that requires fault type

This section presents the fault location method, assuming that the fault type has already been obtained by other modules [4, 9, 10]. The general idea of the method is described as follows. Assuming that a fault occurs at a point on a feeder segment, the transfer impedance and driving point impedance can be obtained, which is a function of the fault location. The voltage at any bus in the system can be written in terms of the transfer and driving point impedance, from which the fault location can be determined.

Derivation of transfer and driving point impedance is provided as follows [8]. Figure 9.2 depicts the one-line diagram of a section of a distribution system, where a three-phase feeder is assumed. Note that the remaining part of the distribution system is not shown.

The following notations are adopted. Note that all the voltage and current quantities are phasors. Without losing generality, the following notations assume that each

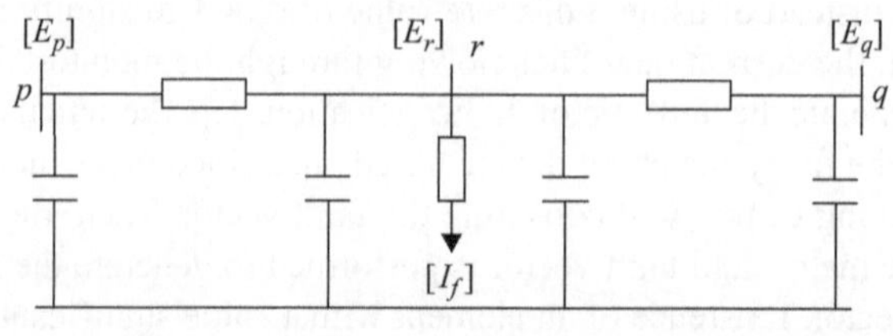

Figure 9.2 A section of a power distribution system

bus consists of three phases; in practices, only those phases present will appear in the vector.

[.]: designation of a matrix or vector;
n: the total number of nodes of the entire distribution system without counting fault nodes r_1, r_2, and r_3; each single phase of a bus is considered to be a node.
p, q: buses of the sample feeder. Bus p comprises nodes p_1, p_2, and p_3 (corresponding to three phases), and bus q includes nodes q_1, q_2, and q_3;
r: fault bus, containing nodes r_1, r_2, and r_3;
$\boldsymbol{E_p}$: node voltage vector, $\boldsymbol{E_p} = [E_{p_1}, E_{p_2}, E_{p_3}]^T$, with T symbolizing vector/matrix transpose. E_{p_1}, E_{p_2}, and E_{p_3} are voltages at nodes p_1, p_2, and p_3, respectively;
$\boldsymbol{E_q}$: node voltage vector, $\boldsymbol{E_q} = [E_{q_1}, E_{q_2}, E_{q_3}]^T$. E_{q_1}, E_{q_2}, and E_{q_3} are voltages at nodes q_1, q_2, and q_3, respectively;
$\boldsymbol{E_r}$: node voltage vector, $\boldsymbol{E_r} = [E_{r_1}, E_{r_2}, E_{r_3}]^T$. E_{r_1}, E_{r_2} and E_{r_3} are voltages at nodes r_1, r_2, and r_3, respectively;
$\boldsymbol{I_f}$: fault current through fault resistances, $\boldsymbol{I_f} = [I_{f1}, I_{f2}, I_{f3}]^T$. I_{f1}, I_{f2}, and I_{f3} are fault currents for phase 1, 2, and 3, respectively;
$\boldsymbol{z}$: the total series impedance matrix of the feeder;
$\boldsymbol{y}$: the total shunt admittance matrix of the feeder due to shunt capacitances;
m: per-unit fault distance from bus p to the fault point;
$\boldsymbol{Z}_0$: the bus impedance matrix of the entire prefault distribution system in phase domain, excluding fictitious nodes r_1, r_2, and r_3. $\boldsymbol{Z}_0$ will be of size n by n;
$Z_{0,kl}$: the element in the kth row and l^{th} column of $\boldsymbol{Z}_0$;
$\boldsymbol{Z}$: the bus impedance matrix in phase domain of the entire distribution system including the fictitious fault nodes. $\boldsymbol{Z}$ will be of size $(n+3)$ by $(n+3)$;
Z_{kl}: the element in the kth row and l^{th} column of $\boldsymbol{Z}$;

In implementation, the fault nodes are numbered as follows: $r_1 = n+1, r_2 = n+2$, and $r_3 = n+3$.

Matrix $\boldsymbol{Z}_0$ can be readily developed. It can be shown that the first n rows and n columns of $\boldsymbol{Z}$ are identical to $\boldsymbol{Z}_0$, and the other rows and columns of $\boldsymbol{Z}$ consist of transfer and driving point impedances related to the fault nodes. The transfer and driving point impedance of $\boldsymbol{Z}$ related to the fault nodes can be obtained as [8]

$$\boldsymbol{Z}_{k_1 r} = \boldsymbol{w}^{-1}\left(\frac{\boldsymbol{Z}_{k_1 p}}{m} + \frac{\boldsymbol{Z}_{k_1 q}}{1-m}\right) \tag{9.1}$$

$$\boldsymbol{Z}_{rr_i} = \boldsymbol{w}^{-1}\left(\frac{\boldsymbol{Z}_{pr_i}}{m} + \frac{\boldsymbol{Z}_{qr_i}}{1-m} + \boldsymbol{z}\boldsymbol{u}_i\right), \quad i = 1, 2, \text{ or } 3. \tag{9.2}$$

$$\boldsymbol{w} = \frac{\boldsymbol{z}\boldsymbol{y}}{2} + \frac{\boldsymbol{u}}{m(1-m)} \tag{9.3}$$

where, $\boldsymbol{Z}_{k_1 r}$: transfer impedance between node k_1 and fault nodes; and $\boldsymbol{Z}_{rr_i}$: driving point and transfer impedance related to fault nodes, $\boldsymbol{Z}_{rr_i} = [Z_{r_1 r_i}, Z_{r_2 r_i}, Z_{r_3 r_i}]^T$.

Varying the value of i will yield relevant driving point and transfer impedances. For example, for $i = 1$, (9.2) will yield $Z_{r_1r_1}$, $Z_{r_2r_1}$, and $Z_{r_3r_1}$ that are the driving point impedance at node r_1, transfer impedances between node r_1 and r_2, and r_1 and r_3, respectively; $\boldsymbol{u}$: a 3×3 identity matrix, whose i^{th} column is denoted by $\boldsymbol{u}_i$.

It is to be noted that the above equations hold for a single-phase, two-phase, or three-phase feeder, where applicable. The key point is that the transfer and driving point impedances are expressed as functions of the fault locations. The fault location algorithm is presented as follows.

The voltage change due to fault at node k_1 is

$$\Delta E_{k_1} = E_{k_1} - E_{k_10} = -Z_{k_1r_1}I_{f_1} - Z_{k_1r_2}I_{f_2} - Z_{k_1r_3}I_{f_3} \tag{9.4}$$

where, ΔE_{k_1}: voltage change, or superimposed voltage, due to the fault; E_{k_1}: voltage at node k_1 during the fault; E_{k_10}: prefault voltage at node k_1; I_{f_i}: fault current flowing through the fault resistance out of the fault node r_i.

Note again that in (9.4), a three-phase fault is considered. In practice, only the terms involving relevant phases will appear in the equations. Fault location will be derived based on (9.5). Methods for single line to ground faults (LG) and three phase to ground faults (LLLG) will be described here, while methods for other types, line to line faults (LL), line to line to ground faults (LLG), and three-phase faults (LLL), are referred to [4, 8].

For an LG fault that occurs at fault node r_1, the fault current through the fault resistance is

$$I_{f_1} = \frac{E_{r_10}}{Z_{r_1r_1} + R_{f_1}} \tag{9.5}$$

where E_{r_10}: prefault voltage at fault node r_1; R_{f_1}: fault resistance.

Prefault voltages at fault nodes $\boldsymbol{E}_{r0}$ can be expressed in terms of the fault location and the prefault node voltages at bus p and q as follows:

$$\boldsymbol{E}_{r0} = \boldsymbol{w}^{-1}\left(\frac{\boldsymbol{E}_{p0}}{m} + \frac{\boldsymbol{E}_{q0}}{1-m}\right) \tag{9.6}$$

where $\boldsymbol{E}_{p0}$ and $\boldsymbol{E}_{q0}$ are prefault voltage vector at respective nodes, which can be estimated utilizing prefault voltages and currents at the substation and the feeder and load impedance.

The voltage change due to the fault at node k_1 is expressed as

$$\Delta E_{k_1} = \frac{-Z_{k_1r_1}E_{r_10}}{Z_{r_1r_1} + R_{f_1}} \tag{9.7}$$

This equation contains two unknown variables, i.e., the fault location and fault resistance. Equation (9.7) can be separated into two real equations, solution to which results in the unknown variables. To simplify computation, E_{r_10} may be approximated by the prefault voltage at bus p or q or k_1.

For an LLLG fault, the fault currents through the fault resistances are given by [4],

$$\begin{bmatrix} I_{f1} \\ I_{f2} \\ I_{f3} \end{bmatrix} = \begin{bmatrix} Z_{r_1r_1} + R_{f1} + R_g & Z_{r_1r_2} + R_g & Z_{r_1r_3} + R_g \\ Z_{r_1r_2} + R_g & Z_{r_2r_2} + R_{f2} + R_g & Z_{r_2r_3} + R_g \\ Z_{r_1r_3} + R_g & Z_{r_2r_3} + R_g & Z_{r_3r_3} + R_{f3} + R_g \end{bmatrix}^{-1} \begin{bmatrix} E_{r_10} \\ E_{r_20} \\ E_{r_30} \end{bmatrix} \tag{9.8}$$

where $E_{r_10}, E_{r_20}, E_{r_30}$: prefault voltages at fault node r_1, r_2, and r_3, respectively; $R_{f1}, R_{f2}, R_{f3}, R_g$: corresponding fault resistances.

The voltage change at the substation nodes can be obtained as

$$\Delta E_{k_1} = -Z_{k_1r_1}I_{f1} - Z_{k_1r_2}I_{f2} - Z_{k_1r_3}I_{f3} \tag{9.9}$$

$$\Delta E_{k_2} = -Z_{k_2r_1}I_{f1} - Z_{k_2r_2}I_{f2} - Z_{k_2r_3}I_{f3} \tag{9.10}$$

$$\Delta E_{k_3} = -Z_{k_3r_1}I_{f1} - Z_{k_3r_2}I_{f2} - Z_{k_3r_3}I_{f3} \tag{9.11}$$

The unknown variable vector is

$$X = [m, R_{f1}, R_{f2}, R_{f3}, R_g]^T \tag{9.12}$$

Equations (9.9–9.11) can be separated into real and imaginary parts to yield six real equations. Define $\boldsymbol{F}(\boldsymbol{X})$ as a function vector composed of these six equations. The unknown vector $\boldsymbol{X}$ is obtained as follows:

$$\boldsymbol{\Delta X} = -(\boldsymbol{H}^T\boldsymbol{H})^{-1}[\boldsymbol{H}^T\boldsymbol{f}(\boldsymbol{X}_i)] \tag{9.13}$$

$$\boldsymbol{H} = \frac{\partial \boldsymbol{F}(\boldsymbol{X}_i)}{\partial \boldsymbol{X}} \tag{9.14}$$

$$\boldsymbol{X}_{i+1} = \boldsymbol{X}_i + \boldsymbol{\Delta X} \tag{9.15}$$

where, i: iteration number starting from 1; $\boldsymbol{\Delta X}$: variable update; $\boldsymbol{X}_i, \boldsymbol{X}_{i+1}$: variable vector at and after i^{th} iteration; $\boldsymbol{H}$: is composed of the derivatives of the functions with respect to the unknown variables.

The feeder shunt capacitances have been modeled based on PI circuit. When the shunt capacitances are ignored, the computational burden will be greatly reduced and the solution is much easier to obtain, especially for LG and LL faults [4]. Therefore, an initial estimate may be obtained by ignoring the shunt capacitances, and then the result can be utilized as the starting point to find out a more accurate estimate considering the shunt capacitances.

It is noted that when measurements from multiple sites are available, an optimization theory can be applied to improve the reliability and accuracy of fault location estimation [11].

When the fault type is provided, the above methods are efficient to derive the fault location. Sometimes, it may be desirable to design fault location methods that obviate the fault type information. One possible approach is illustrated below. Without losing generality, a fault on a three-phase feeder is assumed. By assuming the fault type to be a three phase to ground fault, the method for the LLLG fault can be used to calculate the fault location and all the related fault resistances. The fault resistances should be very large, infinity in theory, for nonfaulted phases. For example, for an AG fault, the calculated values for the fault resistances related to phase B and phase C will be very large. Consequently, this method will find out the fault type as a byproduct. An example is provided below.

The sample nonradial power system shown in figure 4 of [4] is utilized here for illustrating this concept. For example, for an AG fault with fault resistance of 1 Ω, fault location of 0.3 p.u. on a feeder section, the results assuming a fault type of LLLG fault will be: 0.3 p.u. for fault location and 0.1099, 8.8264e4, 2.9349e5, and 0.8901 Ω for the fault resistances. Due to the large value of the second and third fault resistance, it shows that the fault is an AG fault and the fault resistance is $(0.1099 + 0.8901) = 1.0\,\Omega$. In another example, for an AB fault with fault resistance of 1 Ω, fault location of 0.3 p.u., the results assuming a fault type of LLLG fault would be: 0.3 p.u. for fault location, and 0.9589, 0.0410, 2.0003e5, and 9.3062e4 Ω for the fault resistances. This indicates that the fault is actually an AB fault. However, this method imposes more computational burden than if the fault type is provided and if the method specifically designed for each type of fault is utilized.

Another fault location method that does not require fault type is described in the next section.

9.4 Fault location method without requiring fault type

This section presents the fault location algorithm that does not require the fault type information [12]. Fault impedance is usually resistive and therefore only consumes real power. A fault location method can be designed based on this. In the following section, the total complex power consumed by the fault resistances is derived.

Without losing generality, a three-phase fault is assumed as illustrated in Figure 9.3, which is a ground fault. When the impedance z_g is removed, it will be a non-ground fault. In the figure, E_1, E_2, and E_3 are the phase voltages with reference to a specific point, and E_0 is the neutral voltage with reference to the same point. I_1, I_2, I_3, and I_0 are the currents flowing through the fault impedances z_1, z_2, z_3, and z_g, respectively.

For a nonground fault, the total complex power consumed by the fault impedances is calculated as

$$\begin{aligned} S &= (E_1 - E_0)I_1^* + (E_2 - E_0)I_2^* + (E_3 - E_0)I_3^* \\ &= E_1I_1^* + E_2I_2^* + E_3I_3^* - E_0(I_1^* + I_2^* + I_3^*) \\ &= E_1I_1^* + E_2I_2^* + E_3I_3^* \end{aligned} \tag{9.16}$$

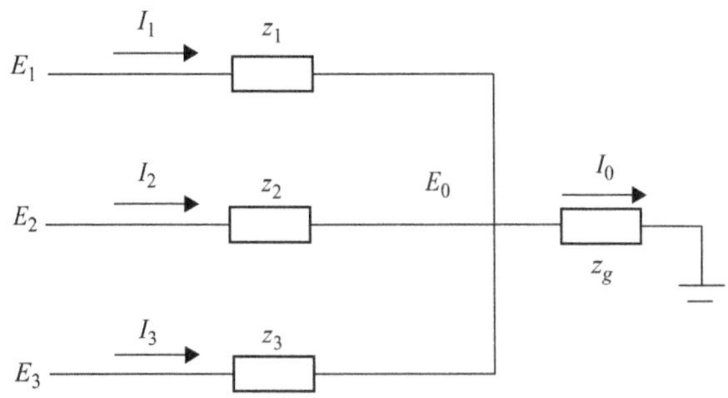

Figure 9.3 Illustration of complex power calculation

In the last step of the simplification, the equality identity $I_1 + I_2 + I_3 = 0$ is used. In (9.16), the symbol "*" denotes complex conjugate.

For a ground fault, the ground will be the voltage reference point. The total complex power consumed by the fault impedances is calculated as

$$\begin{aligned} S &= (E_1 - E_0)I_1^* + (E_2 - E_0)I_2^* + (E_3 - E_0)I_3^* + E_0 I_0^* \\ &= E_1 I_1^* + E_2 I_2^* + E_3 I_3^* - E_0(I_1^* + I_2^* + I_3^* - I_0^*) \\ &= E_1 I_1^* + E_2 I_2^* + E_3 I_3^* \end{aligned} \tag{9.17}$$

In the last step of the simplification, the equality identity $I_1 + I_2 + I_3 = I_0$ is used.

It is thus proved that the general equation for calculating the complex power consumed by the fault impedances for any type of fault can be written as

$$S = E_1 I_1^* + E_2 I_2^* + E_3 I_3^* \tag{9.18}$$

The fault location algorithm that does not require fault type is derived as follows [12]. Without losing generality, a three-phase fault is examined. Based on superimposition theory and the meaning of the transfer impedance, the voltage change due to the fault or superimposed voltage at any bus k can be written as

$$\boldsymbol{\Delta E_k} = -\boldsymbol{Z_{kr} I_f} \tag{9.19}$$

where,

$$\boldsymbol{\Delta E_k} = [\Delta E_{k_1} \quad \Delta E_{k_2} \quad \Delta E_{k_3}]^T \tag{9.20}$$

$$\boldsymbol{Z_{kr}} = \begin{bmatrix} Z_{k_1 r_1} & Z_{k_1 r_2} & Z_{k_1 r_3} \\ Z_{k_2 r_1} & Z_{k_2 r_2} & Z_{k_2 r_3} \\ Z_{k_3 r_1} & Z_{k_3 r_2} & Z_{k_3 r_3} \end{bmatrix} \tag{9.21}$$

$$\boldsymbol{I_f} = [I_{f1} \quad I_{f2} \quad I_{f3}]^T \tag{9.22}$$

$\boldsymbol{\Delta E}_k$: superimposed voltage at bus k, which is assumed to consist of nodes k_1, k_2, and k_3. In practice, only the nodes existing in the system will appear in the equation.

It follows from (9.19) that

$$\boldsymbol{I}_f = -(\boldsymbol{Z}_{kr}^T \boldsymbol{Z}_{kr})^{-1}(\boldsymbol{Z}_{kr}^T \boldsymbol{\Delta E}_k) \tag{9.23}$$

Note that in obtaining (9.23), the pseudo-inverse of matrix $\boldsymbol{Z}_{kr}$ is used because $\boldsymbol{Z}_{kr}$ may not always be a square matrix. For example, for a fault that occurs on a two-phase feeder, $\boldsymbol{Z}_{kr}$ will have a dimension of 3×2, necessitating the use of pseudo-inverse.

The voltage at fault nodes during the fault is given by

$$\boldsymbol{E}_r = \boldsymbol{E}_{r0} - \boldsymbol{Z}_{rr}\boldsymbol{I}_f \tag{9.24}$$

where,

$$\boldsymbol{E}_r = [E_{r_1} \quad E_{r_2} \quad E_{r_3}]^T \tag{9.25}$$

$$\boldsymbol{E}_{r0} = [E_{r_10} \quad E_{r_20} \quad E_{r_30}]^T \tag{9.26}$$

$$\boldsymbol{Z}_{rr} = \begin{bmatrix} Z_{r_1r_1} & Z_{r_1r_2} & Z_{r_1r_3} \\ Z_{r_2r_1} & Z_{r_2r_2} & Z_{r_2r_3} \\ Z_{r_3r_1} & Z_{r_3r_2} & Z_{r_3r_3} \end{bmatrix} \tag{9.27}$$

$\boldsymbol{E}_r, \boldsymbol{E}_{r0}$: voltage at fault nodes r during the fault and preceding the fault, respectively, and T symbolizes matrix and vector transpose.

Based on (9.18), the total complex power consumed by the fault resistances is calculated as

$$S = \boldsymbol{E}_r^T \boldsymbol{I}_f^* \tag{9.28}$$

The fault resistances only consumes real power, so the imaginary part of S is zero

$$\text{Imag}(S) = 0 \tag{9.29}$$

where, Imag(.) yields the imaginary part of its argument.

Then, Newton–Raphson technique can be utilized to solve (9.29) to yield the unknown fault location. An initial value of 0.5 can be used. It can be seen from the derivation that the fault location method is applicable to multisource systems with DGs. The method is certainly applicable to radial systems. The method requires the source impedances to be known, which may be directly provided or can be estimated based on the voltage and current changes due to the fault at the generation sites if such measurements are available. Similarly, optimization theory can be applied based on multiple measurements to improve the reliability and accuracy of fault location estimation.

Alternatively, if the superimposed voltage at another bus l is available, which can be written as $\Delta\boldsymbol{E}_l = -\boldsymbol{Z}_{lr}\boldsymbol{I}_f$, then the fault currents can be eliminated as $(\boldsymbol{Z}_{kr}^T\boldsymbol{Z}_{kr})^{-1}(\boldsymbol{Z}_{kr}^T\Delta\boldsymbol{E}_k) = (\boldsymbol{Z}_{lr}^T\boldsymbol{Z}_{lr})^{-1}(\boldsymbol{Z}_{lr}^T\Delta\boldsymbol{E}_l)$, from which the fault location can be determined.

When the system is radial, the following formulation can be employed to derive the fault location without the need of source impedances. The method is also based on bus impedance matrix. Note that the development of bus impedance matrix in this method will exclude the source impedances.

The fault location algorithm, based on voltages and currents measured at the local substation, is illustrated by examining a fault on a three phase feeder. The method will be applicable to single and two phase feeder as well. No fault type information is required. Without losing generality, the local substation is assumed to have a bus number k, which consists of nodes k_1, k_2, and k_3. The voltage during the fault at the local substation can be written as

$$\boldsymbol{E}_k = \boldsymbol{Z}_{kk}\boldsymbol{I}_k - \boldsymbol{Z}_{kr}\boldsymbol{I}_f \tag{9.30}$$

where, $\boldsymbol{I}_k$ is composed of the phase current measurements at the substation.

The fault current is thus obtained as

$$\boldsymbol{I}_f = (\boldsymbol{Z}_{kr}{}^T\boldsymbol{Z}_{kr})^{-1}\{\boldsymbol{Z}_{kr}{}^T(\boldsymbol{Z}_{kk}\boldsymbol{I}_k - \boldsymbol{E}_k)\} \tag{9.31}$$

The voltage at the fault nodes during the fault is given by

$$\boldsymbol{E}_r = \boldsymbol{Z}_{kr}^T\boldsymbol{I}_k - \boldsymbol{Z}_{rr}\boldsymbol{I}_f \tag{9.32}$$

Based on (9.32), the complex power consumed by the fault resistances is calculated as

$$S = \boldsymbol{E}_r^T((\boldsymbol{Z}_{kr}^T\boldsymbol{Z}_{kr})^{-1}\{\boldsymbol{Z}_{kr}^T(\boldsymbol{Z}_{kk}\boldsymbol{I}_k - \boldsymbol{E}_k)\})^* \tag{9.33}$$

where, "*" denotes complex conjugate.

Because the fault resistances only consume real power, the reactive power consumed is zero as shown below.

$$\text{Imag}(S) = 0 \tag{9.34}$$

Then, Newton–Raphson technique can be utilized to solve (9.29) or (9.34) to yield the unknown fault location. In comparison with the fault location method presented in Section 9.3, this fault-type–free method is general and applicable to any type of fault, and dispenses with the fault-type identification step. However, for certain types of faults such as LG faults, the method in Section 9.3 is more computationally efficient. In practice, both methods may be adopted to corroborate each other.

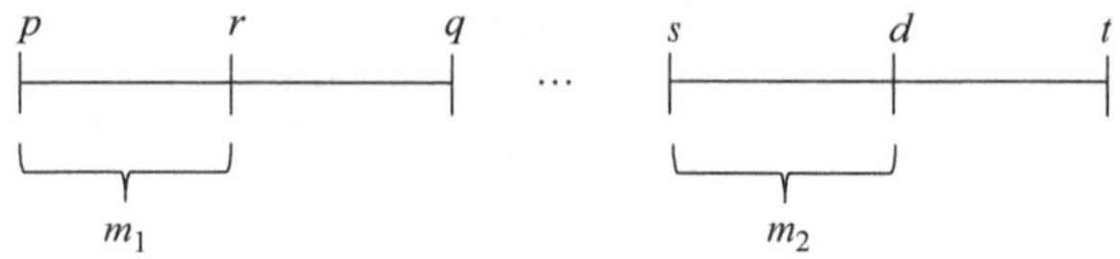

Figure 9.4 Illustration of multiple faults

9.5 Fault location method for multiple simultaneous faults

This section presents fault location methods for multiple simultaneous faults. The described methods will be applicable to unbalanced radial or nonradial meshed networks with DGs and consider shunt capacitances. Figure 9.4 depicts a one-line diagram of a network with two simultaneous faults that occur at different locations. The two faults occur on line section *pq* and *st*, respectively. Fictitious bus *r* is added on line section *pq* and bus *d* is added on line section *st* Suppose that the fault points are m_1 in per unit from bus *p*, and m_2 in per unit from bus *s*.

9.5.1 Fault location method without requiring fault type

This method is applicable to any type of faults and does not require the fault type information. A fault node is added for each phase of the faulted feeder section, regardless of the fault type. In Figure 9.4, bus *r* or *d* may comprise multiple nodes depending on the number of phases of the faulted line segment; there will be one node added for a single-phase line, two nodes for a two-phase line, and three nodes for a three-phase line. The fault currents at the nodes of buses *r* and *d* are denoted as $\boldsymbol{I}_f$ and $\boldsymbol{I}_g$, respectively.

Now assume that measurements from two buses *k* and *l* are available. Then the superimposed voltage quantities for the nodes of the buses are obtained as

$$\begin{bmatrix} \Delta \boldsymbol{E}_k \\ \Delta \boldsymbol{E}_l \end{bmatrix} = - \begin{bmatrix} \boldsymbol{Z}_{kr} & \boldsymbol{Z}_{kd} \\ \boldsymbol{Z}_{lr} & \boldsymbol{Z}_{ld} \end{bmatrix} \begin{bmatrix} \boldsymbol{I}_f \\ \boldsymbol{I}_g \end{bmatrix} \tag{9.35}$$

Equation (9.35) can be rewritten as

$$\Delta \boldsymbol{E}_{kl} = -\boldsymbol{Z}_{klrd} \boldsymbol{I}_{fg} \tag{9.36-a}$$

Now assume that measurements from another two buses *i* and *j* are available,

$$\Delta \boldsymbol{E}_{ij} = -\boldsymbol{Z}_{ijrd} \boldsymbol{I}_{fg} \tag{9.36-b}$$

Eliminating the fault current $\boldsymbol{I}_{fg}$ from (9.36-a) and (9.36-b) yields

$$(\boldsymbol{Z}_{klrd}^T \boldsymbol{Z}_{klrd})^{-1} (\boldsymbol{Z}_{klrd}^T \Delta \boldsymbol{E}_{kl}) = (\boldsymbol{Z}_{ijrd}^T \boldsymbol{Z}_{ijrd})^{-1} (\boldsymbol{Z}_{ijrd}^T \Delta \boldsymbol{E}_{ij}) \tag{9.36-c}$$

Then, the two unknown fault locations can be derived by separating the above equation into real and imaginary parts and solving them.

If the measurements from only two buses k and l are available, the fault location method is discussed below. It follows from (9.36-a) that

$$\boldsymbol{I}_{fg} = -(\boldsymbol{Z}_{klrd}^{T}\boldsymbol{Z}_{klrd})^{-1}(\boldsymbol{Z}_{klrd}^{T}\Delta\boldsymbol{E}_{kl}) \tag{9.37}$$

The voltages at the fault nodes during the fault are

$$\begin{bmatrix} \boldsymbol{E}_r \\ \boldsymbol{E}_d \end{bmatrix} = \begin{bmatrix} \boldsymbol{E}_{r0} \\ \boldsymbol{E}_{d0} \end{bmatrix} - \begin{bmatrix} \boldsymbol{Z}_{rr} & \boldsymbol{Z}_{rd} \\ \boldsymbol{Z}_{rd} & \boldsymbol{Z}_{dd} \end{bmatrix} \begin{bmatrix} \boldsymbol{I}_f \\ \boldsymbol{I}_g \end{bmatrix} \tag{9.38}$$

$\boldsymbol{E}_{r0}$ and $\boldsymbol{E}_{d0}$ can be expressed as functions of the fault locations similar to (9.6). Equation (9.38) are rewritten as

$$\boldsymbol{E}_{rd} = \boldsymbol{E}_{rd0} - \boldsymbol{Z}_{rdrd}\boldsymbol{I}_{fg} \tag{9.39}$$

The complex powers consumed by the fault resistances at the two fault locations are

$$S_r = \boldsymbol{E}_r^T\boldsymbol{I}_f^* \tag{9.40}$$

$$S_d = \boldsymbol{E}_d^T\boldsymbol{I}_g^* \tag{9.41}$$

The imaginary part of the complex power is zero

$$\text{Imag}(S_r) = 0 \tag{9.42}$$

$$\text{Imag}(S_d) = 0 \tag{9.43}$$

Equations (9.42) and (9.43) are thus utilized to find the two unknown fault locations. This method needs a sufficient number of measurements. It is noted that the voltage, current, and impedance matrix have appropriate sizes in the above equations. For example, $\Delta\boldsymbol{E}_k$ is a column vector consisting of voltage changes for all the nodes of bus k. $\boldsymbol{I}_f$ is a column vector consisting of fault currents of all the phases through the fault resistances at bus r; since no fault type is assumed, all the phases are included. $\boldsymbol{Z}_{kr}$ is a matrix consisting of transfer impedance between nodes of bus k and nodes of bus r; The first row corresponds to the first node of bus k, the second row for the second node of bus k, etc. The elements in $\boldsymbol{Z}_{kr}$ and $\boldsymbol{Z}_{rr}$ are functions of m_1, and the elements in $\boldsymbol{Z}_{kd}$ and $\boldsymbol{Z}_{dd}$ are functions of m_2, with similar expressions to (9.1–9.3). The elements in $\boldsymbol{Z}_{rd}$ are functions of m_1 and m_2, which can be derived as

$$\boldsymbol{Z}_{rd} = (\boldsymbol{w}_{st})^{-1}\left(\frac{\boldsymbol{Z}_{rs}}{m_2} + \frac{\boldsymbol{Z}_{rt}}{1-m_2}\right) \tag{9.44}$$

where $\boldsymbol{w}_{st}$ is similar to $\boldsymbol{w}$ in (9.3).

It is observed from the derivation that this method obviates the need of fault type information. This method can be easily extended to cases where there are more than two simultaneous faults occurring in the system. After fault locations are obtained, the fault currents through the fault resistances at the fault points can be calculated, from which the fault type can be determined [12]. Since this method does not require fault type, it is very well suited to evolving faults where the fault evolves from one type to another.

9.5.2 Fault location method based on the information of fault type

If the fault type is already obtained, an alternative method that may be more computationally efficient is presented below. Let r_i represent the fault node i Fault nodes are added only on the faulted phases according to the fault type.

Figure 9.5 illustrates several selected fault types of two simultaneous faults such as LG-LG, LLG-LG, LL-LG, and LLLG-LLLG. LLG-LG indicates that the first fault is an LLG fault and the second one is an LG fault. For LLLG-LLLG, both faults are an LLLG fault, and this case will become an LLL-LLL scenario if the two grounding resistances are removed. Fault currents are shown in the figure and are labeled for the LL-LG case.

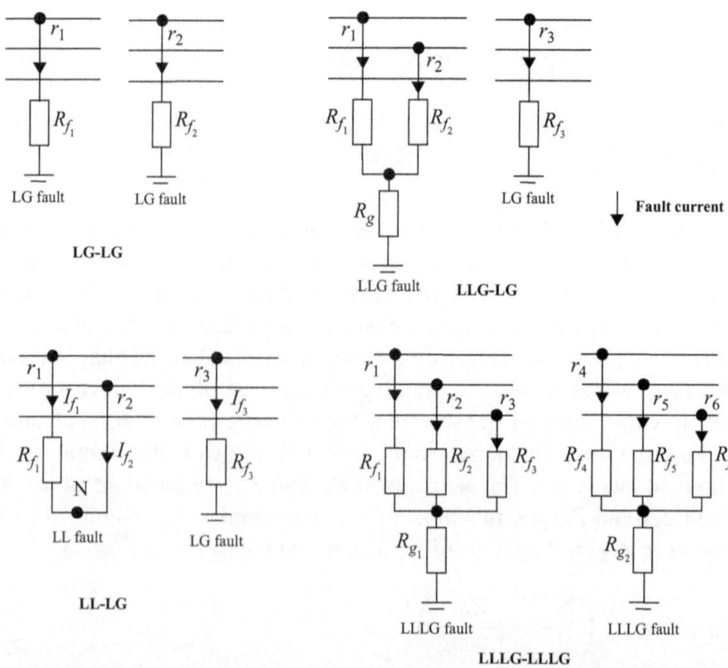

Figure 9.5 Illustration of fault types of multiple faults

For an LG-LG case, where two LG faults occur at nodes r_1 and r_2 at different locations, the fault currents through the fault resistances R_{f_1} and R_{f_2} are determined as

$$\begin{bmatrix} I_{f_1} \\ I_{f_2} \end{bmatrix} = \begin{bmatrix} Z_{r_1r_1} + R_{f_1} & Z_{r_1r_2} \\ Z_{r_1r_2} & Z_{r_2r_2} + R_{f_2} \end{bmatrix}^{-1} \begin{bmatrix} E_{r_10} \\ E_{r_20} \end{bmatrix} \tag{9.45}$$

The voltage change of the nodes at bus k is expressed as

$$\boldsymbol{\Delta E_k} = -\boldsymbol{Z_{kr_1}} I_{f_1} - \boldsymbol{Z_{kr_2}} I_{f_2} \tag{9.46}$$

Similar equations can be derived based on measurements at other buses. Since the transfer and driving-point impedance is a function of fault locations m_1 and m_2, these equations can be used to derive the fault locations. For different types of faults, the general idea for fault location is the same except that the expressions of the fault currents differ as illustrated below.

Suppose that an LLG fault (R_{f_1}, R_{f_2}, R_g) occurs at one location and an LG fault (R_{f_3}) occurs at another location, the fault currents through the fault resistances are obtained as follows.

On one hand, the fault voltages at the fault nodes are

$$\begin{bmatrix} E_{r_1} \\ E_{r_2} \\ E_{r_3} \end{bmatrix} = \begin{bmatrix} E_{r_10} \\ E_{r_20} \\ E_{r_30} \end{bmatrix} - \begin{bmatrix} Z_{r_1r_1} & Z_{r_1r_2} & Z_{r_1r_3} \\ Z_{r_1r_2} & Z_{r_2r_2} & Z_{r_2r_3} \\ Z_{r_1r_3} & Z_{r_2r_3} & Z_{r_3r_3} \end{bmatrix} \begin{bmatrix} I_{f_1} \\ I_{f_2} \\ I_{f_3} \end{bmatrix} \tag{9.47}$$

On the other hand, the fault voltages at the fault nodes can also be written as

$$\begin{bmatrix} E_{r_1} \\ E_{r_2} \\ E_{r_3} \end{bmatrix} = \begin{bmatrix} (R_{f_1} + R_g)I_{f_1} + R_g I_{f_2} \\ R_g I_{f_1} + (R_{f_2} + R_g)I_{f_2} \\ R_{f_3} I_{f_3} \end{bmatrix} = \begin{bmatrix} R_{f_1} + R_g & R_g & 0 \\ R_g & R_{f_2} + R_g & 0 \\ 0 & 0 & R_{f_3} \end{bmatrix} \begin{bmatrix} I_{f_1} \\ I_{f_2} \\ I_{f_3} \end{bmatrix} \tag{9.48}$$

It follows from (9.47) and (9.48) that

$$\begin{bmatrix} I_{f_1} \\ I_{f_2} \\ I_{f_3} \end{bmatrix} = \begin{bmatrix} Z_{r_1r_1} + R_{f_1} + R_g & Z_{r_1r_2} + R_g & Z_{r_1r_3} \\ Z_{r_1r_2} + R_g & Z_{r_2r_2} + R_{f_2} + R_g & Z_{r_2r_3} \\ Z_{r_1r_3} & Z_{r_2r_3} & Z_{r_3r_3} + R_{f_3} \end{bmatrix}^{-1} \begin{bmatrix} E_{r_10} \\ E_{r_20} \\ E_{r_30} \end{bmatrix} \tag{9.49}$$

Suppose that an LL fault (R_{f_1}) occurs at one location and an LG fault (R_{f_3}) occurs at another location, the fault currents through the fault resistances are derived as follows.

The fault voltages at the fault nodes are

$$\begin{bmatrix} E_{r_1} \\ E_{r_2} \\ E_{r_3} \end{bmatrix} = \begin{bmatrix} E_{r_10} \\ E_{r_20} \\ E_{r_30} \end{bmatrix} - \begin{bmatrix} Z_{r_1r_1} & Z_{r_1r_2} & Z_{r_1r_3} \\ Z_{r_2r_1} & Z_{r_2r_2} & Z_{r_2r_3} \\ Z_{r_3r_1} & Z_{r_3r_2} & Z_{r_3r_3} \end{bmatrix} \begin{bmatrix} I_{f1} \\ I_{f2} \\ I_{f3} \end{bmatrix} \tag{9.50}$$

On the other hand, the fault voltages at the fault nodes can also be written as

$$\begin{bmatrix} E_{r_1} \\ E_{r_2} \\ E_{r_3} \end{bmatrix} = \begin{bmatrix} R_{f1} & 0 & 0 \\ 0 & 0 & 0 \\ 0 & 0 & R_{f3} \end{bmatrix} \begin{bmatrix} I_{f1} \\ I_{f2} \\ I_{f3} \end{bmatrix} + \begin{bmatrix} E_N \\ E_N \\ 0 \end{bmatrix} \tag{9.51}$$

where E_N represents the voltage of node N.

Combining (9.50) and (9.51) results in

$$\begin{bmatrix} E_{r_10} \\ E_{r_20} \\ E_{r_30} \end{bmatrix} = \begin{bmatrix} Z_{r_1r_1} + R_{f1} & Z_{r_1r_2} & Z_{r_1r_3} \\ Z_{r_2r_1} & Z_{r_2r_2} & Z_{r_2r_3} \\ Z_{r_3r_1} & Z_{r_3r_2} & Z_{r_3r_3} + R_{f3} \end{bmatrix} \begin{bmatrix} I_{f1} \\ I_{f2} \\ I_{f3} \end{bmatrix} + \begin{bmatrix} E_N \\ E_N \\ 0 \end{bmatrix} \tag{9.52}$$

Subtracting the first row from the second row in (9.52) gives

$$\begin{bmatrix} E_{r_20} - E_{r_10} \\ E_{r_30} \end{bmatrix} = \begin{bmatrix} Z_{r_2r_1} - Z_{r_1r_1} - R_{f1} & Z_{r_2r_2} - Z_{r_1r_2} & Z_{r_2r_3} - Z_{r_1r_3} \\ Z_{r_3r_1} & Z_{r_3r_2} & Z_{r_3r_3} + R_{f3} \end{bmatrix} \begin{bmatrix} I_{f1} \\ I_{f2} \\ I_{f3} \end{bmatrix} \tag{9.53}$$

Since the first fault is an LL fault, the following equation holds

$$I_{f1} + I_{f2} = 0 \tag{9.54}$$

Combining (9.53) and (9.54) leads to the following expression of the fault currents:

$$\begin{bmatrix} I_{f1} \\ I_{f2} \\ I_{f3} \end{bmatrix} = \begin{bmatrix} Z_{r_2r_1} - Z_{r_1r_1} - R_{f1} & Z_{r_2r_2} - Z_{r_1r_2} & Z_{r_2r_3} - Z_{r_1r_3} \\ Z_{r_3r_1} & Z_{r_3r_2} & Z_{r_3r_3} + R_{f3} \\ 1 & 1 & 0 \end{bmatrix}^{-1} \begin{bmatrix} E_{r_20} - E_{r_10} \\ E_{r_30} \\ 0 \end{bmatrix} \tag{9.55}$$

Suppose that an LLL fault (R_{f1}, R_{f2}, R_{f3}) occurs at one location and an LLL fault (R_{f4}, R_{f5}, R_{f6}) occurs at another location. The fault currents through the fault resistances are obtained as

$$\begin{bmatrix} I_{f_1} \\ I_{f_2} \\ I_{f_3} \\ I_{f_4} \\ I_{f_5} \\ I_{f_6} \end{bmatrix} = \begin{bmatrix} Z_{r_2r_1} - Z_{r_1r_1} - R_{f_1} & Z_{r_2r_2} - Z_{r_1r_2} + R_{f_2} & Z_{r_2r_3} - Z_{r_1r_3} & Z_{r_2r_4} - Z_{r_1r_4} & Z_{r_2r_5} - Z_{r_1r_5} & Z_{r_2r_6} - Z_{r_1r_6} \\ Z_{r_3r_1} - Z_{r_1r_1} - R_{f_1} & Z_{r_3r_2} - Z_{r_1r_2} & Z_{r_3r_3} - Z_{r_1r_3} + R_{f_3} & Z_{r_3r_4} - Z_{r_1r_4} & Z_{r_3r_5} - Z_{r_1r_5} & Z_{r_3r_6} - Z_{r_1r_6} \\ 1 & 1 & 1 & 0 & 0 & 0 \\ Z_{r_5r_1} - Z_{r_4r_1} & Z_{r_5r_2} - Z_{r_4r_2} & Z_{r_5r_3} - Z_{r_4r_3} & Z_{r_5r_4} - Z_{r_4r_4} - R_{f_4} & Z_{r_5r_5} - Z_{r_4r_5} + R_{f_5} & Z_{r_5r_6} - Z_{r_4r_6} \\ Z_{r_6r_1} - Z_{r_4r_1} & Z_{r_6r_2} - Z_{r_4r_2} & Z_{r_6r_3} - Z_{r_4r_3} & Z_{r_6r_4} - Z_{r_4r_4} - R_{f_4} & Z_{r_6r_5} - Z_{r_4r_5} & Z_{r_6r_6} - Z_{r_4r_6} + R_{f_6} \\ 0 & 0 & 0 & 1 & 1 & 1 \end{bmatrix}^{-1} \begin{bmatrix} E_{r_2}0 - E_{r_1}0 \\ E_{r_3}0 - E_{r_1}0 \\ 0 \\ E_{r_5}0 - E_{r_4}0 \\ E_{r_6}0 - E_{r_4}0 \\ 0 \end{bmatrix} \tag{9.56}$$

Note that the sum of I_{f_1}, I_{f_2}, and I_{f_3} is zero, and the sum of I_{f_4}, I_{f_5}, and I_{f_6} is zero for LLL-LLL fault, as seen from (9.56).

Suppose that an LLLG fault $(R_{f_1}, R_{f_2}, R_{f_3}, R_{g_1})$ occurs at one location and an LLLG fault $(R_{f_4}, R_{f_5}, R_{f_6}, R_{g_2})$ occurs at another location. The fault currents through the fault resistances are given by

$$\begin{bmatrix} I_{f_1} \\ I_{f_2} \\ I_{f_3} \\ I_{f_4} \\ I_{f_5} \\ I_{f_6} \end{bmatrix} = \begin{bmatrix} Z_{r_1r_1} + R_{f_1} + R_{g_1} & Z_{r_1r_2} + R_{g_1} & Z_{r_1r_3} + R_{g_1} & Z_{r_1r_4} & Z_{r_1r_5} & Z_{r_1r_6} \\ Z_{r_2r_1} + R_{g_1} & Z_{r_2r_2} + R_{f_2} + R_{g_1} & Z_{r_2r_3} + R_{g_1} & Z_{r_2r_4} & Z_{r_2r_5} & Z_{r_2r_6} \\ Z_{r_3r_1} + R_{g_1} & Z_{r_3r_2} + R_{g_1} & Z_{r_3r_3} + R_{f_3} + R_{g_1} & Z_{r_3r_4} & Z_{r_3r_5} & Z_{r_3r_6} \\ Z_{r_4r_1} & Z_{r_4r_2} & Z_{r_4r_3} & Z_{r_4r_4} + R_{f_4} + R_{g_2} & Z_{r_4r_5} + R_{g_2} & Z_{r_4r_6} + R_{g_2} \\ Z_{r_5r_1} & Z_{r_5r_2} & Z_{r_5r_3} & Z_{r_5r_4} + R_{g_2} & Z_{r_5r_5} + R_{f_5} + R_{g_2} & Z_{r_5r_6} + R_{g_2} \\ Z_{r_6r_1} & Z_{r_6r_2} & Z_{r_6r_3} & Z_{r_6r_4} + R_{g_2} & Z_{r_6r_5} + R_{g_2} & Z_{r_6r_6} + R_{f_6} + R_{g_2} \end{bmatrix}^{-1} \begin{bmatrix} E_{r_1}0 \\ E_{r_2}0 \\ E_{r_3}0 \\ E_{r_4}0 \\ E_{r_5}0 \\ E_{r_6}0 \end{bmatrix} \tag{9.57}$$

Fault scenarios with other fault type combinations can be dealt with similarly. It is understood that the methods presented in Section 9.5 can be readily extended to double-circuit lines or multiple phase lines with any number of phases without restrictions.

9.6 Sample calculations and discussions on the presented fault location algorithms

This section first presents sample calculations based on a three-bus system and then presents discussions of the described fault location algorithms.

9.6.1 Sample calculations

This sample system as depicted in Figure 9.6 is used for illustrating the described methods. The system is a 12.47 kV system consisting of a three-phase feeder between bus 1 and 2, and a single-phase lateral between bus 2 and 3. A phase-A load of 120 kVA with a power factor 0.9 is present at bus 3. The lengths of feeders in miles are labeled. The impedance parameters are shown as follows:

Source impedance of source 1:
Positive-sequence: $0.23 + j2.10\,\Omega$, zero-sequence: $0.15 + j1.47\,\Omega$.
Source impedance of source 2:
Positive-sequence: $1.26 + j12.7\,\Omega$, zero-sequence: $1.15 + j11.9\,\Omega$.

The feeder series impedance matrices in ohms per mile and shunt admittance matrix in siemens per mile are given as follows [4].

For the main feeder, the impedance matrix is

$$\begin{bmatrix} 0.7982 + j0.4463 & 0.3192 + j0.0328 & 0.2849 - j0.0143 \\ 0.3192 + j0.0328 & 0.7891 + j0.4041 & 0.3192 + j0.0328 \\ 0.2849 - j0.0143 & 0.3192 + j0.0328 & 0.7982 + j0.4463 \end{bmatrix}$$

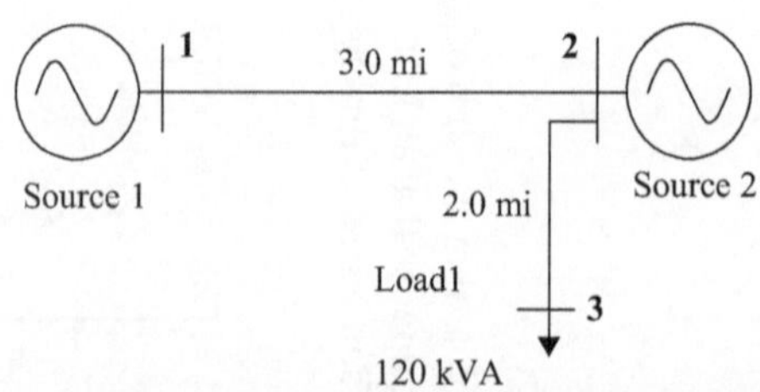

Figure 9.6 A sample three-bus system for fault location illustration

and the admittance matrix is

$$\begin{bmatrix} j96.8897 & 0.0000 & 0.0000 \\ 0 & j96.8897 & 0.0000 \\ 0 & 0 & j96.8897 \end{bmatrix} \times 1\text{e-}6$$

For the single-phase lateral, the impedance matrix is [1.3425 + j0.5124] and the admittance matrix is [j88.9912]*1e-6.

Suppose that an LG fault occurs on the lateral 2-3, with a fault distance of 0.3 p.u. from bus 2 and fault resistance of 5 Ω.

Bus 1 and bus 2 consist of three nodes, respectively. Bus 3 contains 1 node. The nodes are numbered as follows: bus 1 (nodes 1, 2, 3), bus 2 (nodes 4, 5, 6), and bus 3 (node 7). The fictitious fault node is numbered as node 8. We have $k = 1, p = 4, q = 7$, and $r = 8$. The voltage change at node 1 due to the fault, designated as ΔE_1, is utilized to locate the fault. Let $E_{4,0}$, $E_{7,0}$, and $E_{8,0}$ denote the prefault voltage of nodes 4, 7, and 8, respectively. The following expressions are obtained. Note that the unit for voltage is volt and the unit for impedance is ohm.

$$Z_{1,8} = w^{-1}\left(\frac{Z_{1,4}}{m} + \frac{Z_{1,7}}{1-m}\right) \tag{9.58}$$

$$Z_{8,8} = w^{-1}\left(\frac{Z_{4,8}}{m} + \frac{Z_{7,8}}{1-m} + 2.6850 + \mathrm{j}1.0248\right) \tag{9.59}$$

$$w = -9.1198 \times 10^{-5} + \mathrm{j}2.3894 \times 10^{-4} + \frac{1}{m(1-m)} \tag{9.60}$$

$$Z_{4,8} = w^{-1}\left(\frac{Z_{4,4}}{m} + \frac{Z_{4,7}}{1-m}\right) \tag{9.61}$$

$$Z_{7,8} = w^{-1}\left(\frac{Z_{4,7}}{m} + \frac{Z_{7,7}}{1-m}\right) \tag{9.62}$$

$$E_{8,0} = w^{-1}\left(\frac{E_{4,0}}{m} + \frac{E_{7,0}}{1-m}\right) \tag{9.63}$$

$$\Delta E_1 + \frac{Z_{1,8}E_{8,0}}{Z_{8,8} + R_{f1}} = 0 \tag{9.64}$$

where,
$Z_{1,4} = -2.4416 \times 10^{-2} + \mathrm{j}1.4586$
$Z_{1,7} = -2.4740 \times 10^{-2} + \mathrm{j}1.4491$
$Z_{4,4} = 1.6229 + \mathrm{j}2.8012$
$Z_{4,7} = 1.6114 + \mathrm{j}2.7835$
$Z_{7,7} = 4.2673 + \mathrm{j}3.7850$
$E_{4,0} = 7163.28 - \mathrm{j}33.41$
$E_{7,0} = 7116.77 - \mathrm{j}30.83$
$\Delta E_1 = -484.11 - \mathrm{j}1200.66$

Note that a comma is added between the row and column index of the bus impedance matrix for clarity. For instance, $Z_{1,4}$ signifies the first row and fourth column of the bus impedance matrix.

Equation (9.64) is a nonlinear complex equation involving two unknown variables, the fault location m and fault resistance R_{f_1}. Separating it into real and imaginary parts forms two real equations, based on which the unknowns can be obtained by adopting the Newton–Raphson technique. In this example, the result $m = 0.3$ and $R_{f_1} = 5.0$ is reached after five iterations.

It is worth noting that similar equations to those shown above will be obtained regardless of the size of the network, and certainly the impedance constants may be different. The key point is that the computational complexity of fault location will not increase due to the increasing size of the network. For a large network, the main computational requirement is the bus impedance matrix construction, which can be readily derived by applying the well-established procedure.

9.6.2 Discussions on the presented fault location algorithms

The discussions presented in this section are applicable to fault location algorithms that require or do not require fault-type information.

To mitigate the impacts of load variations, a method similar to that discussed in [2,4] can be utilized to compensate the load variations. The basic idea is to calculate the load level based on the measured prefault voltages and currents at the substation, and then calculate the static load impedance accordingly. In the future smart distribution systems with more AMI deployment, the AMI data can be harnessed to derive a more accurate real-time load model.

Moreover, current measurements, when available, can also be exploited for fault location. For the feeder or lateral section between bus l_1 and l_2, its currents are

$$\boldsymbol{I}_{l_1 l_2} = z_{l_1 l_2}^{-1}(\boldsymbol{E}_{l_1} - \boldsymbol{E}_{l_2}) \tag{9.65}$$

where, $\boldsymbol{I}_{l_1 l_2}$ is vector of currents flowing from bus l_1 to l_2, $z_{l_1 l_2}$is impedance matrix of the considered feeder or lateral section, and $\boldsymbol{E}_{l_1}$ and $\boldsymbol{E}_{l_2}$ are bus voltage vectors.

Since bus voltages can be expressed as a function of the fault location, the currents can be formulated as a function of the fault location. Thus, the fault location can be solved by using currents.

In addition, the above discussions utilize the nominal PI circuit to model the feeder shunt capacitances. If the system has longer feeders and thus requires higher modeling accuracy, distributed parameter line model may be adopted. The overall fault location method remains similar.

When measurements from multiple sites are available, optimal estimation theory may be applied to obtain an optimal estimate [11, 13]. The general idea is that multiple equations will be written based on available measurements. The measurement vector (consisting of measurements) and the function vector (consisting of equations involving fault locations) are related by

$$\boldsymbol{S} = \boldsymbol{F}(\boldsymbol{X}) + \boldsymbol{\mu} \tag{9.66}$$

where, $\boldsymbol{\mu}$ is characterized by:

$$\boldsymbol{R} = E(\boldsymbol{\mu}\boldsymbol{\mu}^T) = diag(\sigma_1^2, \ldots, \sigma_N^2) \tag{9.67}$$

$E(.)$ represents the expected value and $diag(.)$ symbols a diagonal matrix with diagonal elements reflecting the error variances of the measurement meters. N is the total number of measurements. Elements of R can be specified according to the accuracy of the meters, a smaller value of which indicates a more accurate measurement. If $\boldsymbol{R}$ is not known, the measurements can be assumed as equally accurate. The optimal estimate of $\boldsymbol{X}$ is obtained by minimizing the cost function defined as:

$$J = (\boldsymbol{S} - \boldsymbol{F}(\boldsymbol{X}))^T \boldsymbol{R}^{-1} (\boldsymbol{S} - \boldsymbol{F}(\boldsymbol{X})) \tag{9.68}$$

The solution can be derived following an iterative procedure [11, 13]. State estimation–based techniques that are capable of detecting and removing bad measurement may be utilized to enhance fault location accuracy.

By taking advantage of optimization theory, the algorithms will be capable of utilizing the data from the main substation, DG sites, automation equipment sites, remote substation, etc., and at the same time picking up the bad data and thus provide the optimal estimate of the fault location. As a byproduct, the optimal estimate of the measurements will also be generated.

When developing the bus impedance matrix, the system models suggested in [14, 15] will be adopted. Generally, a constant impedance model will be utilized for the load. A voltage source in series with a subtransient impedance is commonly used to represent a DG.

It is worth pointing out that when a meter is placed at a generator location to measure voltages and currents, the equivalent source impedance can be accurately estimated based on superimposed voltages and currents. The following equation holds

$$\boldsymbol{z_s} \boldsymbol{\Delta I_s} = \boldsymbol{\Delta E_s} \tag{9.69}$$

where, $\boldsymbol{z_s}$ is source impedance matrix, $\boldsymbol{\Delta I_s}$ is vector of superimposed currents at source location, and $\boldsymbol{\Delta E_s}$ is vector of superimposed voltages at source location.

Usually a source has equal self-impedance and equal mutual impedance for three phases. Hence $[z_s]$ contains two unknown variables, which can be readily obtained. Estimation of source impedance using measurement data may improve fault location accuracy.

Moreover, information such as distributed generator status, recloser status, feeder switching status, as well as loading information from SCADA/DMS and other monitoring systems can be exploited to update and maintain an accurate system topology and model. The developed methods may be integrated into a utility's SCADA/DMS system.

For ungrounded distribution systems, to form the bus impedance matrix, special measures need to be taken. One possible method is to add a fictitious impedance of known value at the measurement bus. The other method is to treat one phase at the measurement site as the reference phase or ground phase. Details are referred to Reference 16 and 17.

9.7 Fault location observability analysis

When measurements are available at only a limited number of sites such as the local substation, some points on the feeder may be unobservable in terms of fault location. In other words, using the available voltage and/or current measurements, it may not be possible to distinguish among some points where faults may occur. In other words, faults occurred on these points will generate the same voltage and current measurements at the recording sites. Sometimes, considering potential measurement errors such as 0.5% or 1%, the unobservable area would be larger than the unobservable area with exact measurements. As a result, the fault location algorithm may yield a list of possible fault location estimates.

When unobservable areas exist, more meters can be added to certain locations until the entire system becomes observable. An optimal meter placement scheme is presented based on network topology and impedance parameters in [18, 19]. The method is able to provide a set of buses where meters are installed so that each fault that occurs in the system can be uniquely determined, while the number of meters needed to be placed in the system is minimized.

9.8 Conclusions

This chapter discusses the challenges of distribution system fault location and presents possible algorithms to overcome some of the challenges. One type of fault location method assumes the fault type to be known and the other dispenses with fault type information. Fault location methods for a single fault and methods for simultaneous multiple faults are presented. When there are only limited measurements available, a portion or portions of the network may be unobservable in terms of fault location. More meters will be needed, and design of an optimal meter placement scheme is possible based on network topology and parameters. It is noted that a great deal of valuable literature on fault location has been published in the past. Interested readers are suggested to read more on this intriguing topic.

References

[1] R. Das, M.S. Sachdev, and T.S. Sidhu, "A fault locator for radial subtransmission and distribution lines," IEEE Power Engineering Society Summer Meeting, Seattle, WA, USA, July 16–20, 2000.
[2] R. Aggarwal, Y. Aslan, and A. Johns, "New concept in fault location for overhead distribution systems using superimposed components," IEE Proceedings—Generation, Transmission and Distribution, vol. 144, no. 3, pp. 309–316, 1997.
[3] A.A. Girgis, C.M. Fallon, and D.L. Lubkeman, "A fault location technique for rural distribution feeders," IEEE Transactions on Industrial Applications, vol. 29, no. 6, pp. 1170–1175, 1993.

[4] Y. Liao, "Generalized fault location methods for overhead electric distribution systems," IEEE Transactions on Power Delivery, vol. 26, no. 1, pp. 53–64. 2011.
[5] R. Krishnathevar and E.E. Ngu, "Generalized impedance-based fault location for distribution systems," IEEE Transactions on Power Delivery, vol. 27, no. 1, pp. 449–451, 2012.
[6] M.E. Baran, "Challenges in state estimation on distribution systems," IEEE Power Engineering Society Summer Meeting, Vancouver, BC, Canada, July 15–19, 2001.
[7] Yuan Liao, "Identification of faulted feeder section in distribution systems using overcurrent information from switching devices," International Journal of Emerging Electric Power Systems vol. 15, no. 2, 2014.
[8] Y. Liao, "Electric distribution system fault location considering shunt capacitances," Electric Power Components and Systems vol. 41, no. 5, pp. 519–536, 2013.
[9] T. Nguyen and Y. Liao, "Transmission line fault type classification based on novel features and neuro-fuzzy system," Electric Power Components and Systems, vol. 38, no. 6, pp. 695–709, 2010.
[10] B. Das, "Fuzzy logic-based fault-type identification in unbalanced radial power distribution system," IEEE Transactions on Power Delivery, vol. 21, no. 1, pp. 278–285, 2006.
[11] Y. Liao and M. Kezunovic, "Optimal estimate of transmission line fault location considering measurement errors," IEEE Transactions on Power Delivery, vol. 22, no. 3, pp. 1335–1341, 2007.
[12] Y. Liao, "A novel method for locating faults on distribution systems," Electric Power Systems Research Journal, vol. 117, pp. 21–26, 2014.
[13] J. Grainger and W. Stevenson, Power System Analysis, McGraw-Hill, Inc., New York, NY, USA, 1994.
[14] W.H. Kersting, "Radial distribution test feeders," IEEE Power Engineering Society Winter Meeting, Columbus, OH, USA, January 28–February 1, 2001.
[15] W.H. Kersting, Distribution System Modeling and Analysis, CRC Press, Taylor & Francis Group, Boca Raton, FL, USA, 2012.
[16] W. Xiu and Y. Liao, "Fault location methods for ungrounded distribution systems using local measurements," International Journal of Emerging Electric Power Systems, vol. 14, no. 5, pp. 467–476, 2013.
[17] W. Xiu and Y. Liao, "Novel fault location methods for ungrounded radial distribution systems using measurements at substation," Electric Power Systems Research, vol. 106, pp. 95–100, 2014.
[18] Y. Liao, "Fault location observability analysis and optimal meter placement based on voltage measurements," Electric Power Systems Research, vol. 79, no. 7, pp. 1062–1068, 2009.
[19] W. Xiu and Y. Liao, "Fault location observability analysis on power distribution systems," Electric Power Components and Systems, vol. 42, no. 16, pp. 1862–1871, 2014.

Chapter 10

Fault detection and classification in distribution network

Zhengyou He[1] *and Jianwei Yang*[1]

10.1 Introduction

Generally speaking, fault classification defines ten types of short-circuit faults: AG, BG, CG, ABG, ACG, BCG, ABC/ABCG, AB, AC, and BC, while fault identification is to identify the faulty feeder that should be cut off the distribution network. Significant amount of research works have been devoted to the classification and selection problem in transmission networks. This is because of the fact that accurate fault classification and selection are very important for relays to operate properly in transmission networks. As the power level of distribution networks increases gradually, effective fault classification and selection become increasingly necessary to facilitate fault location, proper operation of protection system and so forth for system operators.

Various fault identification methods are adopted in different countries. In the Former Soviet Union, most devices used for selecting the faulty feeder employed zero-sequence power direction and the first half-wave principle because most distribution networks adopted small current grounding which means that the distribution systems are ungrounded or grounded via arc-suppression coil. In Japan, power direction and zero-sequence overcurrent protection are major options to cut off faulty feeders in power, steel, and chemical industries, in which ungrounded systems and resistance-grounding systems are used in their power systems.

In Germany and France, resonant grounding is widely adopted to replace traditional grounding methods, such as the resistance neutral grounding method that has been used for decades. Early in the 1930s, German engineers had proposed the application of single-phase grounding protection scheme for transient faults, while lately in France, use of Prony method and wavelet method for extracting the transient components have also been developed for fault classification and selection.

In America, solid neutral grounding systems have been generally used (72% adoption rate) in 22 $\sim$ 70 kV power systems from the middle of 1920s to the middle

[1]School of Electrical Engineering, Southwest Jiaotong University, Chengdu, China

of 1940s. This technique developed rapidly and gradually replaced the ungrounded system in the past few decades.

In China, small current grounding systems, which include ungrounded system, neutral grounding system via arc-suppression coils and high resistance grounding system are widely deployed. Various kinds of high-performance fault-type identification techniques have been devised since the 1960s, aiming to solve the fault-type identification problem. With the rapid development and the widespread application of computer technology in recent years, some small-scale fault type identification and protection devices came out in succession: (i) a micro device for selection of faulty feeder which is based on the amplitude and phase angle of the fifth order harmonics of zero-sequence current [1,2]; (ii) the fault classification device based on the *S* signal injection method [3,4]; (iii) the microcomputer-based identification device using residual current increment method [5]; and (iv) the fault identification device based on wavelet analysis [6], etc. Lately, integrated methods employing transient states, steady states, and harmonics have been gradually applied to fault-type identification devices.

The strict operating condition of fault identification devices makes it essential to carry out more thorough and careful researches on the fault identification problem. Since hardly any single method can solve the problem completely, a comprehensive method that uses both transient and steady state information, and combines different fault features will be the focus of future research.

10.2 Fault type classification and fault data sources in distribution network

10.2.1 Information source of fault classification

When fault happens in a distribution network, e.g., some changes in the node voltage or the branch current can be detected. Then, the protection equipment identifies the faults and triggers protection actions or sends back alarm signals. Eventually, the circuit breaker will be activated to isolate the faulted branch. With the ever-developing science and technology, many automatic measuring facilities can record the failure information, thus providing enough data for fault classification in the distribution network. Considering different conditions of the worksites, these facilities first send the measured information up to the substation, then to the main station in the dispatching layer after preliminary processing of the data. The data source mainly consists of three types: the automatic information (received from the measurement and control substations), protection information and recorder information. The information flow is shown in Figure 10.1 [7].

10.2.1.1 Measurement and control information

Measurement and control devices submit the measurement information and the information regarding the status of the breaker to the integrated automation system

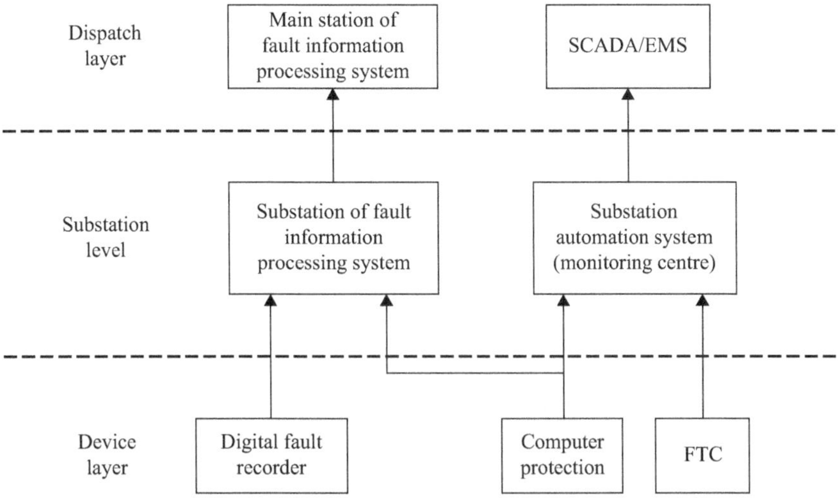

Figure 10.1 Information source for fault diagnosis of distribution systems

Subsequently the system preliminarily processes the information and submits it to the supervisory control and data acquisition (SCADA) system in the dispatching layer.

SCADA mainly stores three kinds of information: status information, measurement information and the information regarding the electrical quantities. The status information can be used for the fault classification system, including the breaker status, the switch status, status of the protection device and the alarm signal of feeder terminal unit (FTU).

Out of this information, the breaker information and the alarm signal of FTU are used for fault classification. These status variables are presented in binary format (using 0 and 1) These status variables can be divided into three groups for representing the facilities' locations, the switch status and the reset actions separately. The updating of SCADA system is processed on the second level, i.e., the substation level.

10.2.1.2 Protection information

The protection device submits the protection information to the integrated automation system and substation of fault information processing system.

The integrated automation system filters the information and submits it to the SCADA system in the dispatch layer. The protection and action information are sent to the onsite relay node.

Generally, fault information substations collect event reports broadcasted by the microprocessor-based protection devices, including protection action information, protection reports, etc. The substations first integrate and process the information, then submit it to the fault information main station so that fault analysis could be conducted in the dispatching layer with the received information. The time interval of the submission of protection information to the dispatching center is measured in seconds.

10.2.1.3 Recorded information

Fault recorders include concentrated recorders and distributed recorders. Concentrated recorders are employed where devices are mostly concentrated. On the other hand, distributed recorders record data on each distributed device and then process all the recorded data in a main control computer. Distributed recorders usually communicate with the main station through the recording network via RS232/485, dial-up network or Ethernet.

Substations could work as recording substations or fault information substations for different requirements. Whichever substation is connected, the information is eventually submitted to the dispatching center.

The recorded data of transient signals under fault condition are used for fault analysis and the evaluation of the behaviors of various protection systems. Thanks to the realization of technologies such as the recorder network, distant data transportation and integrated data processing, recorders from different manufacturers are supposed to be compatible with the unified data format – COMTRADE – for the convenience of analysis and processing. It is notable that the transmission delay of the recorded information to the dispatching center is about 10 minutes.

10.2.2 Scope of fault processing

Fault processing aims to determine the fault type and the fault location (faulty feeder identification) through collected data from measurement, control protection and recording devices. The principle of this mechanism is illustrated in Figure 10.2.

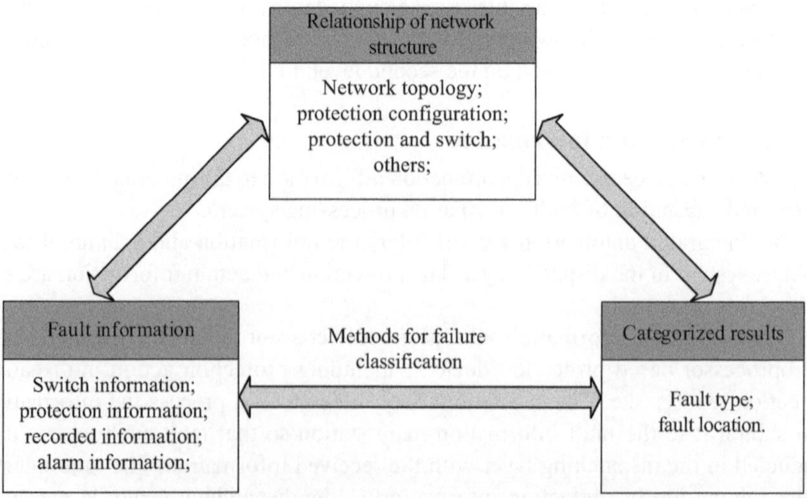

Figure 10.2 Diagram of distribution network fault diagnosis

As observed from Figure 10.2, the mechanism also determines the fault location (identification of faulty feeder) along with fault classification.

10.2.2.1 Fault classification

The objective of fault classification is to identify the type of fault (including AG, BG, CG, ABG, ACG, BCG, ABC/ABCG, AB, AC, and BC) as quickly as possible with the fault information provided by SCADA, protection devices and recording devices once fault happens. As a result, protection measures could be taken timely and the probability of different fault types could be calculated statistically.

10.2.2.2 Faulty feeder identification

Small current grounding (ungrounded or arc-suppression coil grounding) is employed in 6 $\sim$ 66 kV distribution systems in China. Most of the faults in small current grounding systems are single-phase grounding faults. Generally low-impedance short circuits do not happen. In other words, the short-circuit current is so small that the line voltage remains symmetrical.

As a result, the electric power will continue to be delivered for 1 or 2 hours thus the system reliability is improved. Even when instantaneous short circuit happens, the short-circuit arc can be suppressed automatically by itself and the insulation strength is recovered. This is a valuable feature for power system safety. However, higher requirements for system capacity, feeder and cable numbers, current increase in cables, and prolonged operation of the system lead to higher probability of malfunctions (such as two-phase short circuits, serious intermittent arc overvoltage or power system damage). To avoid those damages, faulty feeders should be identified and isolated immediately. To sum up, fault location identification means to identify the faulty feeder as soon as possible with the collected fault data from the SCADA system and recorders, so as to cut off the faulty feeders as soon as possible.

10.3 Methods of fault classification

10.3.1 Abstract

In the past decades, various methods have been developed for fault classification. The determination of correct fault type is a key step in the fault analysis of distribution network. At present, the most widely used fault classification techniques are threshold method, inference rule method, clustering method and AI classifier method.

In the threshold method, the extracted fault characteristic is compared with a threshold to determine the fault type. The principle of threshold method is easy and its physical meaning is clear. Classification can be effectively achieved by setting an appropriate threshold value, but the threshold values require careful selection.

In the inference rule method, logic reasoning, decision tree and fuzzy reasoning are mostly used. Fault classification techniques based on clustering method (using support vector machine) and AI classifier (using neural networks) have drawn a lot of attention in recent years. A lot of research efforts have also been devoted to the improvement of support vector machine and neural network, respectively. Especially in the study of neural network, different structures and types of neural network are employed and well-suited for fault classification studies. Usually, a comprehensive method combining neural network and other theories can lead to better classification results.

Fault classification based on support vector machine and neural network have the ability of learning and generalization, but they must be trained before classifying the fault, and the classification accuracy is directly determined by the size and completeness of the training set.

10.3.2 Methods of fault processing

Some widely used methods of fault processing in recent years are listed as follows:

10.3.2.1 Fault classification method based on ANFIS

Adaptive network-based fuzzy inference system (ANFIS) is a fuzzy reasoning system combining fuzzy reasoning and artificial neural network. This system has the advantages of both methods: the ability of learning from historical database of artificial neural network and the ability of judgment with fuzzy reasoning [8–10]. When the distribution network fails, the excessive transient signals will be a pattern recognition problem for fault classification based on transient signal.

The ANFIS method begins by using wavelet transform to extract the signal information from the fault feature band to examine the statistical properties of different fault types and then constructs the fault classification feature vector based on the statistical properties. Finally, it realizes fault classification of small current grounding system by integrating neural networks based on adaptive fuzzy inference system. The input to the neural network is the fault classification vector constructed above. This method has relatively high accuracy and good adaptability for distribution network fault classification.

10.3.2.2 Fault classification method based on Dempster–Shafer (DS) evidence theory

DS evidence theory provides a way to describe the function of evidence. Through Dempster evidence theory, the reliability of compound proposition will improve if the propositions it consists are all supported by the evidence [11–13]. DS evidence theory could reduce the ambiguity of judgment and improve the accuracy of pattern recognition. In the fault classification scheme based on DS evidence theory, initially the amplitude and angle relationships among symmetric components of post-fault currents and voltages are analyzed theoretically. Subsequently, characteristic quantities for fault classification are identified and based on the characteristic quantities, the basic probability assignments (BPAs) of each identifier is defined. Finally through establishing decision tree and fusing the value of characteristic quantities by DS evidence theory, the fault type is identified. DS evidence theory applied to fault classification in neutral noneffectively grounded distribution network is relatively accurate, fast, and adaptable.

10.3.2.3 Faulty feeder identification method based on *S* Transform

As a time-frequency analysis method, *S* transform is an expanded method of wavelet transform and the short-time Fourier transform. It has the ability of time-frequency analysis of wavelet transform and its computation speed improves dramatically by means of fast Fourier Transform (FFT) [3, 4, 14]. Compared with real-valued wavelet, it can not only extract the amplitude information but also phase angle information of the origin signal and as a result, it is quite a useful tool for identification of faulty feeder identification.

The fault line identification approach based on *S* transform is described as follows: Firstly, through applying the modulus and phase information of the electrical signals at each frequency point obtained from *S* transform, the characteristic frequency points are found by modulus comparison. Secondly, the fault-line vote mechanism is constructed by comparing the phase information. In the next stage, through utilizing the modulus information to reflect the reliability of phase information, the vote confidence degree is defined. Lastly, through multisampling point vote mechanism, the vote statistic diagram was formed; and the fault line is identified by vote statistic diagram integrated with fault line identification confidence degree, which is computed based on the amplitude confidence. The proposed method overcomes the problem of low accuracy when phase angle information of fault signal alone is applied to fault line identification under high impedance fault and strong noises.

10.4 Fault classification in distribution network based on ANFIS

10.4.1 Classification rules based on ANFIS

ANFIS can serve as a basis for constructing a set of fuzzy if–then rules with appropriate membership functions to generate the stipulated input–output pairs [14]. The ANFIS structure is established by embedding the fuzzy inference system into the framework of adaptive networks. It takes advantage of the learning mechanism of neural network and the reasoning ability of fuzzy system to m ake up for the deficiency existing in a single method.

Fuzzy if–then rules or fuzzy conditional statements are expressions of the form if A then B, where A and B are labels of fuzzy sets characterized by appropriate membership functions [16,17]. Assume the system is composed of two inputs (x, y) and one output (f). It has two *if–then* rules by adopting first order Sugeno model which can be described as follows. The structure of ANFIS is presented below in Figure 10.3 [18,19]:

Rule 1: if x is A_1 and y is B_1 then $f_1 = p_1x + q_1y + r_1$;
Rule 2: if x is A_2 and y is B_2 then $f_2 = p_2x + q_2y + r_2$;

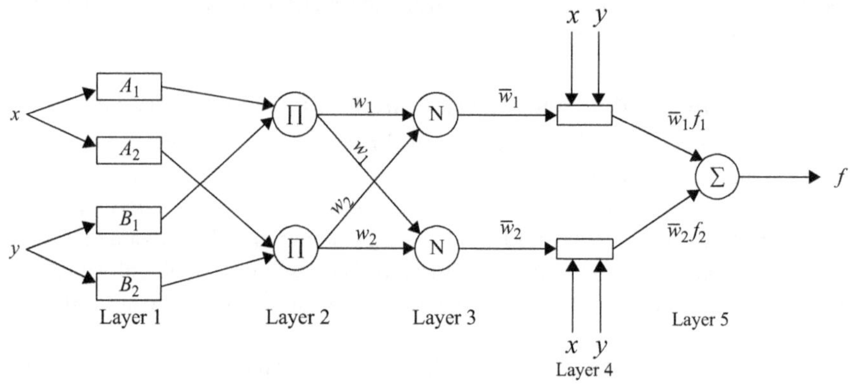

Figure 10.3 Structure frame of ANFIS

ANFIS can be divided into five layers and the specific content can be described as follows:

Layer 1: Output corresponding membership of fuzzy set

$$O_{1i} = \begin{cases} \mu_{A_i}(x) & i = 1,2 \\ \mu_{B_i}(y) & i = 1,2 \end{cases} \tag{10.1}$$

where O_{1i} is the output of i node in the first layer, x and y are input, μ is the membership functions.

Layer 2: Formation the fuzzy rules. The output signals are the product of the corresponding input signal and it can be presented by Π in the figure. w_i is incentive intensity of a rule.

Layer 3: Normalized calculation and it can be expressed by N in Figure 10.3:

$$\overline{w_1} = \frac{w_1}{w_1 + w_2} \tag{10.2}$$

$$\overline{w_2} = \frac{w_2}{w_1 + w_2} \tag{10.3}$$

Layer 4: Calculate the corresponding output signal generated by each rule, and the output signal can be described as follows:

$$O_{4i} = \bar{w}_i f_i = \bar{w}_i\,(p_i x + q_i y + r_i) \quad i = 1,2 \tag{10.4}$$

where O_{4i} is the output of i node in the fourth layer, p_i, q_i, r_i are named posteriori argument which will attain by training ANFIS.

Layer 5: Summing all signals to gain the system output.

10.4.2 ANFIS classifiers

Firstly, measure the three-phase bus voltage in the substation and the three-phase current at the secondary side of transformer. Zero-sequence voltage and fault component of three-phase current after fault can be calculated according to (10.5) and (10.6):

$$u_0(t) = u_a(t) + u_b(t) + u_c(t) \tag{10.5}$$

$$i_p^*(t) = i_p(t) - i_p(t-T) \tag{10.6}$$

where $i_p(t)$ is the current in phase "p" in the first cycle after fault ($p = a, b, c$); $u_a(t)$, $u_b(t)$, $u_c(t)$ are bus voltages of phase a, b and c respectively; T is the fundamental frequency time period.

Wavelet transform is carried out for i_p in a power cycle after fault using spline wavelet. Subsequently, the relative coefficients $\rho_{a,b}$, $\rho_{b,c}$, $\rho_{c,a}$ between the phases and the normalization standard deviation of each phase s_a^*, s_b^*, s_c^* are selected as the inputs of ANFIS. These quantities are described in (10.7)–(10.12)

$$s_p = \left(\frac{1}{n-1} \sum_{i=1}^{n} (i_p^*(n) - E(i_p^*))^2 \right)^{\frac{1}{2}} \tag{10.7}$$

$$s_{\max} = \max(s_p) \quad p = a, b, c \tag{10.8}$$

$$s_p^* = \frac{s_p}{s_{\max}} \tag{10.9}$$

$$\rho_{b,c} = \left| \frac{E(i_b^* i_c^*) - E(i_b^*)E(i_c^*)}{\sqrt{E(i_b^*)^2 - E^2(i_b^*)}\sqrt{E(i_c^*)^2 - E^2(i_c^*)}} \right| \tag{10.10}$$

$$\rho_{a,b} = \left| \frac{E(i_a^* i_b^*) - E(i_a^*)E(i_b^*)}{\sqrt{E(i_a^*)^2 - E^2(i_a^*)}\sqrt{E(i_b^*)^2 - E^2(i_b^*)}} \right| \tag{10.11}$$

$$\rho_{a,c} = \left| \frac{E(i_a^* i_c^*) - E(i_a^*)E(i_c^*)}{\sqrt{E(i_a^*)^2 - E^2(i_a^*)}\sqrt{E(i_c^*)^2 - E^2(i_c^*)}} \right| \tag{10.12}$$

where n is the number of samples during the first fundamental frequency period after the occurrence of the fault; $E(x)$ is the mathematical expectation of variable x in a period of n

FFT has been used to extract the 50 Hz component U_0 of $u_0(t)$ (which is used as the metric for deciding whether the fault involves ground or not).

Utilizing the seven calculated quantities above, we establish three input feature vectors of ANFIS as shown in (10.13) and (10.14)

$$C_1 = \begin{bmatrix} s_a^* & s_b^* & s_c^* & U_0 \end{bmatrix} \tag{10.13}$$

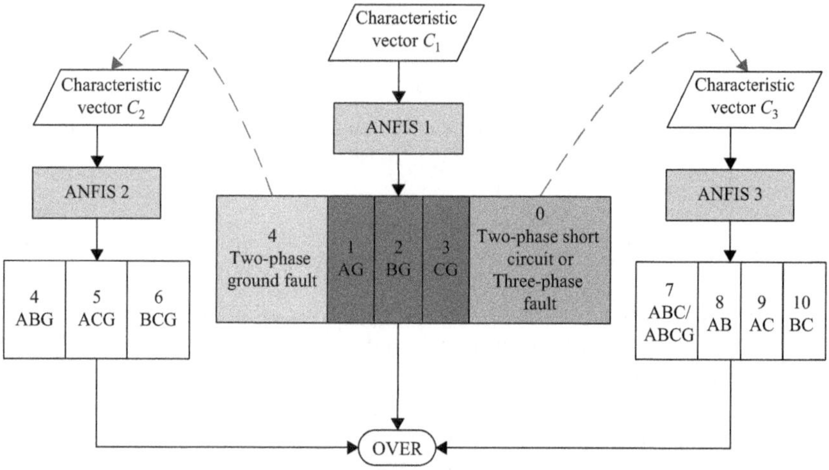

Figure 10.4 Schematic diagram of fault classification

$$C_2 = C_3 = \begin{bmatrix} \rho_{a,b} & \rho_{a,c} & \rho_{b,c} & s_a^* & s_b^* & s_c^* \end{bmatrix} \tag{10.14}$$

There are three ANFIS classifiers of fault classification which have different functions, shown in Figure 10.4.

In Figure 10.4, three ANFISs correspond to different functions. First, ANFIS 1 discriminates among three types of single-phase to ground fault and other types of fault. If the system is diagnosed to have two-phase to ground fault, ANFIS 2 would be used to further determine the faulty phases. If ANFIS 1 estimates that two-phase short circuit or three-phase fault occurred, ANFIS 3 would be needed to make further diagnosis. Because that outputs of ANFISs are digitized, different fault type is represented as different number, which is shown in Figure 10.4. ANFIS 1 has four inputs, and each input has two membership functions which are classed as Gaussian functions. Thus, there are $2^4 = 16$ rules in total. ANFIS 2 and ANFIS 3 both have six inputs, and each input has two membership functions so that both of them have $2^6 = 64$ rules. The details of these rules can be found from [20].

10.4.3 Test simulation and conclusion

The above method has been implemented in the system shown in Figure 10.5. In this system, a total of 3600 fault samples with different fault conditions have been generated. The details of these fault samples are shown in Table 10.1 [21]. The summary of classification results are presented in Table 10.2.

From Table 10.2, it is observed that the results of fault classification method are quite accurate.

The accuracy of classification of fault type in distribution network based on ANFIS can be influenced by noise, load variation and equivalent impedance. So, it is necessary to investigate the effects of these factors on the accuracy of classification.

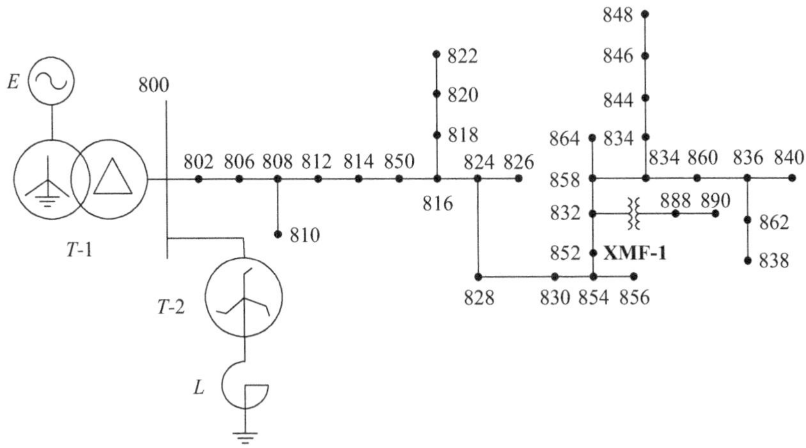

Figure 10.5 Schematic diagram of the test system

Table 10.1 Test cases

Fault location	Fault resistance/Ω	Initial fault angle/(°)	Number of samples
806, 816, 818, 830	0, 5, 10, 25, 40, 50	0, 18, 54, 72, 108, 144	1440
812	1, 50, 100, 200, 500, 1000	0, 18, 36, 90, 126, 162	360
823, 834	20, 70, 150, 300, 600	18, 54, 72, 90, 108, 144	720
848, 860, 862	20, 70, 150, 300, 1200, 1500	18, 54, 72, 90, 108, 144	1080
Summary			3600

Table 10.2 Fault classification results

Classifiers	Outputs(fault type)	Number of samples	Accuracy rate/%
ANFIS1	1 (AG)	360	100
	2 (BG)	360	100
	3 (CG)	360	99.4
	4 (Two-phase grounding)	1080	99.8
	(no grounding)	1440	100
ANFIS2	4 (ABG)	360	99.4
	5 (ACG)	360	100
	6 (BCG)	360	99.7
ANFIS3	7 (ABC/ABCG)	360	100
	8 (AB)	360	100
	9 (AC)	360	100
	10 (BC)	360	100

Table 10.3 Test results with SNR = 40 dB

Classifiers	Outputs (fault type)	Number of samples	Accuracy rate/%
ANFIS1	1 (AG)	360	78.7
	2 (BG)	360	83.6
	3 (CG)	360	80.3
	4 (Two-phase grounding)	1080	93.4
	(no grounding)	1440	100
ANFIS2	4 (ABG)	360	95
	5 (ACG)	360	93.3
	6 (BCG)	360	91.4
ANFIS3	7 (ABC/ABCG)	360	100
	8 (AB)	360	100
	9 (AC)	360	100
	10 (BC)	360	100

Table 10.4 Test results with SNR = 50 dB

Classifiers	Outputs (fault type)	Number of samples	Accuracy rate/%
ANFIS1	1 (AG)	360	98.3
	2 (BG)	360	98.1
	3 (CG)	360	93.1
	4 (Two-phase grounding)	1080	99.7
	(no grounding)	1440	100
ANFIS2	4 (ABG)	360	99.4
	5 (ACG)	360	100
	6 (BCG)	360	99.7
ANFIS3	7 (ABC/ABCG)	360	100
	8 (AB)	360	100
	9 (AC)	360	100
	10 (BC)	360	100

A. Noise interference

Gaussian white noise with different signal to noise ratio (SNR) was added to the test samples of Table 10.1 to study the accuracy of classification method. The summary of the results obtained are shown in Tables 10.3–10.5.

It can be found from the above tables that reduction of SNR results into the reduction of accuracy rate. When SNR = 40 dB, ground fault classification accuracy rate decreases rapidly. Therefore, in practical application, filter can be added to improve the accuracy of classification methods.

B. Variable load

The actual level of load in the distribution system is changing constantly. The effectiveness of the classification method is tested for two cases: (i) actual loads in

Table 10.5 Test results with SNR = 60 dB

Classifiers	Outputs (fault type)	Number of samples	Accuracy rate/%
ANFIS1	1 (AG)	360	100
	2 (BG)	360	100
	3 (CG)	360	99.4
	4 (Two-phase grounding)	1080	99.9
	(no grounding)	1440	100
ANFIS2	4 (ABG)	360	99.7
	5 (ACG)	360	100
	6 (BCG)	360	99.7
ANFIS3	7 (ABC/ABCG)	360	100
	8 (AB)	360	100
	9 (AC)	360	100
	10 (BC)	360	100

Table 10.6 Test results when loads are 50% of the normal load

Classifiers	Outputs (fault type)	Number of samples	Accuracy rate/%
ANFIS1	1 (AG)	360	100
	2 (BG)	360	100
	3 (CG)	360	100
	4 (Two-phase grounding)	1080	99.7
	(no grounding)	1440	100
ANFIS2	4 (ABG)	360	99.2
	5 (ACG)	360	100
	6 (BCG)	360	99.7
ANFIS3	7 (ABC/ABCG)	360	100
	8 (AB)	360	100
	9 (AC)	360	100
	10 (BC)	360	100

the system are 50% of the normal load and (ii) actual loads in the system are twice the normal load. The test results are shown in Tables 10.6 and 10.7.

It can be seen from Tables 10.6 and 10.7 that the accuracy of classification is hardly affected when the loads are decreased. However, increasing load levels will result in the reduction of the accuracy rate. Therefore, the number of training samples needs to be increased with heavy load conditions in order to improve the accuracy of classification methods.

C. Variance conditions of source equivalent resistance

Studies have also been made carried out to investigate the performance of the classification algorithms under variable source equivalent impedance. The various

Table 10.7 Test results when loads twice the normal load

Classifiers	Outputs (fault type)	Number of samples	Accuracy rate/%
ANFIS1	1 (AG)	360	89.2
	2 (BG)	360	88.9
	3 (CG)	360	83.3
	4 (Two-phase grounding)	1080	99.3
	(no grounding)	1440	100
ANFIS2	4 (ABG)	360	99.7
	5 (ACG)	360	100
	6 (BCG)	360	99.7
ANFIS3	7 (ABC/ABCG)	360	100
	8 (AB)	360	100
	9 (AC)	360	100
	10 (BC)	360	100

Table 10.8 Variance conditions of source equivalent impedance

No	Equivalent impedance of system side
1	70% reduction in amplitude
2	50% reduction in amplitude
3	Impedance angle becomes 20°
4	Impedance angle becomes 50°
5	Amplitude increased 100%

Table 10.9 Test results under condition number 1

Classifiers	Outputs (fault type)	Number of samples	Accuracy rate/%
ANFIS1	1 (AG)	360	100
	2 (BG)	360	100
	3 (CG)	360	100
	4 (Two-phase grounding)	1080	99.9
	(no grounding)	1440	100
ANFIS2	4 (ABG)	360	100
	5 (ACG)	360	100
	6 (BCG)	360	99.7
ANFIS3	7 (ABC/ABCG)	360	100
	8 (AB)	360	100
	9 (AC)	360	100
	10 (BC)	360	100

Table 10.10 Test results under condition number 2

Classifiers	Outputs (fault type)	Number of samples	Accuracy rate/%
ANFIS1	1 (AG)	360	100
	2 (BG)	360	100
	3 (CG)	360	99.7
	4 (Two-phase grounding)	1080	99.7
	(no grounding)	1440	100
ANFIS2	4 (ABG)	360	100
	5 (ACG)	360	100
	6 (BCG)	360	99.4
ANFIS3	7 (ABC/ABCG)	360	100
	8 (AB)	360	100
	9 (AC)	360	100
	10 (BC)	360	100

source equivalent impedances considered here are shown in Table 10.8. The results of classification accuracy are shown in Tables 10.9 and 10.10.

It can be concluded that equivalent impedance changes in the system will not reduce the accuracy of the classification of fault type based on ANFIS.

Due to the high flexibility of adaptive networks, the ANFIS can have a large number of variants. The fault classification in distribution network based on ANFIS can rapidly and accurately identify and classify different types of faults and is not susceptible to fault inception angle, fault distance and fault resistance. Furthermore, the fault classification method based on ANFIS is nearly unaffected by noise, harmonics, transfer characteristics of CT or sampling frequency. So it has good adaptability and prospects for engineering application.

10.5 Fault classification using DS evidence fusion

10.5.1 DS evidence fusion

In an information processing system, fusion can take place at three levels: signal level, feature level and decision level. Signal-level fusion is often used to reduce the measurement uncertainty of a single sensor [10]. Feature-level fusion can effectively utilize the complementary information from the signals which are extracted by different signal processing techniques. One of the practical limitations is that the tremendous size of the feature space results in heavy computational burden. To alleviate this problem, the decisions are made individually based on different feature vectors extracted by different methodologies and are then combined together. This approach is known as decision-level fusion. DS evidence theory is an uncertain reasoning theory which has good information fusion capabilities and can reflect the uncertainty objectively. Therefore, DS evidence theory is mainly considered for decision-level information fusion in the classification scheme.

Evidence theory was proposed by Dempster in 1967, and subsequently, Shafer had made extension and further development of this theory in 1976 [11]. Hence, it is also called as DS theory. The theory includes two important definitions which are described as follows [12]

Definition 10.1. *Let U be the framework of discernment. ∅ expresses the set of empty proposition A. If set function* $m : 2^U \rightarrow [0, 1]$ *(*2^U *is the power set of U) and*

$$m(\emptyset) = 0 \tag{10.15}$$

$$\sum_{A \subseteq U} m(A) = 1 \tag{10.16}$$

where m is called as BPA on the discernment framework U. $\forall A \subseteq U$, $m(A)$ *is called as basic probability number which reflects the credibility of A itself.*

Definition 10.2. *Let* $m_1, m_2, \ldots, m_n$ *be the basic probability on the discernment framework U, the orthogonal sum of the multiple distribution function* $m = m_1 \oplus m_2 \oplus \ldots \oplus m_n$ *is expressed in (10.17) and (10.18)*

$$\left.\begin{array}{l} m(\emptyset) = 0 \\ m(A) = T \sum\limits_{\cap A_i = A} \prod\limits_{i \leq j \leq n} m_j(A_i), A \neq \emptyset \end{array}\right\} \tag{10.17}$$

$$T^{-1} = 1 - \sum_{\cap A_i = \emptyset} \sum_{1 \leq j \geq n} m_j(A_i) \tag{10.18}$$

Equation (10.17) is the Dempster's rule of combination, and T is the conflict probability. When $T = 1$, *the contradiction of evidence will happen, and (10.17) cannot be used correspondingly. If* $T \rightarrow 1$, *a serious conflict of evidence takes place which will cause an unfavorable result. As for multiple evidences, the combination results can be obtained by recursion algorithm based on (10.17), and the equation satisfies the commutative law and associative law.*

10.5.2 Fault classification scheme

In this section, a Dempster–Shafer (DS evidence) fusion method based on dynamic absolute probability (DAP) is discussed. A fusion method for selecting the faulty feeder based on DAP is described for neutral noneffectively grounded distribution system. This method measures the fault resistances relying on the variation of the transient factor to assign the absolute probability of the evidences dynamically. After the fusion of the evidences, the faulty feeder is selected [22,23].

The fault classification scheme of neutral noneffectively grounded system which identifies ten types of short circuit fault (including AG, BG, CG, ABG, ACG, BCG, ABC/ABCG, AB, AC, and BC) is discussed. Assume that $\bar{I}_{A1f}$, $\bar{I}_{A2f}$, and $\bar{I}_{A0f}$ are

Table 10.11 Typical value of R_{0f}, R_{2f}, φ_A, and φ_B

Fault types	$\varphi_A/(°)$	$\varphi_B/(°)$	R_{0f}	R_{2f}
AG	0	120	>1	1
BG	120	0	>1	1
CG	120	120	>1	1
ABG	60	60	>1	1
ACG	60	180	>1	1
BCG	180	60	>1	1
AB	60	60	0	1
AC	60	180	0	1
BC	180	60	0	1
ABC/ABCG	–	–	0	0

positive-, negative-, and zero-sequence fundamental fault currents relative to phase "a" at fault point, respectively [24,25]

Define

$$\varphi_A = \left|\arg(\bar{I}_{A2f}) - \arg(\bar{I}_{A1f})\right| \tag{10.19}$$

where, arg(x) denotes the angle of the complex variable x.

Similarly,

$$\varphi_B = \left|\arg(\bar{I}_{B2f}) - \arg(\bar{I}_{B1f}))\right| \tag{10.20}$$

where, $\bar{I}_{B1f}$ and $\bar{I}_{B2f}$ denotes positive- and negative-sequence components of fundamental fault current relative to phase "b".

Define

$$R_{2f} = \left|\bar{I}_{A2f}/\bar{I}_{A1f}\right| \tag{10.21}$$

Assume that $\bar{U}_{S1}$, $\bar{U}_{S2}$, and $\bar{U}_{S0}$ are positive-, negative-, and zero-sequence component of the bus voltage.

Define

$$R_{0f} = \left|\bar{U}_{so}/\bar{U}_{s1}\right| \tag{10.22}$$

R_{0f}, R_{2f}, φ_A, and φ_B are decided as identifiers for fault classification and their typical values for different fault types are listed in Table 10.11 [26]:

Based on Table 10.11, the decision tree of fault classification can be established as Figure 10.6.

It can be seen from Figure 10.6, the fault classification can be seen as an inference problem. Each identifier can be considered as a hypothesis and the D-S evidence fusion rule is employed. Before applying D-S evidence theory, the BPA should be

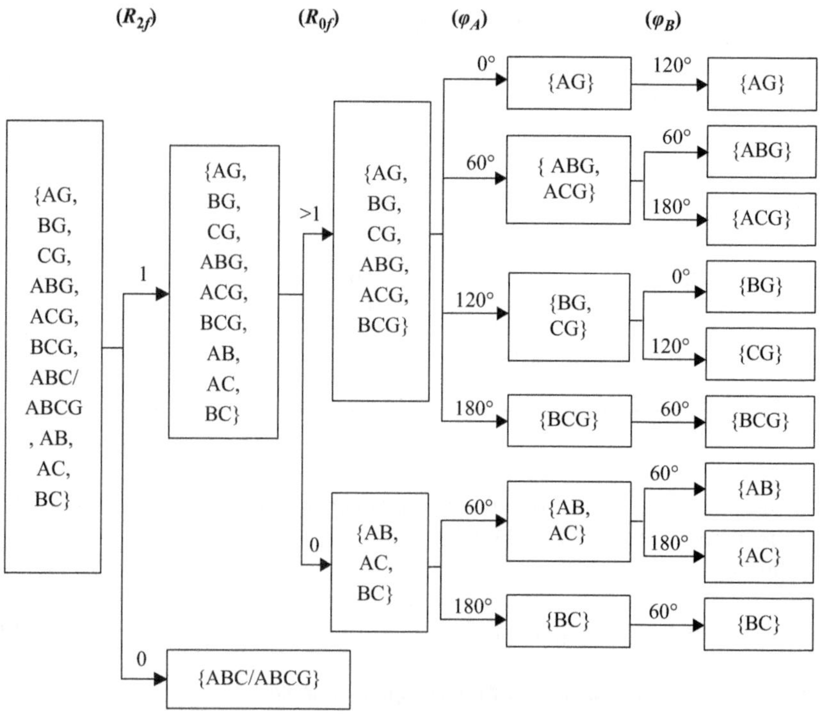

Figure 10.6 Decision tree of fault classification

assigned for each identifier. They are introduced according to the sequence in which they appear in inference process in Figure 10.6.

A trapezoid function is defined by the points O, P, R and S as shown in Figure 10.7.

The BPA of R_{2f} is denoted as m_{R2f} and is defined in Table 10.12.

In Table 10.12, {Others}={non-ABC/ABCG}. The m_{R2F} is plotted in Figure 10.8 as shown below.

The BPA of R_{0f} is denoted as m_{R0f} and is defined as follows.

$$m_{R0f}(\{\text{Ungrounded}\}) = \begin{cases} -1/2 \times (R_{0f} - 2), 0 \leq R_{0f} \leq 2 \\ 0, \qquad R_{0f} > 2 \end{cases} \text{ and}$$

$$m_{R0f}(\{\text{Grounded}\}) = \begin{cases} 1/2 \times R_{0f}, 0 \leq R_{0f} \leq 2 \\ 1, \qquad R_{0f} > 2 \end{cases},$$

where, {Ungrounded} = {ABC/ABCG, AB, AC, BC}, {Grounded} = {AG, BG, CG, ABG, ACG BCG}. The plot of m_{R0f} is shown in Figure 10.9.

The BPAs of φ_A is denoted as $m_{\varphi A}$ and is again defined by a trapezoidal function shown in Figure 10.10. The points O, P, R, and S corresponding to $m_{\varphi A}$ are given in

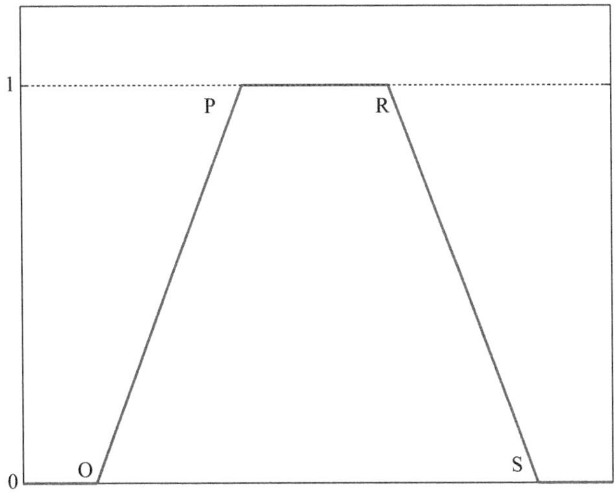

Figure 10.7 Trapezoid function defined by the points O, P, R, and S

Table 10.12 BPA of R_{2f}, m_{R2f}

Fault type	**O**	**P**	**R**	**S**
{ABC/ABCG}	0	0	0.1	0.2
{Others}	0.1	0.2	1.2	1.2

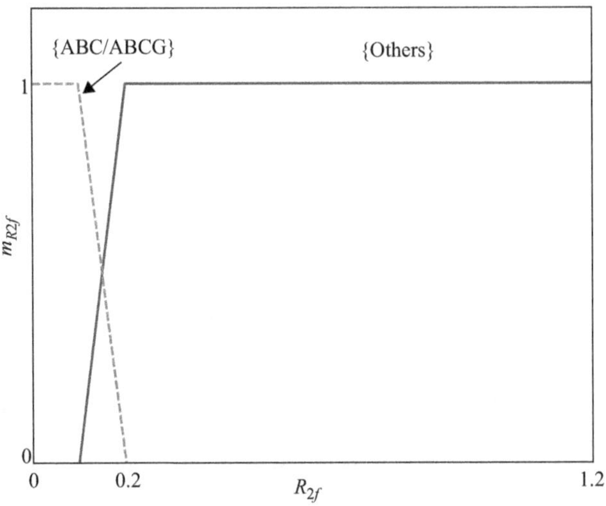

Figure 10.8 m_{R2f}, BPA of R_{2f}

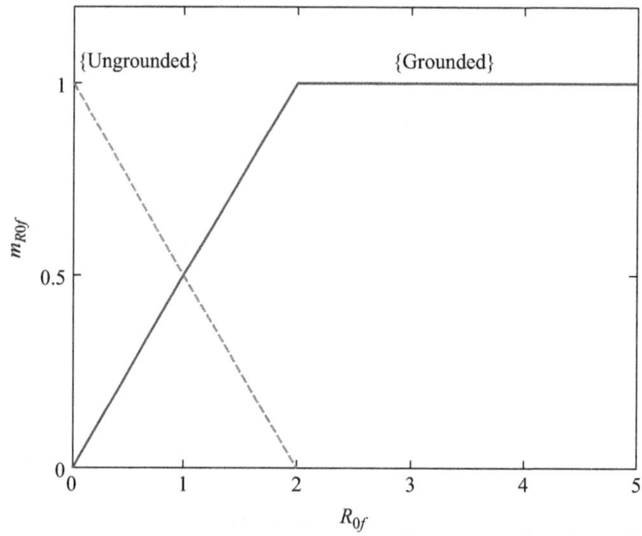

Figure 10.9 m_{r0f}, *BPA of* R_{0f}

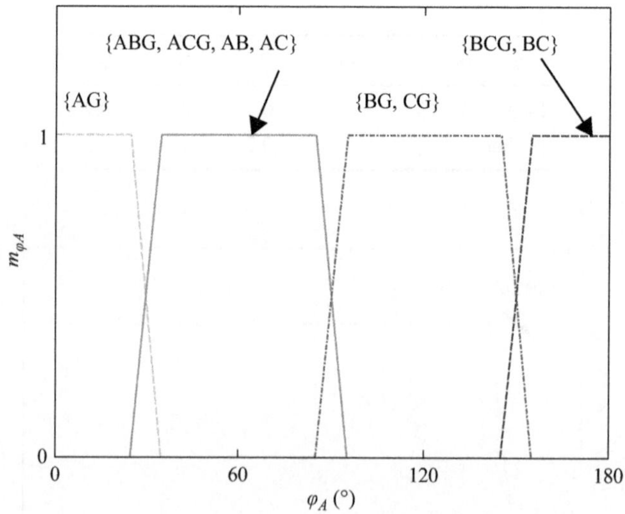

Figure 10.10 $m_{\varphi A}$, *BPA of* φ_A

Table 10.13. The BPAs of φ_B are denoted as $m_{\varphi B}$ and are given in Table 10.14. Its plot is shown in Figure 10.11.

After BPA of each identifier is determined, the flowchart of fault classification is shown in Figure 10.12.

In Figure 10.12, the operator ⊕ denotes the Dempster's rule of combination, and distinct bodies of evidence can be combined by means of Dempster's rule of

Table 10.13 BPA of φ_A, $m_{\varphi A}$

Fault type	**O**	**P**	**R**	**S**
{AG}	0	0	25	35
{ABG, ACG AB, AC}	25	35	85	95
{BG, CG}	85	95	145	155
{BCG, BC}	145	155	180	180

Table 10.14 BPA of φ_B, $m_{\varphi B}$

Fault type	**O**	**P**	**R**	**S**
{BG}	0	0	25	35
{ABG, BCG AB, BC}	25	35	85	95
{AG, CG}	85	95	145	155
{ACG, AC}	145	155	180	180

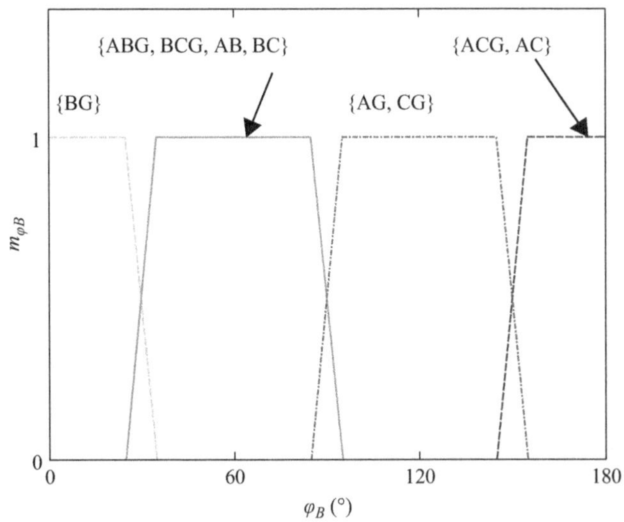

Figure 10.11 $m_{\varphi B}$, BPA of ϕ_B

combination. With two pieces of evidence m_x and m_y Dempster's rule of combination is defined as follows:

$$m_z(A) = \frac{\sum_{B,C \subset U; B \cap C = A} m_x(B) \times m_y(C)}{1 - K} \tag{10.23}$$

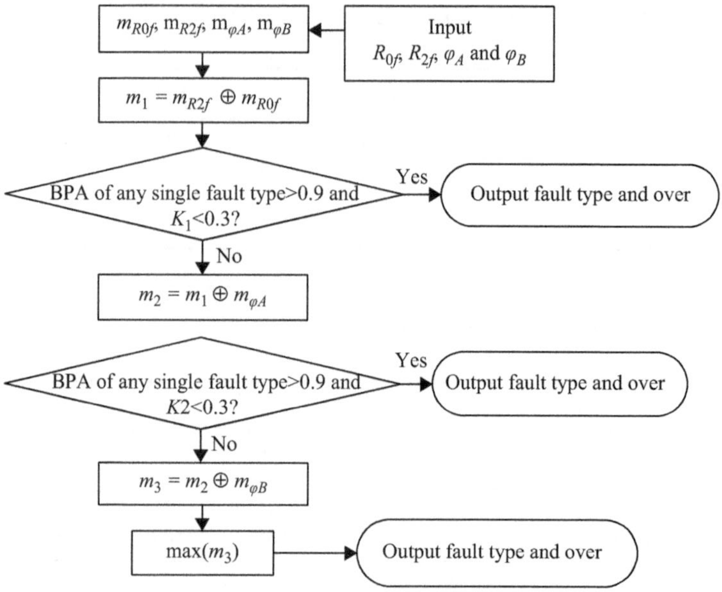

Figure 10.12 Flowchart of fault classification

where, m_x and m_y are two BPAs of universal set U, m_z is the resulting assignment through combining m_x and m_y. $A, B, C \subset U$.

$$K = \sum_{B,C \subset U; B \cap C = \Phi} m_x(B) \times m_y(C) \tag{10.24}$$

where, K represents an degree of the inconsistency which is given by m_x and m_y.

10.5.3 Simulation test and conclusion

The above scheme is verified after establishing a noneffectively grounded distribution network based on traditional IEEE-34 model as shown in Figure 10.13. The system has been simulated in PSCD/EMTDC software [26].

Assume an ABG fault occurs at bus 806, with fault inception angle 18° and fault resistance 0 Ω. Through FFT (sample frequency is 5000 Hz), one can obtain $R_{2f} = 0.99, R_{0f} = 1.89, \varphi_A = 59.9°$ and $\varphi_B = 60.15°$. According to predefined BPAs of each identifiers, the various probabilities can be calculated as $m_{R2f}(\{\text{Others}\}) = 1$, $m_{R0f}(\{\text{Ungrounded}\}) = 0.06$, $m_{R0f}(\{\text{Grounded}\}) = 0.94$, $m_{\phi A}(\{\text{ABG, ACG, AB, AC}\}) = 1$, $m_{\phi B}(\{\text{ABG, BCG, AB, BC}\}) = 1$. According to the flowchart in Figure 10.13, the fusion result of each step is listed in Tables 10.15–10.17.

From Table 10.15 function {Ungrounded} is very low with respect to function {grounded}, so $K_1 = 0$. Now, $m_1(\{\text{AG, BG, CG, ABG, ACG, BCG}\}) = m_{R2f}(\{\text{Others}\}) \times m_{R0f}\{\text{Grounded}\})$; $m_1(\{\text{AB, AC, BC}\}) = m_{R2f}\{\text{Others}\} \times m_{R0f}$

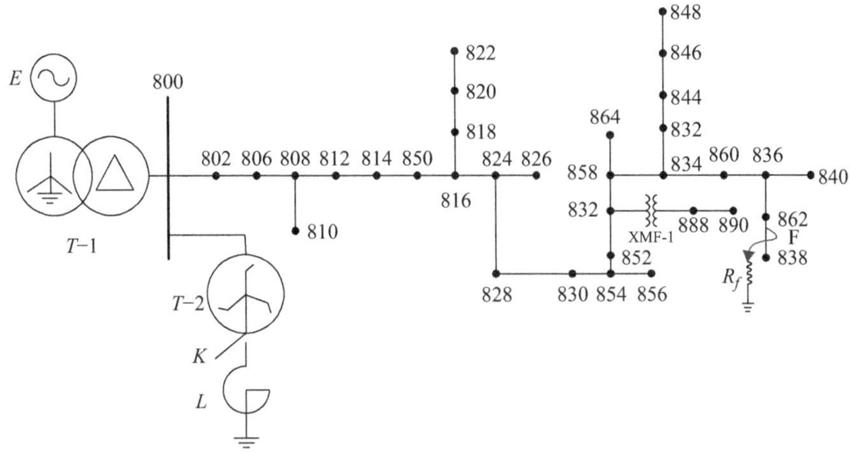

Figure 10.13 Simulating model in PSCAD/EMTDC

Table 10.15 $m_1 = m_{R2f} \oplus m_{R0f}$ *in ABG fault at bus 806, FIA* = 18°, $R_f = 0\,\Omega$

m_{R2f}	m_{R0f}	
	{Ungrounded} (0.06)	**{Grounded} (0.94)**
{Others} (1)	{AB, AC, BC} (0.06)	{AG, BG, CG, ABG, ACG, BCG} (0.94)

Table 10.16 $m_2 = m_1 \oplus m_{\varphi A}$ *in ABG fault at bus 806, FIA* = 18°, $R_f = 0\,\Omega$

m_1	$m_{\varphi A}$ **{ABG, ACG, AB, AC} (1)**
{AG, BG, CG, ABG, ACG, BCG} (0.94)	{ABG, ACG} (0.94)
{AB, AC, BC} (0.06)	{AB, AC} (0.06)

Table 10.17 $m_3 = m_2 \oplus m_{\varphi B}$ *in ABG fault at bus 806, FIA* = 18°, $R_f = 0\,\Omega$

m_2	$m_{\varphi B}$ **{ABG, BCG, AB, BC} (1)**
{ABG, ACG} (0.94)	{ABG} (0.94)
{AB, AC} (0.06)	{AB} (0.06)

Table 10.18 Numerical results of some representative cases

Faulted bus		**806**	**830**	**812**	**832**	**848**
FIA (°)				18	90	144
R_f (Ω)			50	500	1200	1500
	AG	1	1	1	1	1
	BG	1	1	1	1	1
	CG	1	1	1	1	1
Fault types	ABG	0.945	1	1	1	1
	ACG	0.95	1	1	1	1
	BCG	0.94	1	1	1	1
	ABC/ABCG	1	1	1	1	1
	AB	1	1	1	1	1
	AC	1	1	1	1	1
	BC	1	1	1	1	1

{Ungrounded}. Hence m_1({AG, BG, CG, ABG, ACG, BCG}) = 0.94, m_1({AB, AC, BC}) = 0.06.

From Table 10.16, m_2({ABG, ACG}) = 0.94, m_2({AB, AC}) = 0.06 and $K_2 = 0$.

From Table 10.17, m_3({ABG}) = 0.94, m_3({AB}) = 0.06 and $K_3 = 0$. So, fault is identified as ABG type.

Using similar fusion process as above, ten types of fault under 3600 different fault conditions have been simulated. The results show that the above scheme is quite effective and all 3600 test cases are identified correctly. But due to the space constraint, it is impossible to list all the results here. Hence, some representative cases and their numerical results obtained from fusion process are given in Table 10.18.

Through theoretical analysis on amplitude and angle relationship among sequence electric quantities of fault component, a fault classification scheme based on D-S evidence theory is discussed. The described scheme is applied in a neutral noneffectively grounded distribution network. High sampling rate is not required and the scheme can be adopted in resonant grounded distribution network without any modification.

10.6 Faulty feeder selection in distribution network based on *S* transform

10.6.1 Introduction to S transform

S transform is an extension of the idea of continuous wavelet transform (CWT) [6] and is based on a moving and scalable localizing Gaussian window. There are several methods of arriving at the *S* transform. We consider it illuminating to derive the *S* transform as the "phase correction" of the CWT. The CWT $W(\tau, d)$ of a function $h(t)$ is defined by [2]

$$W(\tau, d) = \int_{-\infty}^{\infty} h(t)w(t-\tau, d)dt \tag{10.25}$$

where $w(t, d)$ is a scaled replica of the fundamental mother wavelet; τ is time. The dilation d determines the "width" of the wavelet $w(t, d)$ and thus controls the resolution. Along with (10.25), there exists an admissibility condition on the mother wavelet $w(t, d)$ which says that $w(t, d)$ must have zero mean. Refer to [26] and Young [27] for further details.

The S transform of a function $h(t)$ is defined as a CWT with a specific mother wavelet multiplied by the phase factor

$$S(\tau, f) = e^{i2\pi f\tau} W(\tau, d) \tag{10.26}$$

where the mother wavelet is defined as

$$w(t, f) = \frac{|f|}{\sqrt{2\pi}} e^{-\frac{t^2 f^2}{2}} e^{-i2\pi ft} \tag{10.27}$$

Note that the dilation factor d is the inverse of the frequency f.

The wavelet in (10.27) does not satisfy the condition of zero mean for an admissible wavelet; therefore, (10.26) is not strictly a CWT. Written out explicitly, the S transform is

$$S(\tau, f) = \int_{-\infty}^{\infty} h(t) \frac{|f|}{\sqrt{2\pi}} e^{-\frac{(\tau - t)^2 f^2}{2}} e^{-i2\pi ft} dt \tag{10.28}$$

If the S transform is indeed a representation of the local spectrum, one would expect a simple operation of averaging the local spectra over time to give the Fourier spectrum. It is easy to show that

$$\int_{-\infty}^{\infty} S(\tau, f) d\tau = H(f) \tag{10.29}$$

where $H(f)$ is the Fourier transform of $h(t)$. It follows that $h(t)$ is exactly recoverable from $S(\tau, f)$. Thus

$$h(t) = \int_{-\infty}^{\infty} \left\{ \int_{-\infty}^{\infty} S(\tau, f) d\tau \right\} e^{i2\pi ft} df \tag{10.30}$$

This is clearly distinct from the concepts of the CWT.

The S transform provides an extension of instantaneous frequency (IF) [28] to broadband signals. The 1-D function of the variable τ and fixed parameter f_1 defined by $S(\tau, f_1)$ is called a voice (as with the CWT). The voice for a particular frequency f_1 can be written as

$$S(\tau, f_1) = A(\tau, f_1) e^{i\Phi(\tau, f_1)} \tag{10.31}$$

where A is the magnitude of S; and $\Phi(\tau, f_1)$ is the argument of S of the variable τ for the frequency f_1.

Since a voice isolates a specific component, one may use the phase in (10.31) to determine the IF as defined in [29].

$$\mathrm{IF}(\tau,f_1) = \frac{1}{2\pi}\frac{\partial}{\partial t}\{2\pi f_1 \tau + \Phi(\tau,f_1)\} \tag{10.32}$$

Thus, the absolutely referenced phase information leads to a generalization of the IF of (33) to broadband signals. The validity of (10.32) can easily be seen for the simple case of $h(t) = \cos(2\pi wt)$, where the function $\Phi(\tau,f) = 2\pi(w - f)\tau$.

The S transform can be written as operations on the Fourier spectrum $H(f)$ of $h(t)$

$$S(\tau,f) = \int_{-\infty}^{\infty} H(\alpha + f)e^{-\frac{2\pi^2\alpha^2}{f^2}} e^{i2\pi\alpha\tau} d\alpha \quad f \neq 0 \tag{10.33}$$

The discrete analog of (10.33) is used to compute the discrete S transform by taking advantage of the efficiency of the FFT and the convolution theorem. By not translating the cosinusoid basis functions, the S transform localizes the real and imaginary components of the spectrum independently, localizing the phase spectrum as well as the amplitude spectrum.

10.6.2 Faulty feeder selection method based on S transform

On the basis of the analysis in Section 10.6.1, when S transform is used to extract the phase angle information in faulty feeder selection, amplitude information can be used to describe the reliability of phase angle information. In addition, if the phase angle information of multiple sample points are used, the influence of error on the algorithm in some sample points can be avoided and thus the reliability of faulty feeder selection method can be improved. Consider a four-feeder power distribution system as shown in Figure 10.14 [30]:

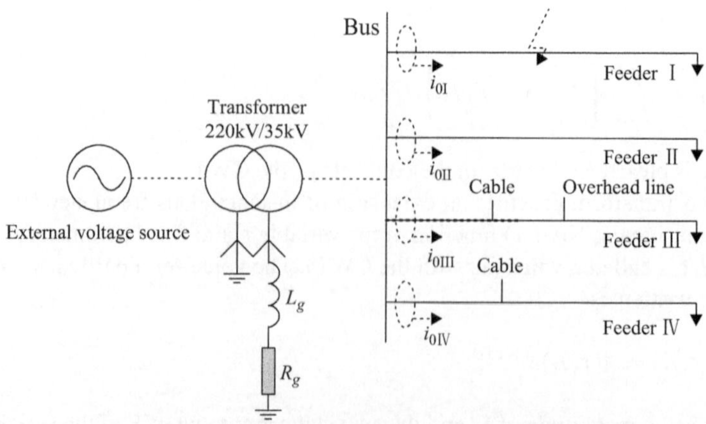

Figure 10.14 Distribution system with 4 feeders

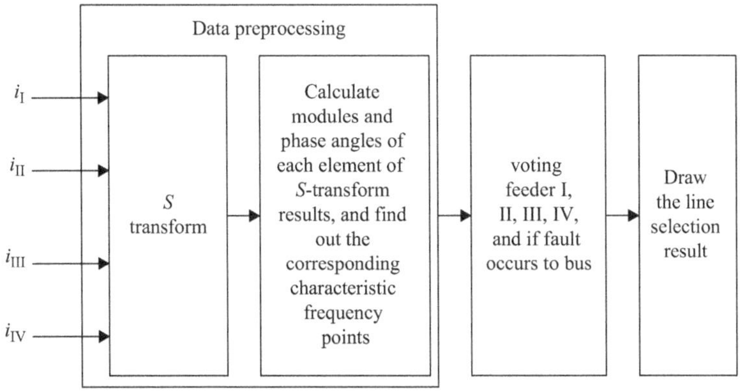

Figure 10.15 Principle block diagram of faulty feeder identification approach base on S transform

In Figure 10.14, R_g and L_g represent the resistance and reactance of the Petersen coil, respectively, and i_{0I}, i_{0II}, i_{0III}, and i_{0IV} are zero-sequence current extracted from the sending ends of the four feeders, respectively. Similarly, i_I, i_{II}, i_{III}, and i_{IV} are currents extracted from the sending ends of the four feeders, respectively. The basic diagram of faulty feeder selection method is shown in Figure 10.15. As shown in Figure 10.15, the faulty feeder selection method includes three steps: data preprocessing, voting for feeders and line selection result.

10.6.2.1 Data preprocessing

Raw data i_{0I}, i_{0II}, i_{0III}, and i_{0IV} are calculated as follows

$$i_{0n}(t) = i_{0n}[(t_f - T/2) : (t_f + T/2)] - i_{0n}[(t_f - 3T/2) : (t_f - T/2)] \quad (10.34)$$

where $n =$ I, II, III, and IV. t_f is the fault instance T is the time period of power frequency. "[a:b]" denotes the time interval between the instant a and the instant b.

Suppose the sampling frequency is 10 kHz and the sampled version of $i_{0n}(t)$ is represented as $i_{0n}(k)$, in which for one power frequency cycle can be obtained as follows

$$i_{0n}(k) = [\, i_{0n}(1) \; i_{0n}(2) \; \ldots \; i_{0n}(200)\,] \quad (10.35)$$

Time-frequency matrix Si_{0n} can be obtained by performing S transform to $i_{0n}(k)$, which is described as follows

$$Si_{0n} = \begin{bmatrix} Si_{0n}(1,1) & Si_{0n}(1,2) & \ldots & Si_{0n}(1,50) \\ Si_{0n}(2,1) & Si_{0n}(2,2) & \ldots & Si_{0n}(2,50) \\ \ldots & \ldots & \ldots & \ldots \\ Si_{0n}(101,1) & Si_{0n}(101,2) & \ldots & Si_{0n}(101,50) \end{bmatrix} \quad (10.36)$$

Calculate module and phase angle of each element:

$$\begin{aligned} \rho(Si_{0n}(m,k)) &= \sqrt{x_n(m,k)^2 + y_n(m,k)^2} \\ \theta(Si_{0n}(m,k)) &= arc\ \tan\left(y_n(m,k)/x_n(m,k)\right) \end{aligned} \tag{10.37}$$

where $Si_{0n}(m,k) = x_n(m,k) + jy_n(m,k)$, in which j is the imaginary operator; $m = 1, 2, \ldots, 101, k = 1, 2, \ldots, 50$; "·" represents a vector, so $\rho(\cdot)$ is the modulus of vector "·", while $\theta(\cdot)$ is the phase angle of vector "·". ρ_{Si0n} and θ_{Si0n} denote the modulus matrix and the phase angle matrix respectively.

The elements in each column of ρ_{Si0n} are sorted in the order of their modulus, and the corresponding frequency points of the first three elements are called feature frequency points. $\rho_{1\_Si0n(k)}$, $\rho_{2\_Si0n(k)}$, and $\rho_{3\_Si0n(k)}$ denote their corresponding magnitudes and $\theta_{1\_Si0n(k)}$, $\theta_{2\_Si0n(k)}$, and $\theta_{3\_Si0n(k)}$ denote their phase angles, respectively, $k = 1, 2, 3, \ldots 50$ ("k" denotes the kth column of the matrix Si_{0n}.

10.6.2.2 Voting mechanism

As previously described, the feature bands of different feeders may not be the same, causing dissimilarity in the selected feature frequency points. Generally the most reliable way to judge whether a fault exists in a feeder or not is the use of feature band corresponding to this feeder. Based on this principle we suppose the corresponding modules of three feature frequency points selected for feeder I are $\rho_{1\_Si0I(k0)}$, $\rho_{2\_Si0I(k0)}$, and $\rho_{3\_Si0I(k0)}$ and corresponding phase angles are $\theta_{1\_Si0I(k0)}$, $\theta_{2\_Si0I(k0)}$, and $\theta_{3\_Si0I(k0)}$. Based on the feature points corresponding with $\rho_{1\_Si0I(k0)}$ the mechanism for deciding whether feeder I is fault or not is as follows:

- If conditions in (10.38) are satisfied, feeder I is considered to be faulty.

$$\begin{cases} \left|\theta_{1\_Si0I(k_0)} - \theta_{1\_Si0II(k_0)}\right| > 90^\circ \\ \left|\theta_{1\_Si0I(k_0)} - \theta_{1\_Si0III(k_0)}\right| > 90^\circ \\ \left|\theta_{1\_Si0I(k_0)} - \theta_{1\_Si0IV(k_0)}\right| > 90^\circ \end{cases} \tag{10.38}$$

where $\theta_{1\_Si0II(k_0)}$, $\theta_{1\_Si0III(k_0)}$, and $\theta_{1\_Si0IV(k_0)}$ denote the phase angles in feeder II, III, and IV of the feature frequency point respectively. Considering the module of the point reflects the accuracy of its phase angle, the voting confidence degree is defined as follows:

$$\lambda_{11}(k_0) = \frac{\rho_{1\_Si0I(k0)}}{\rho_{1\_Si0I(k0)} + \rho_{2\_Si0I(k0)} + \rho_{3\_Si0I(k0)}} \tag{10.39}$$

where the voting confidence degree increases with increasing value of $\lambda_{11}(k_0)$.

- If conditions in (10.40) are satisfied, the fault is considered to exist on the bus.

$$\begin{cases} \left|\theta_{1\_Si0I(k_0)} - \theta_{1\_Si0II(k_0)}\right| \le 90^\circ \\ \left|\theta_{1\_Si0I(k_0)} - \theta_{1\_Si0III(k_0)}\right| \le 90^\circ \\ \left|\theta_{1\_Si0I(k_0)} - \theta_{1\_Si0IV(k_0)}\right| \le 90^\circ \end{cases} \tag{10.40}$$

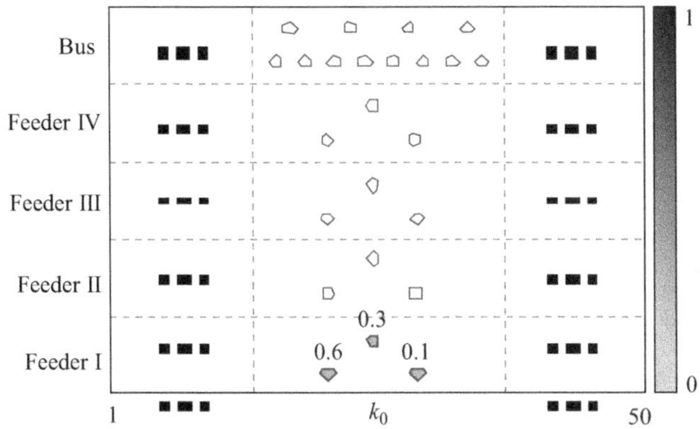

Figure 10.16 Sketch of vote results

where the voting confidence degree is defined the same way as in (10.39).

- If neither condition (1) nor (2) is satisfied, other feeder faults or information errors may have happened and no vote is conducted or the vote is abstained.

The three steps above are results of voting for feeder I according to the corresponding feature frequency with $\rho_{1\_Si0I(k0)}$. Votes according to the corresponding feature frequency with $\rho_{2\_Si0I(k0)}$ and $\rho_{3\_Si0I(k0)}$ should also be carried out The definition of degree of voting confidence is same as that of (10.39) except that the numerator is replaced accordingly.

The final result is showed in Figure 10.16, in which the three circles in a box represent the votes of three feature frequency points, and the gray level denotes the voting confidence degree where a higher level of gray manifests a higher voting confidence degree. A hollow circle is used to denote an absence of a vote in order to make it clear. It can be seen in Figure 10.16 that the three feature frequency points chosen by feeder I indicate fault to feeder I, and the voting confidences are 0.6, 0.3 and 0.1; all the three feature frequency points chosen by feeder II abstain from voting. As four feeders select 12 feature frequency points in total to vote whether the bus is faulty or not, four groups of results are listed in the box corresponding to "bus."

10.6.2.3 Fuse/merge the voting results

Firstly, each box with vote in Figure 10.16 are converted to fuzzy set (feeder n fault, $\lambda_n(k)$) and (bus fault, $\lambda_{Bus}(k)$), $\lambda_n(k)$, and $\lambda_{Bus}(k)$ are defined as follows:

$$\lambda_n(k) = \lambda_{1n}(k) + \lambda_{2n}(k) + \lambda_{3n}(k) \tag{10.41}$$

$$\lambda_{Bus\_n}(k) = \lambda_{1n}(k) + \lambda_{2n}(k) + \lambda_{3n}(k) \tag{10.42}$$

where $n =$ I, II, III, and IV. As an example, in Figure 10.16, the voting results of box k_0 are converted to (feeder I fault, $\lambda_{\mathrm{I}}(k_0) = \lambda_{1\mathrm{I}}(k_0) + \lambda_{2\mathrm{I}}(k_0) + \lambda_{3\mathrm{I}}(k_0) = 0.6 + 0.3 + 0.1 = 1$).

As Shannon fuzzy entropy can comprehensively depict the fuzziness and uncertainty of an event, faulty feeder identification confidence degree is defined by Shannon fuzzy entropy and the fuzzy set is handled with fusion calculation. The Shannon fuzzy entropy is defined as follows:

$$S(x) = -\sum_{n=1}^{N}[x_n \log_2 x_n + (1 - x_n)\log_2(1 - x_n)], \forall x = [x_1, x_2, ..., x_N]^{\mathrm{T}} \in [0, 1]^N \tag{10.43}$$

The Shannon fuzzy entropy is nonmonotonic while faulty feeder identification confidence degree should be monotonic. So $\lambda_n(k)$ and $\lambda_{Bus}(k)$ are mapped from range [0, 1] to range [0.5, 1]:

$$\begin{cases} \lambda_n^*(k) = (\lambda_n(k) + 1)/2 \\ \lambda_{Bus}^*(k) = (\lambda_{Bus\_n}(k) + 1)/2 \end{cases} \tag{10.44}$$

The meaning of confidence degree is opposite to Shannon fuzzy entropy. A larger confidence degree proved the event to be more clear and the corresponding fuzzy entropy is supposed to be smaller and vice versa. Moreover, the confidence degree should range in [0, 1]. In view of the above, faulty feeder identification confidence degree α_n and α_{Bus} are defined in (10.45) and (10.47) below.

$$\alpha_n = 1 - S(\lambda_n^*)/50, n = \mathrm{I,II,III,IV} \tag{10.45}$$

where,

$$S(\lambda_n^*) = -\sum_{k=1}^{50}\{\lambda_n^*(k)\log_2\lambda_n^*(k) + [1 - \lambda_n^*(k)]\log_2(1 - \lambda_n^*(k))\} \tag{10.46}$$

$$\alpha_{Bus} = 1 - S(\lambda_{Bus}^*)/(50 \times 4) \tag{10.47}$$

In (10.47),

$$S(\lambda_{Bus}^*) = \sum_{n=1}^{\mathrm{IV}}\sum_{k=1}^{50}\{\lambda_{Bus\_n}^*(k)\log_2\lambda_{Bus\_n}^*(k) + [1 - \lambda_{Bus\_n}^*(k)]\log_2(1 - \lambda_{Bus\_n}^*(k))\} \tag{10.48}$$

The faulty feeder identification confidence degree is obtained by calculation of the vote results in Figure 10.16 through (10.45) and (10.47). The final faulty feeder selection result is shown in the form of the faulty feeder number and the faulty feeder identification confidence degree.

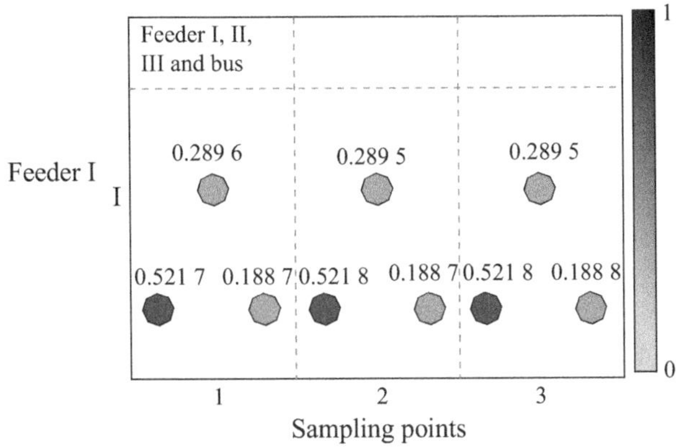

Figure 10.17 Voting results at the first three sampling points when a fault on feeder I occurs

10.6.3 Simulation test and conclusion

The four-feeder power distribution system shown in Figure 10.14 has been modeled in PSCAD/EMTDC. In the system, feeder I and feeder II are overhead lines, and feeder IV is a cable line, while feeder III is a hybrid transmission line.

10.6.3.1 Case 1 (normal fault situation)

Consider the following case: an AG fault occurs on feeder I, at a location of 13 km away from the bus, and the fault inception angle is 72°. The fault resistance is 50 Ω. Figure 10.17 shows the voting results of the first 3 sampling points.

It can be seen in Figure 10.17 that the feature frequency points chosen by feeder I all vote for feeder I and the feature frequency points chosen by other feeders abstain from voting. At sampling point 1, the feature frequency point with the largest modulus value chosen by feeder I would vote for feeder I to be faulty, and the voting confidence is 0.5127; the feature frequency point with the second largest modulus would again vote for feeder I to be faulty, and the voting confidence is 0.2896; the feature frequency point with the third biggest module would again vote feeder I to be faulty, and the voting confidence is 0.1887. Finally, the calculation result indicates that the voting confidence of feeder I is 1.

10.6.3.2 Case 2 (high-impedance fault situation)

In this case, an AG fault occurs on feeder III, the fault location is 12 km away from the bus and the fault inception angle is 18°. The fault resistance is 2.5 kΩ. In this case, the fault resistance is larger and the fault inception angle is smaller. It can be seen from the fault wave that the zero-sequence feeder current is reduced significantly. The peak value of zero-sequence currents of feeder I and II is less than 0.1 A.

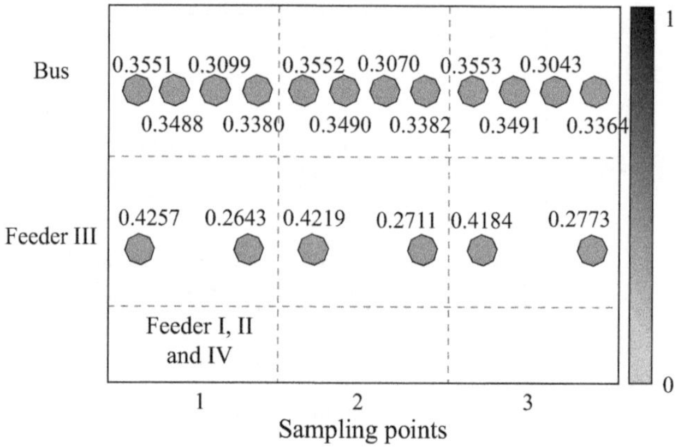

Figure 10.18 Voting results at the first three sampling points when a fault on feeder III occurs

The voting results in this case are shown in Figure 10.18.

It can be seen in Figure 10.18 that the feature frequency points with the largest modulus and the third largest modulus chosen by feeder III all vote fault for feeder III, and the voting confidences are 0.4257 and 0.2643, respectively. But the feature frequency point with the second biggest module chosen by all four feeders vote fault for bus, with the voting confidences of 0.3551, 0.3488, 0.3099, and 0.338, respectively. The calculation result indicates that the fault occurs on feeder III or the bus with the voting confidence of 0.4387 or 0.0798, respectively.

The voting confidence of feeder III is higher than that of the bus. As a result, the fault occurs on feeder III. The data of voting confidence 0.4387 reveals that the method works in this adverse condition also. However, the selection result has low confidence in being correct.

10.6.3.3 Statistical results

Furthermore, to test the validity of the faulty feeder selection method, all of the following cases in Table 10.19 are simulated.

The performance of the faulty feeder selection method has been investigated for all the test cases listed in Table 10.19. It has been found that for the test cases shown in Table 10.19, the faulty feeder selection method has 100% accuracy. Some of the results are listed in Table 10.20.

10.6.3.4 Conclusion

On the basis of the extraction of the amplitude-frequency and phase-frequency characteristics of signals using *S* transform, the faulty feeder selection method is discussed, which merges the voting results of several sampling points.

A voting mechanism for faulty feeder selection is introduced, and the voting confidence is defined with the consideration of the phase angle shift caused by the module at certain frequency points. The phase of zero-sequence feeder current is obtained with *S* transform.

Table 10.19 Fault conditions of test cases

Fault feeder	Fault inception angle/°	Fault resistance/Ω	Fault distance/km	Number
Feeder I	0, 18, 36, 54, 72, 90, 108, 126, 144, 162	0, 5, 50, 100, 200, 500, 700, 1k, 1.5k, 2.5k	10 random round numbers between 0.5 and 15	1000
Feeder II			10 random round numbers between 0.5 and 30	1000
Feeder III			10 random round numbers between 0.5 and 12	1000
Feeder IV			10 random round numbers between 0.5 and 8	1000
Bus				100

Table 10.20 Partial fault line identification results of test cases

Fault feeder	Fault inception angle/°	Fault resistance/Ω	Fault distance/km	Selection results	
				Fault line	Vote confidence
Feeder I		5	3	Feeder I	1
Feeder II	36	500	8	Feeder II	1
Feeder III			5	Feeder III	1
Feeder III	18	2.5 k	9	Feeder II	0.4837
				Bus	0.0798
Feeder IV	90	2.5 k	4	Feeder IV	0.9530
Feeder IV	108	1 k	4	Feeder IV	1
Bus	18	2.5 k		Bus	1
Bus	126	1 k		Bus	0.9612

The results of faulty feeder selection are presented in the form of the faulty feeder number and the faulty feeder identification confidence degree. The faulty feeder identification results and its confidence degrees are provided to related staffs, helping them gain an insight knowledge on the fault condition and the reliability of faulty feeder selection.

References

[1] Zheng Q., Yang Y.-H., "Analysis of technology of fault line selection for single-phase-to-earth faults in neutral point non-effectively grounded system," Automation of Electric Power Systems. vol. 28, no. 14. pp. 1–5, 2004.

[2] Pang Q.-L., Chen S.-Y., "Artificial immune algorithm based faulty line detection method for indirectly grounding power system," Power System Protection and Control. vol. 37, no. 24. pp. 27–31, 2009.
[3] Stockwell, R.G., Mansinha L., and Lowe R.P., "Localization of the complex spectrum: the S transform," IEEE Transactions on Signal Processing, vol.44, no. 4, pp. 998–1001, 1996.
[4] Dash P.K, Panigrahi B.K, and Panda G. "Power quality analysis using S-transform," IEEE Transaction on Power Delivery, vol. 18, no.2, pp. 406–411, 2003.
[5] Liang R., Xin J., Wang C.-L., *et al.*, "Fault line selection in small current grounding system by improved active component method," High Voltage Engineering, vol. 36, no. 2, pp. 375–379, 2010.
[6] Daubechies I., "The wavelet transform, time-frequency localization and signal analysis," IEEE Transactions on Information Theory, vol. 36, no. 5, pp. 961–1005, 1990.
[7] He Z.Y., Yang J.W., and Zang T.L., "Intelligent Information Processing in Power Grid Fault Diagnosis," Chengdu: Southwest Jiaotong University Press, 2012.
[8] Jang, J.-S.R. "ANFIS: adaptive-network-based fuzzy inference system," IEEE Transactions on Systems, Man, and Cybernetics. vol. 23, no. 3. pp. 665–685, 1993.
[9] Reddy M., Mohanta D.K., "Performance evaluation of an adaptive-network-based fuzzy inference system approach for location of faults on transmission lines using monte carlo simulation," IEEE Transaction on Fuzzy Systems, vol. 16, no. 4, pp. 909–919, 2008.
[10] Reddy M.J., Mohanta D.K., "Adaptive-neuro-fuzzy inference system approach for transmission line fault classification and location incorporating effects of power swings," IET Generation, Transmission & Distribution. vol. 2, no. 2, pp. 235–244, 2008.
[11] Zhuting Y., Hongxia P., "Engine fault diagnosis based on the weighted DS evidence theory," Computational Intelligence and Applications (IWCIA), 2014 IEEE 7th International Workshop on Hiroshima, pp. 7–8, 2014.
[12] Cuzzolin F., "Geometry of Dempster's rule of combination," IEEE Transactions on Systems, Man and Cybernetics, Part B (Cybernetics), vol. 34, no.2, pp. 961–977, 2004.
[13] Dewasurendra D.A., Bauer P.H., and Premaratne K., "Evidence filtering," IEEE Transactions on Signal Processing, vol. 55, no. 12, pp. 5796–5805, 2007.
[14] Lee I.W.C., Dash P.K., "S-transform-based intelligent system for classification of power quality disturbance signals," IEEE Transactions on Industrial Electronics, vol. 50, no. 4, pp. 800–805, 2003.
[15] Xiang-ning L., Pei L., and Yan G., "Ultra-high-speed line directional protection based on transient and mathematical morphology," Proceedings of the CSEE, vol. 25, no. 4, pp. 13–18, 2005.
[16] Sun Z.-L., Au K.-F., and Choi T.-M. "A neuro-fuzzy inference system through integration of fuzzy logic and extreme learning machines," IEEE Transactions on Systems, vol. 10, no. 5, pp. 1321–1331, 2007.

[17] Adeh J., Afradi H., "A new and accurate fault location algorithm for combined transmission lines using adaptive network-based fuzzy inference system," Electric Power Systems Research, vol. 79, no. 11, pp. 1538–1545, 2009.
[18] Das B., "Fuzzy logic-based fault-type identification in unbalanced radial power distribution system," IEEE Transactions on Power Delivery, vol. 21, no. 1, pp. 278–285, 2006.
[19] Zhang J., He Z.Y., and Tan X.J., "An ANFIS based fault classification in distribution network and its adaptability analysis," Power System Protection and Control, vol. 39, no. 4, pp. 23–29, 2011.
[20] Zhang J., Li X.P., and HE Z.Y., "Fault classification technique for power distribution network using adaptive network based fuzzy inference system," Proceeding of the CSEE, vol. 30, no. 25, pp. 87–92, 2010.
[21] Borghetti A., Bosetti M., Di Silvestro M., *et al.*, "Continuous-wavelet transform for fault location in distribution power networks: definition of mother wavelets inferred from fault originated transients," IEEE Transactions on Power Systems, vol. 23, no. 2, pp. 380–388, 2008.
[22] Denoeux T., "A k-nearest neighbor classification rule based on Dempster–Shafer theory," IEEE Transactions on Systems, Man and Cybernetics, vol. 25, no. 5, pp. 804–813, 1995.
[23] Salim R.H., de Oliveira K., Filomena A.D., *et al.*, "Hybrid fault diagnosis scheme implementation for power distribution systems automation," IEEE Transactions on Power Delivery, vol. 23, no. 4, pp. 1846–1856, 2008.
[24] Valsan S.P., Swarup K.S., "High-speed fault classification in power lines: theory and FPGA-based implementation," IEEE Transactions on Industrial Electronics, vol. 56, no. 5, pp.1793–1800, 2009.
[25] Liu J.F., Wilson P.L., and Jayasinghe R.P., "A PSCAD/EMTDC based simulation study of protective relay," in Proc. 8th Inst. Elect. Eng. Int. Conf. IEE Devices in Power System Protection, vol. 1, pp. 264–267, 2004.
[26] Zhang J., He Z.Y., and Zang T.L., "A fault phase selection approach in neutral non-effectively grounded system based on D-S evidence theory and stationary component," Power System Protection and Control, vol. 39, no. 11, pp. 49–55, 2011.
[27] Young R.K., Wavelet theory and its applications. Dordrecht: Kluwer, 1993.
[28] Bracewell R.N., The Fourier transform and its applications. New York, McGraw-Hill, 1978.
[29] Lin X.N., Kursan I., Sun, J.W., Li, Z.T., Zheng R.J., Bo Z.Q., Xu J.F., and Wei Z.K., "Fault section identification based on per-unit zero sequence current," in 12th IET International Conference on Developments in Power System Protection (DPSP 2014), pp. 1–6, 2014.
[30] Zhang J., He Z.Y., and Jia Y., "Fault line identification approach based on S-transform," Proceedings of the CSEE, vol. 31, no. 10, pp. 109–155, 2011.

Chapter 11

Economic analysis/cost–benefit analysis

Chun-Lien Su[1] *and Jen-Hao Teng*[2]

11.1 Introduction

The distribution network is an important part of the total electrical supply system, as it provides the final link between the bulk transmission system and the customers. It has been reported that 80% of the customer service interruptions are due to failures in the distribution networks. Increasing attention to research and application of Distribution Automation (DA) systems has assisted the utilities in solving these challenges. DA systems have been defined by the IEEE as systems that enable an electric utility to monitor, coordinate, and operate distribution network components in real-time mode from remote control centers [1]. The DA can be implemented in phases to include remote monitoring and control of substation, feeder and consumer devices, and loads [2, 3]. DA systems have been implemented by many utilities around the world to achieve significant and immediate improvement in reliability and hence service to the electricity customers.

DA systems are usually modular and can be implemented in phases to include remote monitoring and control of substations, feeders, and consumer loads. DA systems are built to achieve the goals of operation and maintenance (O&M) cost reduction, capital investment deferment, supply reliability improvement, and operation efficiency enhancement [3]. The installation of a DA system requires a large capital investment; therefore, the economic analysis of DA systems is critical to identify the DA functions that produce significant operational benefits.

Many literatures are relevant to the economic study on various aspects of DA [4–11]. Reference 4 provided a set of general guidelines for evaluating DA. References 5–9 developed some economic formulas for benefit/cost evaluation of various combinations of DA functions. References 10 and 11 used software packages to estimate the present worth of all costs and benefits associated with the DA functions. Reference 12 showed the importance to identify the costs and benefits of the DA project, including the value of improved reliability to the customers.

This chapter presents the methodology and results for the economic evaluation of an extensive DA project. The outage costs and benefits of the DA system are quantified by using standard mathematical expressions. An economic evaluation method based on value-based analysis and present worth analysis is performed to identify the most

[1] Department of Marine Engineering, National Kaohsiung Marine University, Kaohsiung, TAIWAN 805
[2] Department of Electrical Engineering, National Sun Yat-Sen University, Kaohsiung, TAIWAN 804

beneficial functions in the DA projects. Test results obtained in this chapter have indicated that the feeder automation (FA) is the most beneficial function to utilities. Therefore, the candidate feeders and number of switch for further FA extension is also conducted in this chapter. The results and expressions of this chapter can serve as the basis for the utilities' DA system implementation guideline. The results can also be used for cost–benefit analysis, return-of-investment, or possible performance measure of the DA systems.

This study is conducted in four phases: (1) to perform an extensive review of all functions of the DA system, (2) to carry out benefit analysis based on the expected functionality of the DA and assumptions regarding the utilities' operation, (3) to collect field data and evaluate the real return of the DA project, and (4) to find the candidate feeders for further FA extension and determine the optimal number of switches for a feeder.

This chapter is divided into five sections. Section I describes the overview of the study. Section II reviews the implementation of the DA project and describes the implemented functions for the project. The benefits of the DA functions are analyzed and quantified in Section III. The economic evaluation results are also represented in Section III. Section IV describes the methodology to find the candidate feeders for further FA extension and determine the number of switches for a feeder. Finally, Section V summarizes the overall conclusions of the study.

11.2 The DA project review

To illustrate the value-based DA economic analysis, a practical DA system, Tai-Chung DA system, implemented in the Tai-Chung District of Taipower is used for the study. The system consists of the feeder control center, onsite devices for control and data acquisition, and the communication system. There are 16 substations, 139 feeders, 464 automatic switches, 92 capacitor banks in the Tai-Chung DA system, and 1941 and 160 customers participating in the automatic meter reading (AMR) and direct load control programs, respectively. The operation and planning functions in the Tai-Chung DA system are classified as FA system, distribution analysis system, and customer management system. Main objectives of this project are to reduce operating and maintenance (O&M) cost, improve service reliability, and provide better customer reliability with enhanced technology level. The Tai-Chung DA system is designed based on open system architecture, which means that this system can have an unlimited capability through hardware and software upgradations. There are three main parts in the DA system including feeder control center, onsite devices for control and data acquisition, and communication infrastructure. The feeder control center processes the real-time operating data including the digital and analog signals and then issues proper commands to control and coordinate the system operation through the onsite devices and communication system. The onsite devices capture the feeder operating data and customers' energy consumption data, send these data to the control center, and execute the control actions received from the control center. The monitoring data and control action signal between the control center and the onsite devices are

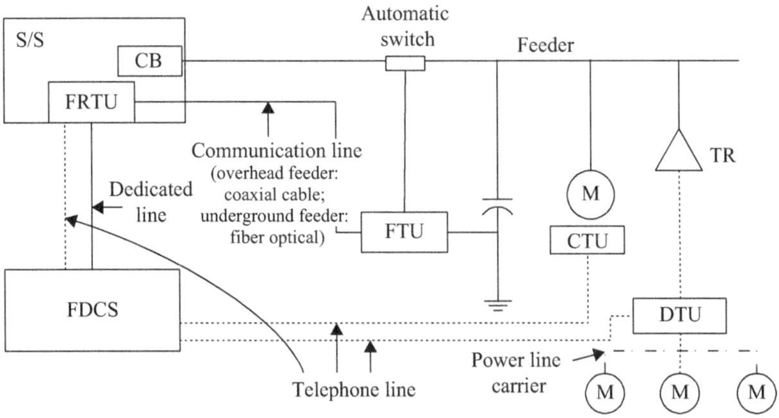

FRTU: Feeder remote terminal unit
FTU: Feeder terminal unit
CTU: Customer terminal unit
DTU: Data terminal unit

FDCS: Feeder control center M: Meter
S/S: Secondary substation
TR: Transfomer

Figure 11.1 The structure of Tai-Chung DA system

transferred via various types of communication medium. The main structure of this system is shown in Figure 11.1.

The functions included in the Tai-Chung DA system are for both operational and planning purposes. These functions can be classified as FA functions, distribution analysis functions, and customer management functions. The FA functions are used to monitor the system status, record system operation data, and control onsite devices according to the operating conditions. The FA functions include

- Supervisory control and data acquisition (SCADA)
- Fault detection, isolation, and service restoration (FDIR)
- Capacitor bank control (CBC)
- Information storage and retrieval (ISR)
- Outage scheduler (OS)
- Trouble call system (TCS)
- Dispatcher training simulator (DTS)

The distribution analysis functions provide both dispatchers and engineers the capability to analyze the Tai-Chung distribution system under specific operating conditions. These functions are

- Distribution power flow (DPF)
- Optimal switching (OSW)
- Short-circuit analysis (SCA)
- Protection coordination
- Optimal capacitor placement (OCP)

The customer management functions are used to monitor and collect the customer data and provide the real-time pricing and direct load controls. The functions cover

- AMR
- Transformer load survey (TLS)
- Load management (LM)

All the above described hardware devices and software functions have been successfully implemented and run on Tai-Chung District. The following section presents the cost and benefit of the DA project and the results of economic analysis.

11.3 The DA system economic analysis

The successful implementation of the DA project results in

- More accurate data and information for distribution system operation and planning
- Deferred capital expenditure
- Reduced operation and maintenance expenses
- Better outage response and restoration
- Enhanced operation efficiency
- Enhanced customer satisfaction
- Positive public image

Although distribution systems and associated operating costs vary among utilities, it is important to identify and quantify the expected benefits in the DA. Generally, the benefits of DA are classified into two parts: the quantifiable benefits and the unquantifiable (intangible) benefits. The quantifiable benefits that can be expressed as dollar values include reduced O&M costs, reduced customer interruption costs (CICs), deferred capital investments, and increased energy revenues, etc. The unquantifiable benefits are improved public image and safety, better quality of system information, and enhanced customer satisfaction, etc. Though not all DA's benefits can be quantified, they are nevertheless valuable to utilities. For example, enhanced public image from shortened restoration time during emergency conditions and better quality of information for engineering and planning are benefits that cannot be quantified but are highly worthwhile for utilities. Table 11.1(a–c) shows the potential benefits of the DA project. In Table 11.1(a–c), the symbol "√" indicates that the DA function could have this benefit. In the table, the other benefits are unquantifiable ones that are described above.

The quantified benefits of each function in Table 11.1 are analyzed and can be formulated as follows. The equations used to calculate the quantified benefits shown

Table 11.1 Benefits of the Tai-Chung DA project

DA functions/ Benefits	Capital deferral	O&M cost reduction	Customer cost reduction	Revenue increase	Others
(a) Feeder automation functions					
SCADA		√			√
FDIR	√	√	√	√	√
CBC	√	√			√
ISR		√			√
OS		√			√
TCS		√			√
DTS		√			√
(b) Distribution analysis functions					
DPF		√			√
OSW		√			√
SCA		√			√
PCN		√			√
OCP		√			√
(c) Customer management functions					
AMR		√		√	
TLS	√	√			√
LM	√	√	√		√

in Table 11.1 are mainly determined by the discussions with the Taipower's staffs in Tai-Chung District and the equations presented in Reference 4.

11.3.1 FA functions

The major benefit of SCADA is to reduce the labor costs to collect system information and operate onsite devices. Its benefit is expressed as

$$\begin{aligned}\mathrm{SCADA}^B = {} & (\mathrm{LC_{SCADA}}(\$/\mathrm{hrs}) \times \mathrm{DC^{NA}}(\mathrm{hr}) \times \mathrm{DT} + \mathrm{LC_{SCADA}}(\$/\mathrm{hr}) \\ & \times \mathrm{SC^{NA}}(\mathrm{hrs}) \times \mathrm{ST}) \times \mathrm{CPV_1}\end{aligned} \tag{11.1}$$

where, SCADA^B is the quantified benefit of the SCADA function. $\mathrm{LC_{SCADA}}(\$/\mathrm{hr})$ is the labor rate for data collection and onsite device operation. $\mathrm{DC^{NA}}(\mathrm{hrs})$ and $\mathrm{SC^{NA}}(\mathrm{hrs})$ are the average time for data collection and remote device operation before DA, respectively. DT and ST are the average number of data collection and remote device operation per year before DA, respectively. $\mathrm{CPV_1}$ is the cumulative present value.

Since most costs and benefits in the economic evaluation are the future annual costs and benefits, it needs the CPV to discount the value of future costs. For the reduced O&M and increased revenues, the CPV takes the interest rate, inflation rate, and economic life of equipment into account. It can be expressed as follows [13, 14]

$$P_1 = \frac{\left(1 + \frac{\mathrm{IF}}{100}\right)}{\left(1 + \frac{\mathrm{IR}}{100}\right)} \tag{11.2a}$$

$$\mathrm{CPV}_1 = \frac{1 - (P_1)^{\mathrm{EL}}}{1 - (P_1)} \tag{11.2b}$$

For CIC reduction, the CPV that needs to incorporate the interest rate, inflation rate, load growth rate, and economic life of equipment is determined by the following equations

$$P_2 = \frac{\left(1 + \frac{\mathrm{IF}}{100}\right)\left(1 + \frac{\mathrm{LG}}{100}\right)}{\left(1 + \frac{\mathrm{IR}}{100}\right)} \tag{11.3a}$$

$$\mathrm{CPV}_2 = \frac{1 - (P_2)^{\mathrm{EL}}}{1 - (P_2)} \tag{11.3b}$$

where, IR, IF, LG, and EL are the interest rate, inflation rate, load growth rate, and economic life of equipment, respectively.

The FDIR function can reduce the time required for feeder fault location, isolation, and restoration. Therefore, the CIC can be reduced, energy revenue can be increased, and labor cost for fault location, isolation, and restoration can be reduced. Its benefit then is calculated by

$$\begin{aligned} \mathrm{FDIR}^{B} = {} & ((\mathrm{LC_{FDIR}}(\$/\mathrm{hr}) * \mathrm{FDIR}^{\mathrm{NA}}(\mathrm{hrs}) - \mathrm{LC_{FDIR}}(\$/\mathrm{hr}) * \mathrm{FDIR}^{A}(\mathrm{hrs})) * \mathrm{FT} \\ & + (\mathrm{LC_{FDIR}}(\$/\mathrm{hr}) * \mathrm{FR}^{\mathrm{NA}}(\mathrm{hrs}) - \mathrm{LC_{FDIR}}(\$/\mathrm{hr}) * \mathrm{FR}^{A}(\mathrm{hrs})) * \mathrm{FT}) * \mathrm{CPV}_1 \\ & + (\mathrm{RR}^{\mathrm{NA}}(\$/\mathrm{year}) - \mathrm{RR}^{A}(\$/\mathrm{year}) + \mathrm{CIC}^{\mathrm{NA}}(\$/\mathrm{year}) - \mathrm{CIC}^{A}(\$/\mathrm{year})) * \mathrm{CPV}_2 \end{aligned} \tag{11.4}$$

where, FDIR^B is the quantified benefit of FDIR function. $\mathrm{LC_{FDIR}}(\$/\mathrm{hr})$ is the labor rate for onsite fault repair. $\mathrm{FDIR}^{\mathrm{NA}}(\mathrm{hrs})$ and $\mathrm{FDIR}^{A}(\mathrm{hrs})$ are the average time for fault detection, isolation and restoration before and after DA, respectively. $\mathrm{FR}^{\mathrm{NA}}(\mathrm{hrs})$ and $\mathrm{FR}^{A}(\mathrm{hrs})$ are the average time for each fault repair before and after DA, respectively. FT is the average number of feeder failure per year. $\mathrm{RR}^{\mathrm{NA}}$ (\$/year) and RR^{A} (\$/year) are the reduced revenues due to the fault per year before and after DA, respectively. $\mathrm{CIC}^{\mathrm{NA}}$ (\$/year) and CIC^{A} (\$/year) are the CIC due to the fault per year before and after DA, respectively.

Using feeder historical reliability data, when an outage occurred in any *section* of a feeder, the corresponding outage duration times of all *sections* in the feeder can be easily determined. The information can be expressed in an outage duration matrix γ. The element γ_{ij} is an average duration of service interruption of Section j for a fault at Section i. Feeder topology, switch location, section failure rate, switch operations (manual or automatic) time, and load transfer capability of the feeder will affect γ_{ij}. Feeder loads and switches are assumed to be evenly distributed on the feeders for simplicity in the estimation of CICs and energy revenue reductions before and after FA. The outage duration matrix γ is expressed as

$$\gamma = \begin{bmatrix} t_{repair} & t_{feeder} & t_{feeder} & \cdots & t_{feeder} \\ t_{ss} & t_{repair} & \ddots & & \vdots \\ t_{ss} & & \ddots & \ddots & \vdots \\ \vdots & & & \ddots & t_{feeder} \\ t_{ss} & \cdots & \cdots & t_{ss} & t_{repair} \end{bmatrix} \tag{11.5}$$

where, t_{ss}, t_{feeder}, and t_{repair} are the average time for restoration from substation, from other feeders, and fault repair for each fault, respectively.

For faults occurred on Section i, the CIC per year can be written as

$$\begin{aligned} \text{CIC}_i = {} & \frac{\lambda \times l}{(n+1)} \times \left(\text{IC}(t_{\text{ss}}) \times \frac{(i-1) \times L}{(n+1)} + \text{IC}(t_{\text{repair}}) \times \frac{L}{(n+1)} + \text{IC}(t_{\text{feeder}}) \right. \\ & \left. \times \frac{(n-i+1) \times L}{(n+1)} \right) \end{aligned} \tag{11.6}$$

where, λ is the average failure rate (failure/mile/year), l is the feeder length (miles), and L is the average feeder load (in kilowatts), n is the number of switch. $IC(t)$ is the CIC per kilowatt (\$/kW).

And the CIC per year for a feeder can be estimated by

$$\text{CIC}(\$/\text{year}) = \sum_{i=1}^{n+1} \text{CIC}_i = \frac{\lambda \times l \times L}{(n+1)} \times \left(\frac{n}{2} \times IC(t_{\text{ss}}) + IC\left(t_{\text{repair}}\right) + \frac{n}{2} \times IC(t_{\text{feeder}}) \right) \tag{11.7}$$

Similarly, the energy revenue reduction per year due to faults occurred on Section i, can be written as

$$\begin{aligned} \text{RR}_i = {} & K_A \times \frac{\lambda \times l}{(n+1)} \times \left(t_{\text{ss}} \times \frac{(i-1) \times L}{(n+1)} + t_{\text{repair}} \times \frac{L}{(n+1)} \right. \\ & \left. + t_{\text{feeder}} \times \frac{(n-i+1) \times L}{(n+1)} \right) \end{aligned} \tag{11.8}$$

where K_A(\$/kWh) is energy cost.

The reduced revenue due to a fault can be expressed as

$$\text{RR}(\$/\text{year}) = \sum_{i=1}^{n+1} \text{RR}_i = \frac{\lambda \times l \times L}{(n+1)} \times \left(\frac{n}{2} \times t_{\text{ss}} + t_{\text{repair}} + \frac{n}{2} \times t_{\text{feeder}} \right) \tag{11.9}$$

The potential benefits of CBC function are capacitor banks repair and maintenance savings and fuel savings through reduced primary losses. They are expressed as

$$\begin{aligned}\mathrm{CBC}^B &= (\mathrm{LC_{CBC}}(\$/\mathrm{hr}) \times ((\mathrm{CFR}^{\mathrm{NA}}(\mathrm{hrs}) \times \mathrm{CFT}^{\mathrm{NA}} + \mathrm{CI}^{\mathrm{NA}}(\mathrm{hrs}) \times \mathrm{CIT}^{\mathrm{NA}}) \\ &\quad - (\mathrm{CFR}^A(\mathrm{hrs}) \times \mathrm{CFT}^A + \mathrm{CI}^A(\mathrm{hrs}) * \mathrm{CIT}^A)) + (\mathrm{GC}(\$/\mathrm{kWh}) \\ &\quad \times \mathrm{LLR}_1^A(\mathrm{kWh/year}) + \mathrm{FGC}(\$/\mathrm{kWyear}) \times \mathrm{LLR}_2^A(\mathrm{kW/year}))) \times \mathrm{CPV}_1\end{aligned} \tag{11.10}$$

where, CBC^B is the quantified benefit for CBC function. $\mathrm{LC_{CBC}}$ (\$/hr) is the labor rate for capacitor bank failure repair and onsite inspection. $\mathrm{CFR}^{\mathrm{NA}}$(hrs) and CFR^A(hrs) are the average time for capacitor bank failure repair before and after DA, respectively. $\mathrm{CI}^{\mathrm{NA}}$(hrs) and CI^A(hrs) are the average time for capacitor bank onsite inspection before and after DA, respectively. $\mathrm{CFT}^{\mathrm{NA}}$ and CFT^A are the average numbers of capacitor bank failure repair per year before and after DA, respectively. CIT^{NA} and CIT^A are the average numbers of capacitor bank onsite inspection per year before and after DA, respectively. GC (\$/kWh) and FGC (\$/kW year) are generation cost and fixed generation cost, respectively. LLR_1^A(kWh/year) and LLR_2^A(kW/year) are the average amounts of reduced energy and power for line losses per year after DA, respectively.

The ISR function can reduce the labor costs for data storage and retrieval, therefore its benefit is calculated by

$$\mathrm{ISR}^B = (\mathrm{LC_{ISR}}(\$/\mathrm{hr}) \times (\mathrm{IS}^{\mathrm{NA}}(\mathrm{hrs}) \times \mathrm{IST}^{\mathrm{NA}} - \mathrm{IS}^A(\mathrm{hrs}) \times \mathrm{IST}^A)) \times \mathrm{CPV}_1 \tag{11.11}$$

where, ISR^B is the quantified benefit for ISR function. $\mathrm{LC_{ISR}}$ (\$/hr) is the labor rate for data storage and retrieval. $\mathrm{IS}^{\mathrm{NA}}$(hrs) and IS^A(hrs) are the average time of each data storage and retrieval before and after DA, respectively. $\mathrm{IST}^{\mathrm{NA}}$ and IST^A are the average number of data storage and retrieval per year before and after DA, respectively.

The benefit of the OS function is to reduce the labor costs to schedule distribution equipment outages. It can be calculated by

$$\mathrm{OS}^B = (\mathrm{LC_{OS}}(\$/\mathrm{hr}) \times (\mathrm{OS}^{\mathrm{NA}}(\mathrm{hrs}) \times \mathrm{OST}^{\mathrm{NA}} - \mathrm{OS}^A(\mathrm{hrs}) \times \mathrm{OST}^A)) \times \mathrm{CPV}_1 \tag{11.12}$$

where, OS^B is the quantified benefit of OS function. LC_{OS} (\$/hr) is the labor rate for scheduling equipment outages. $\mathrm{OS}^{\mathrm{NA}}$(hrs) and OS^A(hrs) are the average time of each equipment outage scheduling work before and after DA, respectively. $\mathrm{OST}^{\mathrm{NA}}$ and OST^A are the average numbers of distribution equipment outage scheduling per year before and after DA, respectively.

TCS function can reduce the labor costs to handle trouble calls and its benefit is shown as follows

$$\mathrm{TCS}^B = (\mathrm{LC_{TCS}}(\$/\mathrm{hr}) \times (\mathrm{TC}^{\mathrm{NA}}(\mathrm{hrs}) \times \mathrm{TCT}^{\mathrm{NA}} - \mathrm{TC}^A(\mathrm{hrs}) \times \mathrm{TCT}^A)) \times \mathrm{CPV}_1 \tag{11.13}$$

where, TCS^B is the quantified benefit of TCS function. $\mathrm{LC_{TCS}}$(\$/hr) is the labor rate for handling trouble call. $\mathrm{TC}^{\mathrm{NA}}$(hrs) and TC^A(hrs) are the average processing time of each trouble call before and after DA, respectively. TCT^{NA} and TCT^{A} are the average numbers of trouble calls per year before and after DA, respectively.

The benefit of DTS function is to reduce the labor cost for training and can be calculated by

$$DTS^{B} = (LC_{DTS}(\$/hr) \times (DT^{NA}(hrs) \times DTT^{NA} - DT^{A}(hrs) \times DTT^{A})) \times CPV_{1} \quad (11.14)$$

where, DTS^{B} is the quantified benefit for DTS function. $LC_{DTS}(\$/hr)$ is the labor rate for dispatcher training. $DT^{NA}(hrs)$ and $DT^{A}(hrs)$ are the average time of each training before and after DA, respectively. DTT^{NA} and DT^{A} are the average numbers of dispatcher training per year before and after DA, respectively.

11.3.2 Distribution analysis functions

The distribution analysis functions including DPF, OSW, SCA, PCN, and OCP functions are to compute and analyze distribution system behaviors for operational and planning purposes. The main benefit of these functions is to reduce the required labor costs to perform these functions. Therefore, the benefit of each distribution analysis function can be expressed as

$$DAF^{B} = (LC_{DAF}(\$/hr) \times (DA^{NA}(hrs) \times DAT^{NA} - DA^{A}(hrs) \times DAT^{A})) \times CPV_{1} \quad (11.15)$$

where DAF^{B} is the quantified benefit of the distribution analysis function. $LC_{DAF}(\$/hr)$ is the labor rate for executing the function. $DA^{NA}(hrs)$ and $DA^{A}(hrs)$ are the average execution time of the function before and after DA, respectively. DAT^{NA} and DAT^{A} are the average execution numbers of the function per year before and after DA, respectively.

11.3.3 Customer management functions

The benefit of the AMR function is to reduce manpower costs for meter reading and reduce revenue losses due to theft of service. Its benefit is expressed as

$$\begin{aligned} AMR^{B} = {} & (LC_{MR}(\$/customer) \times AM(customers) \times AMT + EL(kWh/ \\ & customer\ year) \times AM\ (customers) \times K_{A}(\$/kWh)) \times CPV_{1} \end{aligned} \quad (11.16)$$

where, AMR^{B} is the quantified benefit of AMR function. $LC_{MR}(\$/customer)$ is the labor rate for meter reading per customer. AM (customers) is the number of customers participating in AMR program. AMT is the average number of meter reading per year before DA. EL(kWh/customer-year) is the average amount of yearly energy losses due to theft of service per customer. The parameters $LC_{MR}(\$/customer)$, AM(customers), and AMT can have different values for different types of customers.

The benefit of TLS function can be quantified as

$$TLS^B = (LC_{TLS}(\$/\text{transformer}) \times TL(\text{transformers}) \times TLT) \times CPV_1 \quad (11.17)$$

where, TLS^B is the quantified benefit of TLS function. LC_{TLS}($/transformer) is the labor rate for metering load survey per transformer. TL (transformers) is the number of transformers with automatic load metering function. TLT is the average number of load survey per year.

The benefits of LM function are fuel savings due to load shift and energy savings due to reduced load. They can be calculated by

$$\begin{aligned} LM^B = {} & (LME(\text{kWh/year}) \times GC(\$/\text{kWh}) + LMP(\text{kW/year}) \\ & \times FGC(\$/\text{kW-year})) \times CPV_1 \end{aligned} \quad (11.18)$$

where, LM^B is the quantified benefit of LM function. LME(kWh/year) and LMP(kW/year) are the average amounts of energy and power for LM control per year, respectively.

It is necessary to define economic evaluation parameters before evaluating the benefits and costs of the DA project. The economic parameters used in this study include the capital costs, the annual O&M costs, the benefit evaluation parameters, and the basic economic evaluation parameters such as interest rate, inflation rate, and economic life of equipment. The capital costs include basic DA system hardware and software plus additional costs to implement the DA functions on the basic DA system platform. The O&M costs and benefit evaluation parameters for each DA function are collected and obtained from the field operation data and historical statistics in Tai-Chung District. In this study, the total capital cost and annual O&M cost for the Tai-Chung DA project are about $29,207,481 and $185,840 [15], respectively. The interest rate, inflation rate, economic life of equipment, and energy cost are assumed to be 3%, 5%, 15 years, and $0.0818/kWh, respectively. They are considered as constant throughout the evaluation. Interruption cost functions for various customer types are important for CIC calculation in the above expressions. The interruption cost functions used in this chapter was conducted in 2002 by the Taiwan Economic Research Institute [16], and is shown in Table 11.2.

The quantifiable benefits of the implemented DA functions are evaluated by using the benefit evaluation expressions. The benefit/cost ratio for life of equipment is used as a primary guide for economic evaluation. If this ratio is greater than unity

Table 11.2 Interruption cost functions in Taiwan ($/kW)

Interruption duration (minutes)	Residential	Commercial	Industrial
1	0.054	0.624	2.523
15	0.818	1.976	5.047
30	1.636	5.555	7.723
60	3.273	10.889	19.423

the DA project is economical; on the other hand, if the ratio is less than unity, the DA project cannot obtain economical value based only on quantifiable benefits. The economic evaluation results for the Tai-Chung DA project are shown in Table 11.3. From these results, it is seen that the benefit/cost ratio for the Tai-Chung DA project is about 0.77454. Among all implemented DA functions, the FDIR function is the most economic. Its benefit/cost ratio is 0.7147 that is about 92% of the benefits for the project. The FA functions have the largest benefit among all DA functions. The distribution analysis functions and customer management functions are not justifiable based only on consideration of quantifiable benefits and they may be justified based on intangible benefits. The unquantifiable benefits should be taken into account for making a decision to judge the success of these functions. Note that in Table 11.3, the benefit/cost ratios for the ISR, OS, and DPF functions are negative. This is because before DA the Tai-Chung District has no ISR and DPF functions, these two functions can be seen as additional costs for the project. In addition, the time required for performing the OS function is longer than that for the system without this function, which resulted in a negative value of benefit/cost ratio.

Table 11.3 Economic evaluation results

Function category	**Present value of benefits ($)**	**Annual benefits ($)**	**Benefit/ cost ratio**
FA functions	23899474.745	973984.322	0.73746
SCADA	568511.513	33013.030	0.01754
FDIR	23162007.482	931160.160	0.71470
CBC	78276.463	4545.455	0.00242
ISR	−565.104	−32.815	−0.00002
OS	−75934.692	−4409.470	−0.00234
TCS	154654.852	8980.689	0.00477
DTS	12524.234	727.273	0.00039
Distribution analysis functions	98366.587	5712.073	0.00304
DPF	−3653.632	−212.164	−0.00011
OSW	24265.703	1409.091	0.00075
SCA	4435.614	257.573	0.00014
PCN	2870.085	166.664	0.00009
OCP	70448.817	4090.909	0.00217
Customer management functions	1103480.695	64078.283	0.03405
AMR	34784.808	2019.927	0.00107
LM	279669.142	16240.174	0.00863
TLS	789026.745	45818.182	0.02435
Present value of capital costs ($)		29207481.418	
Present value of O&M costs ($)		3200319.094	
Benefit/cost ratio		0.77454	

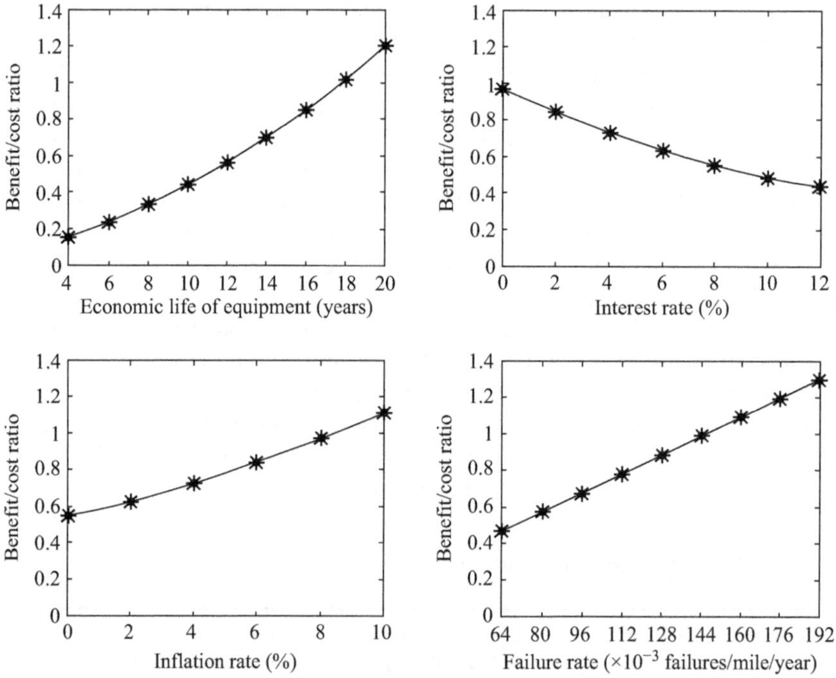

Figure 11.2 Effects of basic economic evaluation parameters on the benefit/cost ratio of the Tai-Chung DA system

To determine the effects of basic economic evaluation parameters on the economic evaluation results, a sensitivity analysis is also performed. Figure 11.2 shows the sensitivity of the benefit/cost ratio of the DA system to economic life of equipment, interest rate, inflation rate, and failure rate, respectively. From the results, it is found that the benefit/cost ratio increases with increases in economic life of equipment, inflation rate, and failure rate. It means that DA systems are suitable for the area with large values of failure rate and inflation rate. The results also indicate that if a small value of the interest rate is considered, there will be a good economic justification of DA systems. The results also show that the basic economic parameters significantly affect the economic analysis results. For better results, accurate economic evaluation parameters should be used in the DA system economic analysis for making a corrective decision.

Since most of the benefits of the DA system come from the reliability improvement measures, it is important to understand the effects of the CICs on the benefit/cost ratio of the DA system. Using the CIC data in Table 11.2 as a basis, the changes in benefit/cost ratio for different CICs are shown in Figure 11.3. It can be seen from the result that the benefit/cost ratio increases as the CIC increases. The implication of the result is that the area with high CICs could have a high priority for the implementation of DA systems.

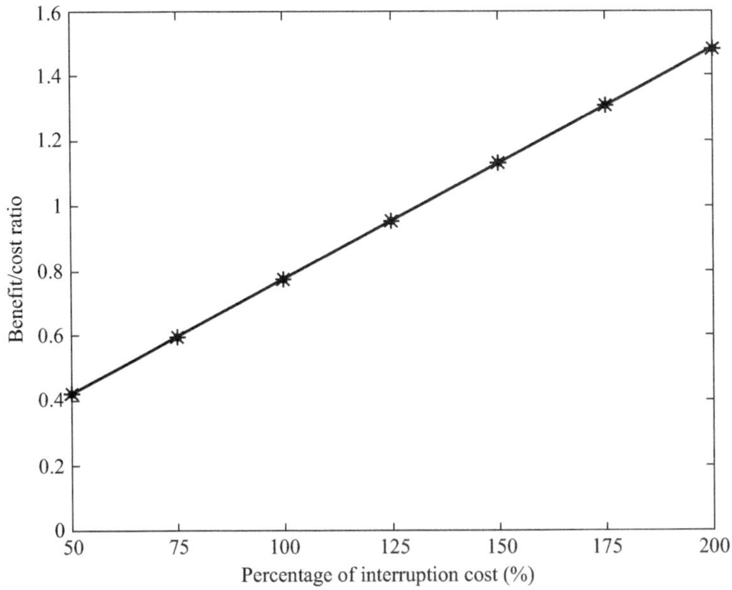

Figure 11.3 Sensitivity of benefit/cost ratio to interruption cost

11.4 Further extension for FA

There are currently 138 feeders with automation functions for the Tai-Chung District. In the foreseeing future, more and more feeders will be automated. For the future FA extension, it is necessary to determine which types of feeders are suitable for FA implementation. From Table 11.3, it can be seen that FDIR function plays a major role on the quantified benefit of FA due to the reduced time required for the feeder restoration; therefore, the CICs are reduced and the utility energy revenues are increased after FA. So, the process of determining the appropriate feeder to be automated should calculate the reduced CICs and increased revenues for each feeder. Table 11.4(a) and 11.4(b) show the average times required for feeder restoration from substation, from other feeders, and fault repair before and after FA, respectively. The data shown in these two tables indicate that the average time required for feeder restoration from substation, from other feeders, and fault repair are 47.17, 76.2, and 76.2 minutes before FA as compared to 2, 5, and 32 minutes after FA, respectively. The interruption durations are reduced due to the implementation of the FA system, which results in the lower interruption costs.

Using the data in Table 11.2 as a reference, the interruption costs for different customer types in the Tai-Chung DA area before and after FA can be calculated. The results are also shown in Tables 11.4(a) and 11.4(b). Since the CICs and energy revenues heavily depend on network operating conditions, the detailed calculations of CIC and reduced revenues due to a fault need to consider the uncertainties on network topology, load distribution, and switch location. Some complicated stochastic

Table 11.4 Interruption costs of different customer types

Interruption duration (minutes)	**CICs ($/kW)**		
	Residential	**Commercial**	**Industrial**
(a) Before FA			
47.17 (feeder restoration from substation)	2.549	8.269	13.680
76.2 (feeder restoration from other feeders)	4.118	14.363	27.671
76.2 (fault repair)	4.118	14.363	27.671
(b) After FA			
2 (feeder restoration from substation)	0.108	0.534	2.768
5 (feeder restoration from other feeders)	0.270	0.982	3.083
32 (fault repair)	1.729	5.434	8.541

methods such as Monte Carlo simulation can be used to more exactly predict the interruption cost; however, they are more time-consuming. From the historical data the average outage data such as outage rate and outage duration can be predicted; therefore, the future interruption cost of a distribution feeder could be predicted by using the average load and average outage data. In this chapter, average feeder load and outage data are used to estimate the CIC and reduced utility energy revenue. In the computations, feeder loads are assumed to be evenly distributed on the feeder. Then the CIC and reduced revenues can be estimated by using (11.7) and (11.9).

After the restoration time before and after FA is determined, the CIC and RR before and after FA can be calculated by (11.7) and (11.9), respectively. Therefore, the benefits for CIC and utility energy revenue can be obtained.

In addition, the system reliability index such as average system interruption duration index (ASIDI) which provides information in system average duration of interruptions can also be calculated. The definition of ASIDI is shown as follows

$$\text{ASIDI} = \frac{\text{Connected kVA duration interrupted}}{\text{Total connected kVA served}} \tag{11.19}$$

The ASIDI per year due to faults occurred on Section i can be determined by

$$\text{ASIDI}_i = \frac{\lambda \times l}{(n+1)} \times \left(t_{ss} \times \frac{(i-1)}{(n+1)} + t_{\text{repair}} \times \frac{1}{(n+1)} + t_{\text{feeder}} \times \frac{(n-i+1)}{(n+1)} \right) \tag{11.20}$$

The ASIDI after the installation of n switches can be expressed as

$$\text{ASIDI} = \frac{\lambda \times l}{n+1} \times \left(\frac{n}{2} \times t_{\text{ss}} + t_{\text{repair}} + \frac{n}{2} \times t_{\text{feeder}} \right) \tag{11.21}$$

Using the feeder GS51 as an example for illustrating the procedure of calculating the CIC and energy revenue for a feeder before and after FA, the specification data for GS51 are

- Average load is 4600 kW
- Feeder length is 1.243 miles
- Switch number is 3
- Residential customers is 28.57% and commercial customers is 71.43%
- Load growth rate is 5.15%, and outage rate is 0.112 failures/mile/year

The CICs for GS51 before and after FA are computed by using (11.6), and they are $6178 and $988, respectively. It means that the annual reduced interruption cost for GS51 is $5190 due to the FA implementation. The annual increased utility energy revenue is $48 after FA by using (11.8). Therefore, the annual total increased benefit for GS51 due to the implemented FDIR function is $5238 that equals the sum of the annual reduced interruption cost and the annual increased revenue. The cumulative benefit during economic life of equipment then can be computed by using present worth analysis. With (11.2a) and (11.2b), the present value and cumulative present value are calculated as 1.0719 and 25.5033, respectively. The total benefit including the reduced interruption costs and energy revenues for GS51 during the economic life of equipment is $133586. Besides, the ASIDI before and after FA can be easily determined by using (11.11), and they are 14.55 minutes and 2.38 minutes, respectively. The computation procedure mentioned above is helpful to find the feeders which have large benefits after FA.

The expressions shown in (11.6) and (11.8) are also used to compute the potential benefits of overhead and underground feeders after FA and further to determine which type of feeder is suitable for FA extension. For this purpose, an economic analysis of overhead and underground feeders for different types of customers including residential, commercial, and industrial customers is performed. Table 11.5 shows the interruption durations of the overhead and underground feeders in the Tai-Chung DA area before FA. These data are used in the economic analysis. It can be seen from Table 11.5 that the time for fault detection and isolation for the underground feeder is longer than that for the overhead feeder. Tables 11.6 and 11.7 show the capital and O&M costs of communication links and automation devices for overhead and underground feeders, respectively. It shows that the line switches (LS) and communication links for the underground FA are more expensive than those for the overhead

Table 11.5 Interruption durations for overhead and underground feeders before FA

Interruption duration (minutes)	Overhead feeder	Underground feeder
Feeder restoration from substation	26	61
Feeder restoration from other feeders	79	69
Failure repair	79	69

Table 11.6 Communication link costs for overhead and underground FAs

Costs	Overhead feeder	Underground feeder
Capital cost ($/mile)	10833	12999
O&M cost ($/mile/year)	31	32

Table 11.7 FA devices costs for overhead and underground FAs

Costs	FTU		LS	
	Overhead feeder	Underground feeder	Overhead feeder	Underground feeder
Capital cost ($/each one)	4114	4736	5421	20370
O&M cost ($/each one/year)	93	93	19	26

feeder. The total outage duration for the overhead feeder is shorter than that for the underground feeder (Table 11.5). It is interesting to know the benefit/cost ratios of these two types of feeder to be automated.

To achieve the economic analysis of FA extension, a spreadsheet file was developed. With the file, the benefit/cost ratios of the overhead and underground feeders for residential, commercial, and industrial customers under different feeder failures, feeder lengths, feeder loads, and numbers of switch are performed. In the analysis, the basic network data are assumed as: average load is 4600 kW, feeder length is 1.24 miles, number of LS is 3, load growth rate is 5.15%, and feeder failure rate is 0.13 failures/mile/year.

Using the CIC and energy revenue estimation formulas expressed in (11.7) and (11.9) and feeder interruption durations data and FA devices costs information shown in Tables 11.5–11.7, the benefit/cost ratios of the overhead and underground feeders for residential, commercial, and industrial customers are computed. The results are shown in Figures 11.4–11.7, which depict the sensitivity of the benefit/cost ratios of the overhead and underground feeders for different customer types with respect to feeder failure rate, feeder length, feeder loads, and number of switch, respectively. In Figures 11.4–11.7, the test cases for residential, commercial, and industrial customers mean that only this customer type is considered in the feeder. From the results, it can be found that the industrial customers can have the largest benefit/cost ratio since they have higher interruption costs than those for the commercial and residential customers. The results also indicate that since the overhead feeder has lower capital costs than the underground feeder, the benefit/cost ratio of the overhead feeder is larger than that for the underground feeder under the same feeder failure rate. The results imply

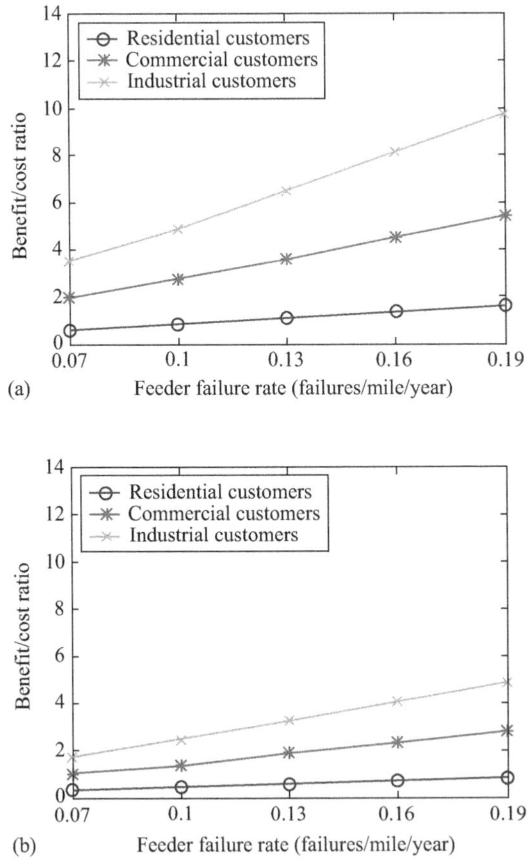

Figure 11.4 Effects of feeder failure rates on benefit/cost ratios of FA extension. (a) Overhead feeder; (b) underground feeder

that the FA extension is preferable for industrial customers and overhead feeders. It can also be seen that the benefit/cost ratio of the automated feeder increases with an increase in the feeder failure rate, feeder length, and feeder load. The benefit/cost ratio of the automated feeder decreases when the number of LS increases.

The analysis can further be used to find the optimum number of switch for a candidate feeder when the service reliability is specified. For example, considering an underground feeder that has the following parameters and objective

- Average load is 5200 kW
- Feeder length is 2.361 miles
- Commercial customers is 36.85% and industrial customers is 63.15%
- Load growth rate is 5.15%, and outage rate is 0.112 failures/mile/year
- The ASIDI should be smaller than 4 minutes

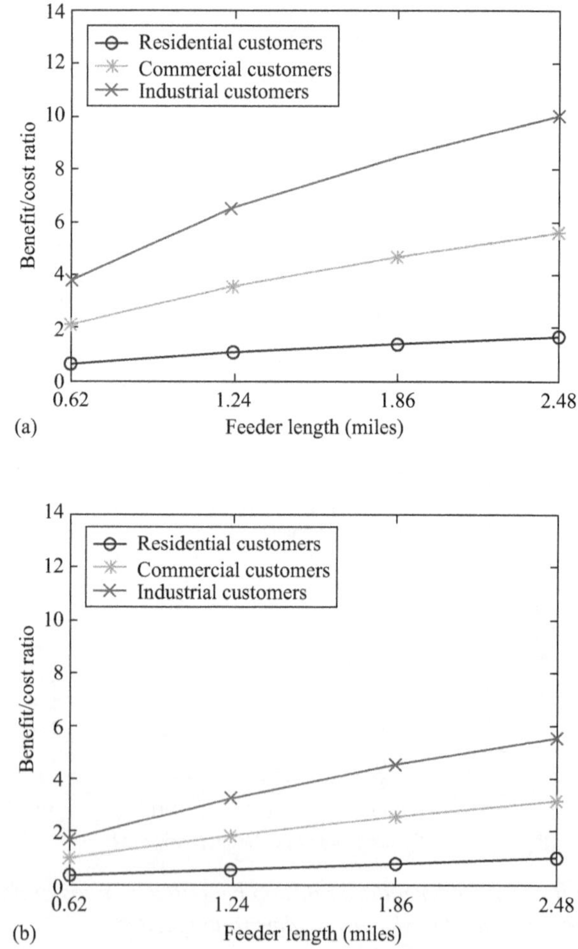

Figure 11.5 Benefit/cost ratios of FA extension under different feeder lengths. (a) Overhead feeder; (b) underground feeder

The ASIDI before FA is 28.05 minutes. The benefit/cost ratio and ASIDI for this feeder with respect to the number of switch is shown in Table 11.8. It can be seen that the benefit/cost ratio and ASIDI decrease with an increase in the number of LS. When four LS are considered to be installed, the benefit/cost ratio is 5.796 and ASIDI is 3.92 minutes. For this case, the obtained ASIDI is smaller than the predefined reliability index value. The result has indicated that the optimum number of switch of this feeder for balancing FA investment and service reliability is four. At this time the benefit/cost ratio of the automated feeder is 5.796. They are bold values given in Table 11.8.

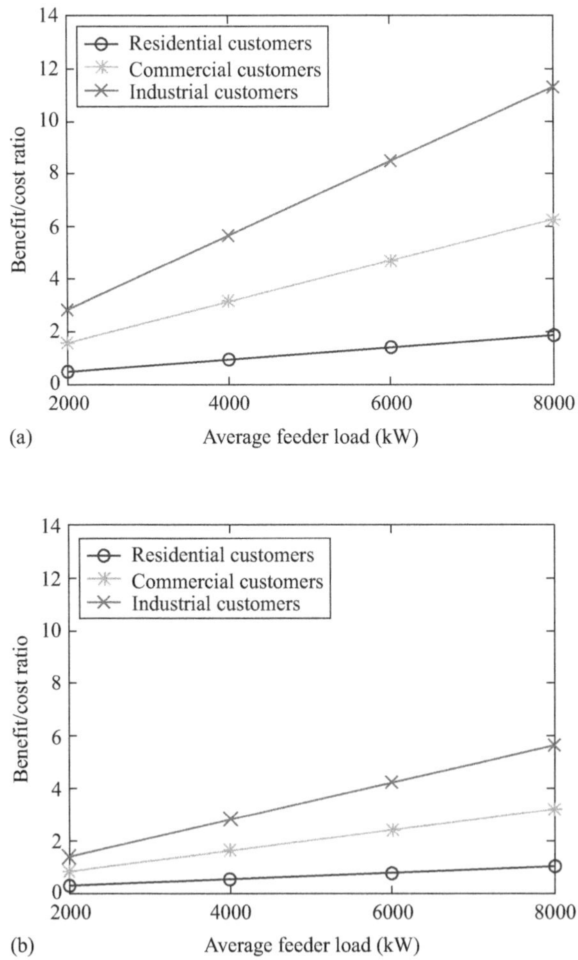

Figure 11.6 Effects of feeder loads on benefit/cost ratios of FA extension. (a) Overhead feeder; (b) underground feeder

11.5 Conclusions and discussions

A value-based economic evaluation method for DA has been described in this chapter to evaluate any DA project. This method takes CICs into account in the benefit analysis and uses the present worth analysis to perform the project economic evaluation. The FDIR function appears to be most cost-effective; the distribution analysis and customer management functions may not be cost-justified based on quantifiable benefits but may be justified if the intangible benefits are of concern. The methodology, used to find the candidate feeders for further FA extension and determine the optimal

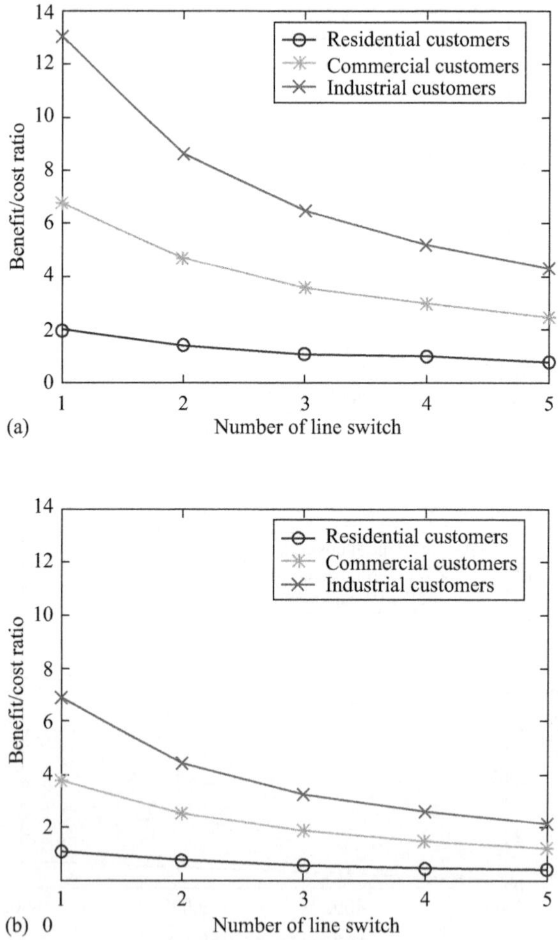

Figure 11.7 Effects of number of LS on benefit/costratios of FA extension. (a) Overhead feeder; (b) underground feeder

Table 11.8 Benefit/cost ratio and ASIDI versus number of switch

Number of Switch	**1**	**2**	**3**	**4**	**5**
B/C ratio*	12.775	9.138	7.097	**5.796**	4.89
ASIDI (minutes)	7.56	5.54	4.53	**3.92**	3.51

*B/C ratio means benefit/cost ratio.

number of switch for a feeder, is also discussed in this chapter. The study results have indicated that the implementation of the FA is suitable for the feeders with the high values of feeder failure rate, feeder load, and feeder length. The industrial customers and overhead feeders are preferable for implementing FA. The results also indicate that the implementation of DA system is suitable for the area with the high values of interruption costs, failure rate, and inflation rate, and can be justified when a small value of interest rate is considered. The methodology and formulas presented in this chapter can assist the utility in assessing the reliability improvement benefits of the DA system. The study results can also serve as the base for the utility's future DA system implementation guideline.

References

[1] *IEEE Tutorial Course on Distribution Automation*, IEEE Power Engineering Society, 88 EH0280-8-PWR, 1988.

[2] E. Gardner, "Distribution automation today: separating tools from toys," *Proceedings of the 37*th *Annual Rural Electric Power Conference*, Kansas, MO, 1993, pp. B2/1–216.

[3] H.L. Smith, "DA/DSM directions, an overview of distribution automation and demand side management with implications of future trends," *IEEE Computer Applications in Power*, Vol. 7, No. 4, 1994, pp. 23–25.

[4] General Electric Company, "Guidelines for evaluating distribution automation," *EPRI EL-3728 Final Report*, November 1984.

[5] J.S. Lawler, L.D. Monteen, J.B. Patton, and D.T. Rizy, "Impact of automation on the reliability of the Athens Utilities Board's distribution system," *IEEE Transanctions on Power Delivery*, Vol. 4, No. 1, 1989, pp. 770–778.

[6] D.L. Brown, J.W. Skeen, P. Daryani, and F.A. Rahimi, "Prospects for distribution automation at Pacific Gas & Electric company," *IEEE Transactions on Power Delivery*, Vol. 6, No. 4, 1991, pp. 1946–1954.

[7] N.S. Markushevich, I.C. Herejk, and R.E. Nielsen, "Functional requirements and cost-benefit study for distribution automation at B. C. Hydro," *IEEE Transactions on Power Systems*, Vol. 9, No. 2, 1994, pp. 772–781.

[8] R. Boateng, L. Nguyen, and S. Agarwal, "Distribution systems reliability—Lakeland electric case study," *Proceedings of the 2003 Annual reliability and maintainability Symposium*, January 27–30, 2003, pp. 546–550.

[9] S. Haacke, S. Border, D. Stevens, and B. Uluski, "Plan ahead for substation automation," *IEEE Power & Energy Magazine*, 2003, pp. 32–41.

[10] A. Pahwa, J.K. Sholtis, D.M. Rucker, and W.N. Dowling, "Field test of distribution automation equipment at Kansas utilities," *IEEE Transactions on Power Delivery*, Vol. 6, No. 4, 1991, pp. 1564–1572.

[11] M. Lehtonen, and S. Kupari, "A method for cost benefit analysis of distribution automation," *Proceedings of the International Conference on Energy Management and Power Delivery*, Vol. 1, 1995, pp. 49–54.

[12] D. Gruenemeyer, "Distribution automation: how should it be evaluated?" *Proceedings of the 35th Annual Rural Electric Power Conference*, Dearborn, MI, 1991, pp. C3/1–C310.
[13] D.G. Newnan, J.P. Lavelle, and T.G. Eschenbach, *Engineering Economic Analysis* (12th ed.), Oxford University Press, New York, NY, USA, 2013.
[14] C.A. Collier and B.L. William, *Engineering Economic and Cost Analysis*, HarperCollins, New York, NY, USA, c1988.
[15] C.N. Lu, C.L. Su, and J.H. Teng, "Economic evaluation of Tai-Chung district distribution automation project," *Final Report Submitted to Taiwan Power Company*, May 2003.
[16] J.L. Chang, Z.Y. Lin, and Z.Y. Wu, "Power system reliability evaluation and reasonable reserve margins analysis," *Final Report Submitted to Taiwan Power Company*, April 2002.

Chapter 12
Distribution automation at Enexis – a case study

J. Morren[1,2], *S. van der Heijden*[1], *and J.G. Slootweg*[1,2]

12.1 Introduction

Enexis is one of the major distribution system operators (DSOs) in the Netherlands, located in the southern and north-eastern of the Netherlands. With a network of approximately 130,000 km of medium- and low-voltage cables, it is supplying electricity to about 2.7 million customers. Besides electricity, natural gas is also supplied to approximately 2.1 million customers. The light grey areas in Figure 12.1 show the areas in which Enexis supplies electricity.

The electricity sector is facing major challenges. The increasing application of renewable energy sources, often of a decentralised and intermittent nature, the use of electric cars and the ever increasing demand for more reliable electricity networks drive DSOs to improve the capacity and reliability of their network. Some of the major challenges will be described shortly in this section.

In the past, networks were passive and the power flow was top-down. More and more small-scale distributed generation is connected to the network, however, making power flows bidirectional and operation of the network became more difficult.

Most of the electricity networks have been built in the 50s and 60s, and thus, they are approaching the end of their lifetime. Meanwhile, society is becoming more and more dependent on a secure supply of electricity. Power outages are no longer accepted as inevitable or inherent to the supply of electricity. Public information regarding power interruptions is getting more important.

In the classical way of solving an interruption, switching is done manually, and as a result, the technicians have to travel between different locations. Especially in large cities (during rush-hour), this takes a long time, leading to long restoration times, as the density of traffic is increasing year by year.

The average age of the employees of most distribution system operators is relatively high, and therefore, in the near future, a lot of them will retire, and new employees have to be found. With the shortage of students with technical education, this will become a problem in the future.

The trends described above imply that in the future, in a society that is becoming more and more dependent on electricity, the network is becoming older and more

[1]Enexis B.V., 's-Hertogenbosch, The Netherlands
[2]Eindhoven University of Technology, Eindhoven, The Netherlands

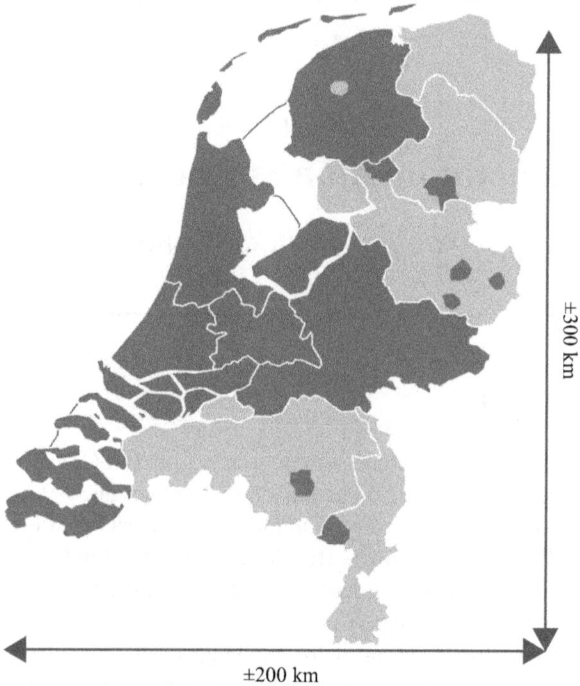

Figure 12.1 Supply area of Enexis

Figure 12.2 Structure of the chapter

complicated, while the number of employees is decreasing. Solutions have to be found to face these challenges.

Aware of these challenges, Enexis started a project in 2008 to investigate the main challenges that had to be expected and to find possible solutions for them. One of the main results of this project was the introduction of distribution automation (DA) in the medium voltage (MV) networks of Enexis.

This chapter will describe in detail the DA concept developed by Enexis, and will explain which steps have been taken. The main lessons learned will be presented. Afterwards the preparation of the large-scale roll-out and the impact of DA on the organisation of the DSO will be described. Finally the findings after automation of more than 1000 substations will be given.

The structure of this chapter is as follows (Figure 12.2): DA concept and functional design; Pilots; Technical design; Implementation; Results.

12.2 DA concept

In Section 12.1, several challenges that the network operators face are mentioned. The question is whether the current design and operation of the MV network is still appropriate for these developments.

12.2.1 Risk analysis

Enexis used its risk-based asset management (RBAM) methodology [1] to answer this question and to find possible solutions. The starting point for this was the risk 'Obsolescence of the design and operation of MV networks to meet new future demands'.

According to the RBAM approach the risk that is defined is confronted with the company values. From this analysis, it becomes clear that mainly the business values 'reliability' and 'reputation' are affected. Reliability is affected as the information in the current MV network is not sufficient to transport the fluctuating load from the distributed generation. Also the network that is becoming older, giving a larger possibility of failures, poses a risk on reliability. The reputation of Enexis can be put at stake because the demands of society concerning reliability and dissemination of public information are not adequately incorporated in the current network operation.

The risk matrix of Enexis is used to determine the risk levels of the affected business values. By combining both the effect and the frequency, it can be determined that the risk level is 'high' for both reliability and reputation [2].

12.2.2 Strategy development

The new developments that must be faced typically ask for a more flexible operation, with a high reliability and more dissemination of public information, while at the same time, the operation should be less labour-intensive. The combination of these requirements asks for more automation in the network.

The MV networks (10 and 20 kV) of Enexis typically consist of a transmission part, that is designed redundantly (n-1 secure), and a distribution part, that has a ring-configuration but is operated as a radial grid (Figure 12.3). In case of a failure, the fault is cleared automatically. However, the isolation of the faulty section and the subsequent restoration of power are done manually. This implies a great potential for improvement of reliability and a reduction of labour-intensity by applying DA. Moreover, DA will also facilitate dissemination of public information. Therefore, a concept of remote switching and fault localisation in the MV-distribution grid was developed. Many variants of remote control are possible: in theory, all circuit breakers and switch disconnectors could be equipped with remote control, but at a certain point the additional reliability improvement is no longer worthwhile, i.e., the additional investment does not pay off. In Ref. [3], it has been shown that the most optimal solution is to automate the switch disconnectors approximately in the middle of both feeders and at the normally open point (NOP) (S1, S3 and S2, respectively, in Figure 12.3).

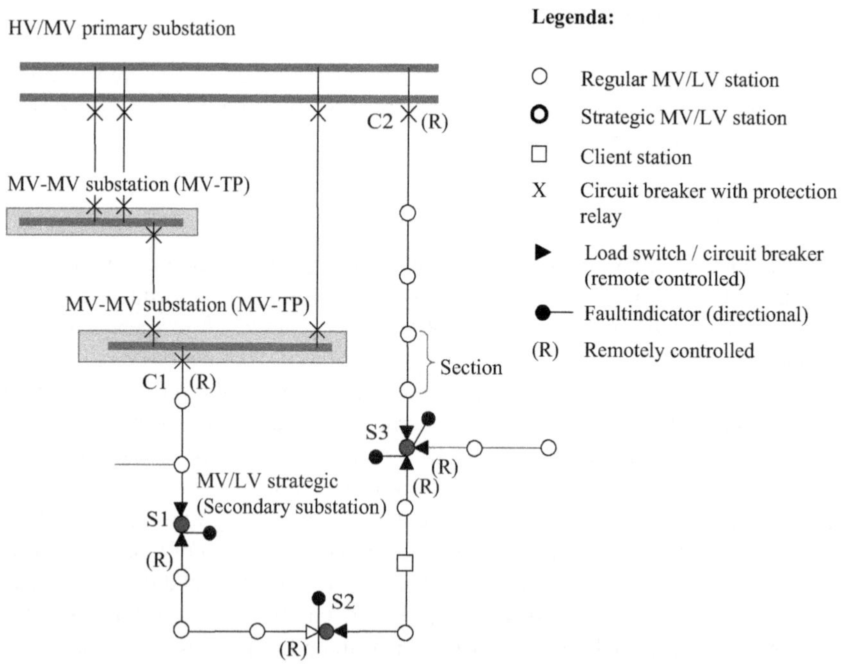

Figure 12.3 Enexis' distribution automation concept

12.2.3 DA concept

The resulting DA concept is shown in Figure 12.3. As soon as a fault occurs in the cable between C1 and S1, the circuit breaker C1 will disconnect the feeder. In the conventional situation the fault had first to be located and isolated manually and then power supply to the feeder could be restored. With DA switch disconnector, S1 can be opened remotely. Immediately afterwards, the NOP S2 will be closed and customers in the second part are re-energised. In case of a fault between S1 and S2, S1 will be opened remotely and C1 closed, such that customers in the first part of the feeder are re-energised immediately. This application of DA results in a significant reduction of the customer minutes lost (CML).

12.3 Pilot project

After defining the functional design, a large pilot project has been launched. This pilot project had several goals. One important goal was to see whether the components that were needed to implement Enexis' DA concept were available on the market. Second, the goal was to investigate which practical problems would be encountered when the concept would be implemented in practice. The last main goal was to prepare for the large-scale roll-out. In this section the above issues will be discussed.

12.3.1 Availability of components

One of the main goals of the pilot project was to test the required components if they were available in the market or to develop them if they were not available yet. The main issues encountered will be described in this section.

A key component in the whole concept is the remotely controlled Ring-Main Unit (RMU). At Enexis, these RMUs are located in small substations or in even smaller kiosk substations. This requires that the RMUs should be very small. It became clear rather soon already that no off-the-shelve solutions were available. The available RMUs were too large, did not have the possibility to motorise them, or required a lot of modifications before they were suited for remote motorised control. Together with a manufacturer, Enexis developed an RMU that could be remotely controlled and did meet the requirements of Enexis (i.e. was small enough).

Another important part is the required automation. This consists of fault indicators, measuring of current at the low-voltage side of the transformer, the control of the motor drive and the gateway to the control room. The goal of Enexis was to have a one-box solution to which only the current measurements and the fault passage indicators had to be connected. However, during the preparation of the pilot project, it became clear that most manufacturers active in the area of DA did not have that kind of solutions yet. Proposed solutions consisted of a lot of different components, all of which have to be connected together in order to obtain the desired functionality. During the pilot project a number of manufacturers were challenged to develop a system that met the requirements of Enexis. That resulted in several dedicated products, and finally, after a tender procedure, one of them was selected by Enexis.

Because of the increasing amount of distributed generation in the MV networks, the fault indicators should be bi-directional. At the start of the pilot project, these were not available (based on proven technology). Together with some manufacturers, they were developed and tested. The result was a good product which also was able to measure the voltage and current in the MV network.

12.3.2 Other issues

Besides the fact that a lot of the required components were not available, some other issues were encountered.

At the start of the project, it was also assumed that it would be possible to automate the existing switchgear. However, during the pilot project, it became clear that in most cases this was not possible. This implies that in those cases the switchgear had to be replaced, resulting in an increase of the cost of DA projects.

Also a lot of kiosk-type substations were not suited for application of DA. They are often too small already for placing an RMU, so finding space for additional DA equipment is even more difficult. Also the climatic conditions in those kiosks are extreme harsh. So, in order to make DA possible, often the complete substation and RMU have to be replaced by a new one.

In discussions on automation of power systems, it is mostly assumed that fast and reliable communication with a high bandwidth (e.g. glass fibre) is available everywhere. However, the reality is quite different. From a research project that has been carried out for Enexis, it became clear that at that moment, wireless communication was the only feasible option for the DA project of Enexis. The used protocols and data to be transferred are not optimised for a mobile medium which has its own characteristics and a data-based tariff structure. In addition, no international security standards were available describing how to implement adequate security measures over mobile communication networks.

12.3.3 Preparation of large-scale roll-out

The third main goal of the pilot phase was the preparation of the large-scale roll-out of the DA programme in the MV networks of Enexis. This section describes the most important issues that have to be taken into account for the large scale roll-out. The most important issues are:

- Different types of switchgear—there is a large diversity of switchgear (>100), both in manufacturer and type, and the same type even can have different versions;
- Different types of substation—substations are different in dimensions, layout and climatic conditions, and as a result, different solutions have to be found;
- Sixty to eighty percentage of the protection relays cannot be used (as these are mechanical relays, they have no possibility for communication with (standardised) protocols, etc.);
- Primary equipment is often not suited for automation by a motor drive, implying that a part of the primary switchgear has to be replaced in order to make DA possible;
- Specific knowledge, experience and skills required—installing and maintaining DA equipment is quite different from the work that is done nowadays, implying that training of people is necessary.

12.4 Technical design

After the pilot phase and the lessons that had been learned in that phase, the technical design could be finalised. The result for the most important parts (the MV/MV substations, the MV/LV substations and the supervisory control and data acquisition (SCADA) system) are presented in this section [4].

12.4.1 MV/LV substations

The remote operation requires that the RMU should have a motor drive for remote opening and closing the load-break switches. Because it has to operate at the moment when there is no power, a battery is required, with enough capacity to operate the load-break switches several times. In order to know whether the fault was in the first or in the second half of the feeder, there should be a short-circuit/earth-fault indicator on at least one of the two feeders connected to the RMU. For grid management and planning, load conditions are needed which requires some additional measurements. For exchange of data and control signals a communication medium is required, for which public available wireless communication showed to be the most cost-effective solution. In order to make quick installation (and replacement, if necessary) possible, it was required that all functionally should be put together in one so-called 'DA box'. An overview of the complete solution is shown in Figure 12.4.

12.4.2 MV/MV substations

In order to be able to disconnect and reconnect quarters of the distribution ring, it is also needed that the circuit breakers at the beginning of each feeder are remotely controllable. In case of distribution feeders directly connected to the primary switchgear in the HV/MV substations, this is in most cases possible, as the circuit breakers are already suited for remote control. However, Enexis has a grid structure in which most distribution feeders are connected to MV/MV substations, and in these substations the circuit breakers are not remotely controllable. In some of the newest substations, motor control of the circuit breakers is already available, but in older substations even this is not possible, and the switchgear has to be retrofitted or replaced.

Fault information and data are acquired by digital protection relays, which also control the circuit breakers. The relays are interconnected by a LAN-based network. For communication with the SCADA-system, GPRS and UMTS/LTE are the preferred medium for the short term (because of the low costs). Based on the functional design described above, a technical design has been developed. An overview of it is given in Figure 12.5.

12.4.3 SCADA/distribution management system

Another important part of the whole DA concept is the new SCADA/distribution management system (DMS) system. Besides a conventional SCADA system, it is a combined geographical and data information system. This means that the operators in the control room always have the latest grid diagrams including the data points from the equipments in the field.

12.5 Implementation

The final stage before the large-scale roll-out could start was to implement the DA concept and everything related to it throughout the organisation, which means:

- Inform the engineers when they have to use which components.
- Train the technicians regarding installations of the systems.
- Introduce new working processes.
- Teach the operators how to use this new functionality.

This section gives an overview of the impact on the whole organisation, all its processes and the colleagues working in it, of the introduction of DA. Subsequently

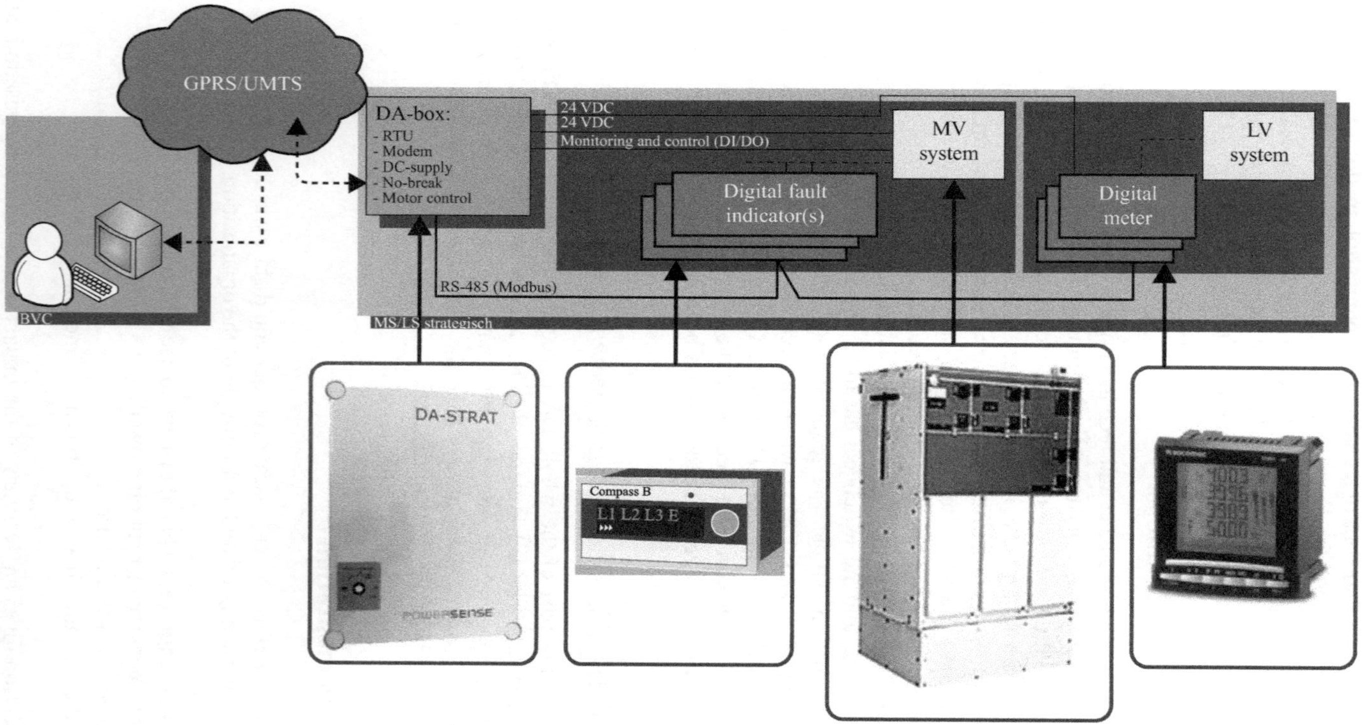

Figure 12.4 Technical design for MV/LV substation

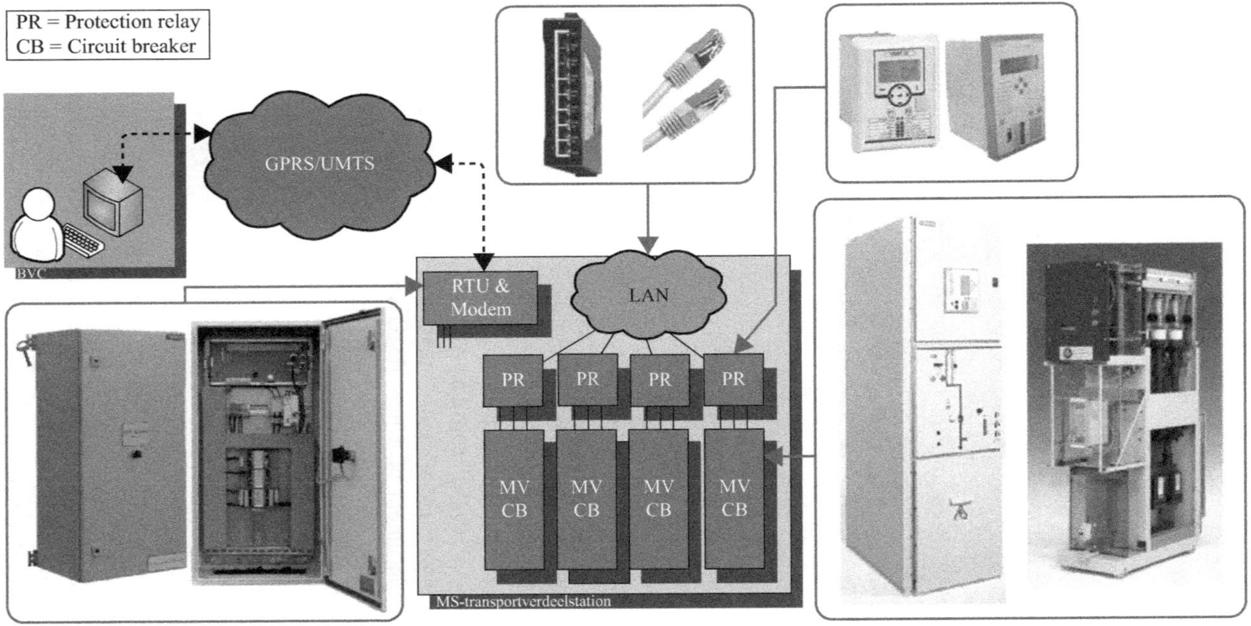

Figure 12.5 Technical design for MV/MV substation

the acceptance by colleagues, the impact on the daily work, the required changes in the company processes and the required knowledge and education will be discussed [3].

12.5.1 Acceptance by workforce

Introduction of DA means introduction of more automation in the network. Two extremes in this regard can be noticed. On one hand, there are colleagues who embrace the new technologies and who believe in new opportunities. On the other hand, there are those who see it as a threat and/or a downfall. This last group has its doubts about the reliability and lifetime of the new products. Besides that, they also see a lot of additional work because of the new equipments. This group has to be taken seriously, and it is important to listen to them very carefully in order to understand the real issues. Lastly, when this scepticism can be changed into enthusiasm, it will have a positive influence on the whole organisation.

12.5.2 Impact on daily work

During the implementation of DA, it was noticed that the impact on the organisation was larger than was expected beforehand. The knowledge that is required was one issue, but not the most important. For most colleagues, in-depth knowledge is not needed, and some basic understanding of the new techniques is enough. However, the new DA equipment has significant influence on the organisation as a whole. Similar to other network operators, Enexis has a number of processes in which the work within the company is organised. Most of these processes could be done almost independently from each other. However, during the introduction of the DA programme within Enexis, it became clear that this resulted in much more dependencies between different departments. In the past, each department could do more or less its own part of the job independently, but now there are more interdependencies between the departments. For example, the placement of a new RMU with DA requires coordination with the SCADA department and the colleagues operating the DMS. The impact will be discussed in more detail in the next section.

12.5.3 Changing company processes

In this sub-section, as an example, a description is given of the process from selection to acceptance of substations, including the additions as a result of DA.

The process starts with the selection for a new substation, then the planning and design starts, after that the design will be realised and finally the new substation will be connected to the grid including acceptance of the grid manager, as shown schematically in Figure 12.6. In the past, all those steps could be done more or less independently from each other.

Intake: According to the policy of Enexis, every year the (asset) engineers have to select a number of substations (mainly RMUs) that should be upgraded to a DA substation. The goal is to find these parts of the distribution grid where the introduction

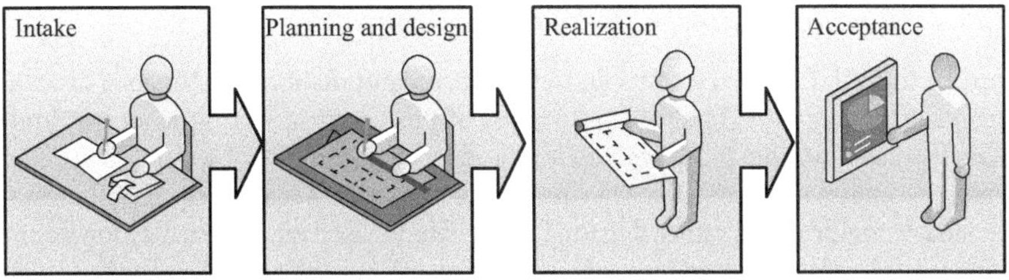

Figure 12.6 The basic elements of the process for building a substation

of the DA equipment will have the largest positive influence on the reliability of the network. Besides that, for every new RMU that is placed for other reasons, they have to decide whether it should be automated or not.

Planning and design: Adding DA requires more cooperation with the SCADA department, because they have to connect the substations to the SCADA system and in some cases they also have to programme the equipment in the substations (remote terminal unit (RTU), modem). Besides, the engineer needs to register the whole project, step by step, in the GIS and ERP systems before realisation already. This is because of the new SCADA/DMS system that is integrated with these enterprise data systems. In the past, all this was done at once, at the end of the whole process, when the substation was placed and commissioned already, and everything was in operation.

Realisation: Both placement and commissioning of new switchgear require more cooperation and time than in the past. Before the introduction of the DA programme, new switchgear could just be placed, the cables connected and put into operation. Now, also all the DA equipment (RTU, modem, short-circuit indicators, motor drives, etc.) have to be commissioned. This requires a close cooperation between those departments that are responsible for placement of the switchgear and the departments that are responsible for, and have the knowledge of, the DA equipment.

Furthermore, when the new substation is placed, also the commissioning of it will take much more time than in the past. All the additional DA equipment that is used has to be tested. This requires knowledge of both the primary equipment (the switchgear) and secondary equipment (all the DA equipment). All this has to be tested together, while in the past, it could mostly be done separately.

Acceptance: Now that the substations can be directly controlled and observed from the control room, the project teams have to prove that all signals and measurements are correct. Technically this is already a challenge but especially for the whole project planning process, as on the same moment people in the substation (both for the primary equipment and the secondary (DA) equipment), people at the SCADA department and people in the control room have to be available and have to cooperate. In the past, all these steps could be done individually, requiring much less coordination and planning. Furthermore, in addition to other substations (without DA), some extra information about the DA equipment needs to be registered.

12.5.4 Required knowledge and education

Introduction of DA also means introduction of automation equipment (so-called secondary equipment). The basic principles of this secondary equipment are fundamentally different from those of the primary equipment such as cables or transformers. Primary components are static. They remain the same for decades. On the other hand, secondary equipment is rather dynamic. It is mainly based on software which implies that a lot of settings have to be made, which have to be determined and tested. It can be expected that the software (firmware) in the equipment will change every couple of years and the software to programme and test it even more often. All this requires a complete new way of thinking for people. It is therefore not obvious that people who have worked with primary equipment for decades already will be able to learn working with DA equipment easily.

So far, only the people who worked with protection relays and with substation automation are more or less familiar with (the working principles of) the new equipment. But this is only a limited number of people, and with the new DA equipment, more of them are needed. Also for them things will change. The largest group are the protection specialists. Most of them are used to work with individual protection relays only. However, in the MV/MV substations, they now have to work with Local Area Networks (LANs), RTUs and so on. The principles of an Ethernet network or of the IEC61850 protocol are new also for the protection specialists. So, also for them the working principles of the new technologies are not obvious.

So, it can be concluded that the number of people that is familiar with DA equipment is rather limited, and that even they do not have enough knowledge in most cases. This implies that an extensive training programme is required to provide the protection specialists with more knowledge and train a significant number of new employees in order to be able to realise the introduction of an increasing amount of DA equipment.

12.6 Results

12.6.1 Realisation

In 2012, Enexis started with the large-scale roll-out of DA. At the end of 2015, more than 1000 RMUs have been automated. Although initially people had to get used to it, DA is now daily business at Enexis, and people are very positive about it and ask for increasing the roll-out speed. Some examples of substations that have been automated are shown in Figure 12.7.

Figure 12.8 gives an overview of all locations with DA in the Enexis supply area (marked with dots).

12.6.2 CML savings

Since the start of the large scale DA roll-out in 2012, it proves to be very successful in reducing the CML. Although still a rather small percentage of the substations is

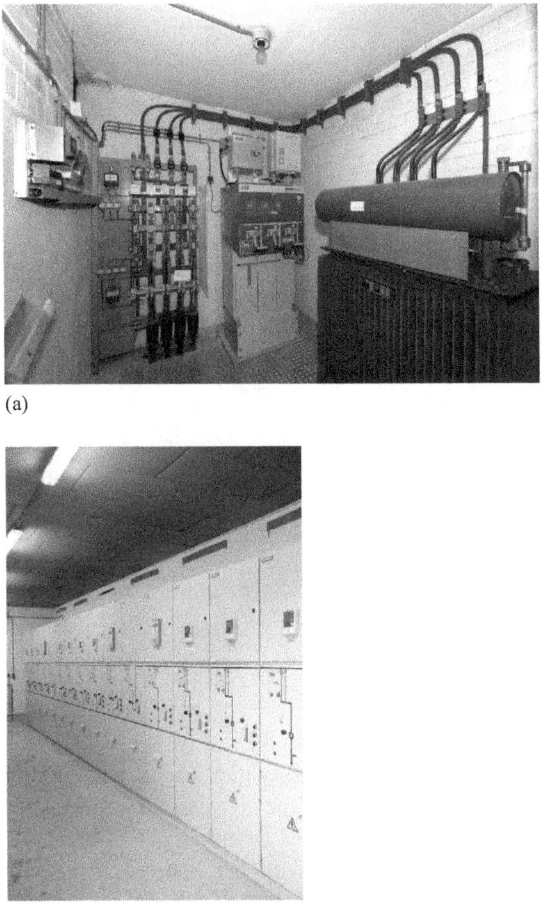

(a)

(b)

Figure 12.7 Example of substations equipped with DA: MV/LV substation (a), MV/MV substation (b)

automated (about 15% of the total number that is expected to be automated), and on average, these substations have been automated less than 2 years ago, until the end of 2014 already about 120 faults have been restored by using DA, saving about 3.2 million CML.

From the interruptions that have been restored by DA, it has become clear that even about 50% of the CML of an interruption can be reduced by applying DA, instead of the 40% that was initially assumed. This makes application of DA even more attractive. The higher reduction is due to the fact that in practice, there is often more than one remotely controlled RMU in a feeder, which means that a larger part can be restored remotely.

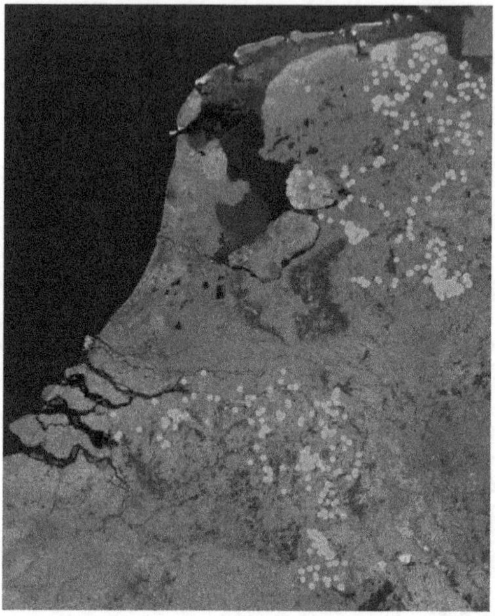

Figure 12.8 Locations of substations with DA

As an example the reduction in CML will be calculated for an interruption that occurred on 27 February, 2014. A part of the network diagram is shown in Figure 12.9. In this figure, two feeders are shown. The first consists of the parts A, B and C, and the second consists of part D. The two feeders can be connected with the NOP. The fault was in the second half of the feeder (the first part of the feeder is marked by A, the second part by B and C). The boxes with the numbers 1 to 5 show the actions that have been performed to restore the power. After occurrence of the fault, first a switch-disconnector in the middle of the feeder could be opened (1) and then the circuit breaker closed (2). Afterwards the faulted cable section had to be disconnected locally. This has to be done at both sides, giving two switching actions (3). Now the last part of the feeder (part C) can be re-energised by closing the normally open point (4) to the other feeder (part D). Also part B, between the switch disconnector, which was opened as the first step and the fault location, can be closed (5) and also part B of the feeder is re-energised. After these actions, power supply is restored for all customers and the fault cable section can be repaired.

The reduction in CML is calculated in Table 12.1. The CML for the situation without DA is based on the average recovery times for restoration of supply. The CML in case of DA are realised in reality. From the numbers, it can be concluded that the reduction in CML due to DA is 49%.

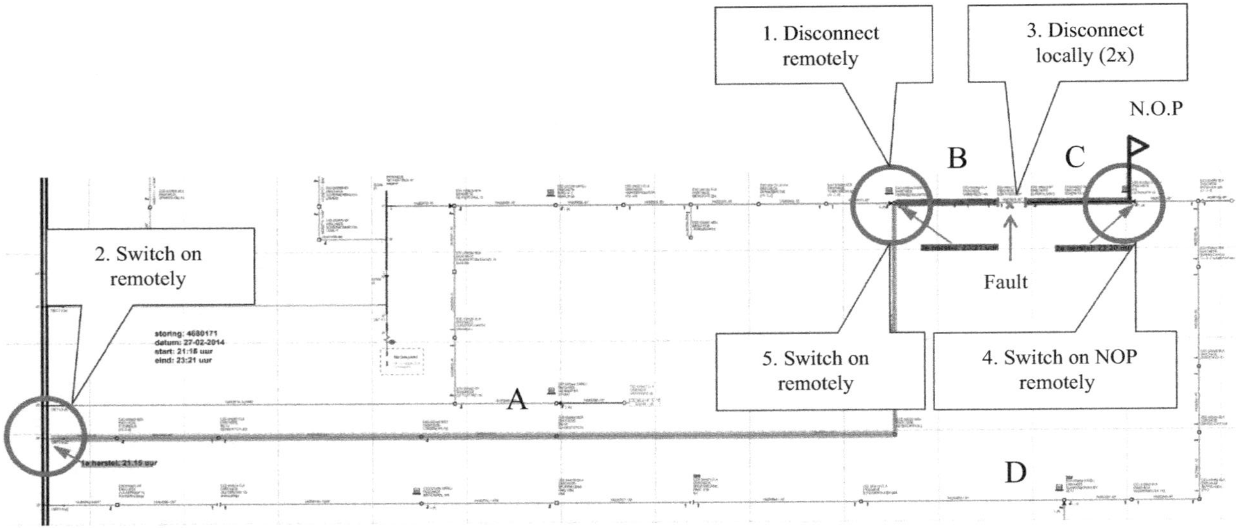

Figure 12.9 Network diagram in which fault was restored by using DA

Table 12.1 Reduction in CML due to DA

	Recovery time No DA	**With DA**	**CML Customers**	**No DA**	**With DA**
First	87	3	882	76734	2646
Second	132	125	441	58212	55125
Third	174	126	294	51156	37044
			Total	186102	94815

12.6.3 Other advantages

One of the main drivers for the roll-out DA is the reduction in CML. Several other advantages have been noticed already. One of the reasons for very positive acceptance of DA by the employees is that it also is used during regular maintenance work. As most of the switching actions can be done from the control centre, this results in less movements between different substations and thus resulting into an efficiency improvement.

Another advantage of remote switching is that it improves the safety of the work, as switching no longer has to be done locally but can be done remotely.

The roll-out of DA also results in more measurements in the distribution network. With the increase of decentralised and fluctuating power sources such as wind and solar, this is useful to have better insight in the power flows, and therefore overloading of the network can be avoided. In the future, it will also be of importance for new developments such as smart charging of electric vehicles and demand side management. The introduction of DA can therefore be considered as a step towards a smarter grid.

12.6.4 Lessons learnt and future developments

After 3 years of large-scale roll-out of DA, it is possible to evaluate the roll-out and investigate what will be the future developments.

One of the assumptions at the start of the DA roll-out was that most of the existing switchgear could easily be modified for application of DA. However, during the roll-out, it was noticed that there is a huge diversity in installed base of MV switchgear. Network operators have a history of over 100 years, and taking into account the typical lifetime of switchgear of 40–60 years, this results in a huge diversity in the installed base. This means that for a significant number of different types of switchgear a solution has to be found to make them suitable for remote switching. Together with manufacturers, solutions have been found for different types of switchgear.

Another important lesson that was learned was that it is very important to involve all relevant departments and colleagues to come to a solution that is accepted by everyone. This results in the best solutions, as all available experience is included in the final solution. In this way, it is avoided that during the roll-out, discussions will start and there is no support.

As mentioned in Section 12.6.2 the results in CML reduction are even better than expected on beforehand. Because of this and other advantages of DA (Section 12.6.3), Enexis will continue the roll-out of DA. The target of Enexis is to keep the CML constant for the coming years although higher failure rates (because of ageing assets) may be expected. From the RBAM methodology, it is clear that DA is one of the most cost-effective measures to compensate for this higher failure rate and reach the goal of constant CML. The investments will even increase in the coming years to a level that is two times as high as in 2014. With this level, in 4 or 5 years all larger cities in the supply area of Enexis can be provided with DA.

It is expected that the cost of DA equipment will decrease in the coming years. When Enexis started its roll-out of DA, some of the required components had still to be developed or were made specifically for Enexis. However, DA is now becoming a more common application and applied on a larger scale, and therefore, more equipment will become available and the cost will decrease. This development will make it even more cost-effective to continue with the roll-out of DA.

References

[1] Slootweg, J.G., Clemens, G.J.M.B., 'Implementing and certifying a state of the art asset management approach', *Proceedings of the 19th CIRED Conference*, June 2007, paper no. 65.

[2] Berende, M.J.C., Slootweg, J.G., Kuiper, J., Peters J.C.F.M., 'Asset management arguments for smart grids', *Proceedings of Cired Seminar 2008*, paper no. 12.

[3] Groot, R.W.J., de Morren, J., Slootweg, J.G., 'Smart integration of distribution automation applications', Proceedings of the 3rd IEEE PES Innovative Smart Grid Technologies (ISGT) Europe Conference, October 2012, p. 1–7.

[4] Morren, J., Heijden S. van der, 'The PAC-men revival – increasing need of PAC qualified colleagues', *Proceedings of the PAC World conference 2012*, paper no. 22.

Index